FAO TRAINING SERIES 16/2

simple methods for aquaculture

TOPOGRAPHY

for freshwater fish culture topographical surveys

Text: A.G. Coche
Graphic design: T. Laughlin

FOOD AND AGRICULTURE ORGANIZATION OF THE UNITED NATIONS
Rome 1989

David Lubin Memorial Library Cataloguing-in-Publication Data

FAO, Rome (Italy)

Topography for freshwater fish culture: topographical surveys.
(FAO Training Series, No. 16/2)

1. Topography 2. Fish culture
I. Title II. Series

FAO code: 44 AGRIS: P31 M12
ISBN 92-5-102591-6

P-41

ISBN 92-5-102591-6

THE AQUACULTURE TRAINING MANUALS

The training manuals on simple methods for aquaculture, published in the FAO Training Series are prepared by the Inland Water Resources and Aquaculture Service of the Fishery Resources and Environment Division, Fisheries Department. They are written in simple language and present methods and equipment useful not only for those responsible for field projects and aquaculture extension in developing countries but also for use in aquaculture training centres.

They concentrate on most aspects of semi-intensive fish culture in fresh waters, from selection of the site and building of the fish farm to the raising and final harvesting of the fish.

The following manuals on simple methods of aquaculture have been published in the FAO Training Series:

Volume 4 — Water for freshwater fish culture
Volume 6 — Soil and freshwater fish culture
Volume 16/1 — Topography for freshwater fish culture
Topographical tools
Volume 16/2 — Topography for freshwater fish culture
Topographical surveys

The following manuals are being prepared:
Pond construction for freshwater fish culture
Management for freshwater fish culture

FAO would like to have readers' reactions to these manuals. Comments, criticism and opinions, as well as contributions, will help to improve future editions. Please send them to the Senior Fishery Resources Officer (Aquaculture), FAO/FIRI, Via delle Terme di Caracalla, 00100 Rome, Italy.

HOW TO USE THIS MANUAL

The material in the two volumes of this manual is presented in sequence, beginning with basic definitions. The reader is then led step by step from the easiest instructions and most basic materials to the more difficult and finally the complex.

The most basic information is presented on white pages, while the more difficult material, which may not be of interest to all readers, is presented on coloured pages.

Some of the more technical words are marked with an asterisk (*) and are defined in the Glossary on page 257.

For more advanced readers who wish to know more about topography, a list of specialized books for further reading is suggested on page 262.

CONTENTS

TABLES AND CHARTS

7 TOPOGRAPHICAL SURVEYS — PLAN SURVEYING

What is a topographical survey?

1. A survey of your fish culture site can help you do one of two things: make a map to help you plan your work; or lay out marks on the ground that will guide you as you work.

Site

2. **Topographical surveys** will help you to make plans or maps of an area that show:

- the main **physical features on the ground**, such as rivers, lakes, reservoirs, roads, forests or large rocks; or the various features of the fish-farm, such as ponds, dams, dikes, drainage ditches or sources of water;
- the difference in height between land forms, such as valleys, plains, hills or slopes; or the difference in height between the features of the fish-farm. These differences are called the **vertical relief**.

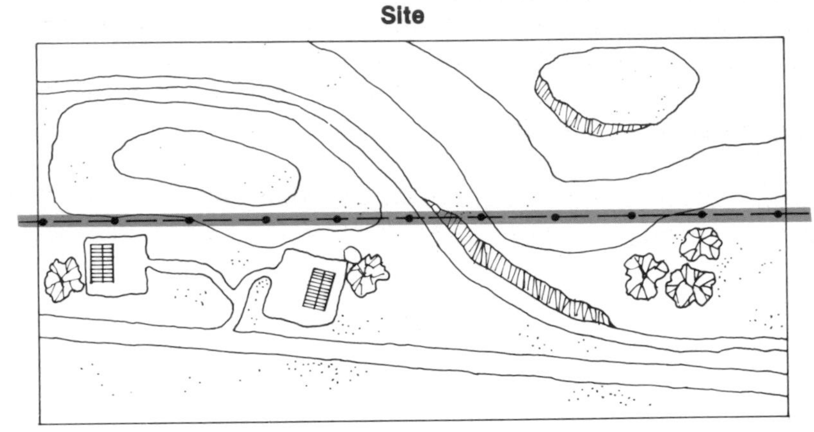

Map

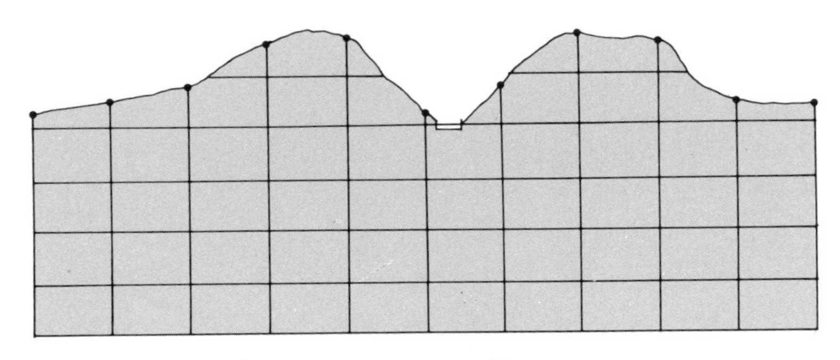

Vertical profile

What do topographical surveys involve?

3. The purpose of the first type of topographical survey is to establish, on a horizontal plane, the position of one or more points in relation to the position of one or more other points. To do this, you will measure **horizontal distances** and **horizontal angles** or **directions**. You will use a method called **plan surveying**, which will be explained in this chapter.

4. The purpose of the second type of topographical survey is to find the elevation (or vertical height) of one or more points above a definite horizontal plane. To do this, you will measure **horizontal distances** and **height differences**; you may also need to lay out contour lines. You will use a method called **direct levelling**, which will be explained in Chapter 8.

5. You will learn how to make **plans and maps** based on the results of plan surveying and direct levelling in Chapter 9.

Site

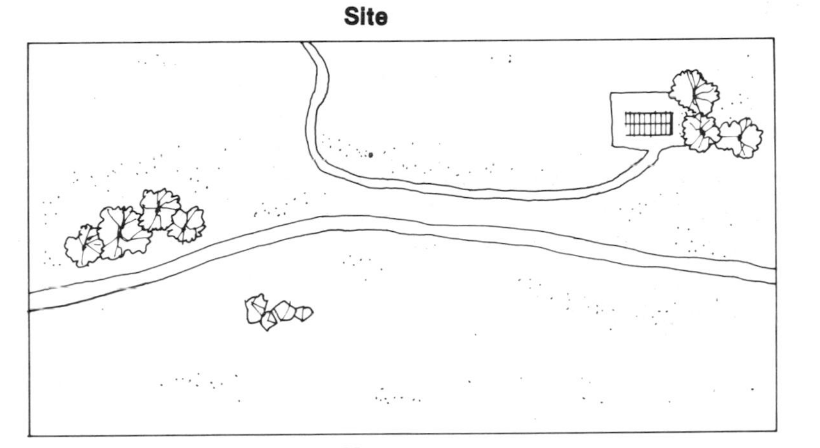

Map

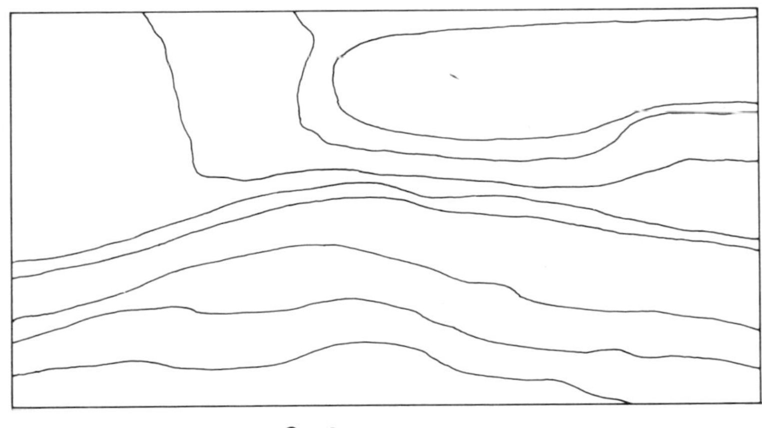

Contour map

6. When you plan a topographical study, the most important rule to remember is that you must work **from the whole to the part**, keeping in mind all of the work you will need to do as you begin the first steps. Different types of survey require different levels of accuracy, but you should lay down the first points of each survey as accurately as possible. You will adjust all the work you do later to agree with these first points.

Example

You need to plan survey a fish-farm site.

(a) First, you must make **a perimeter survey ABCDEA**. Besides these summits and boundaries, add several **major points and lines**, such as AJ and EO. They run across the interior to create right angles, which will help you in your calculations. This survey gives the **primary survey points**, which you should determine and plot very accurately.

(b) Then, lay out **minor lines** such as FP and TN. They go between the major lines to divide the area into **blocks**. This gives you the **secondary survey points**, which you may determine less accurately.

(c) Finally, survey **details** in each block using **tertiary points**, for which less accuracy is also acceptable.

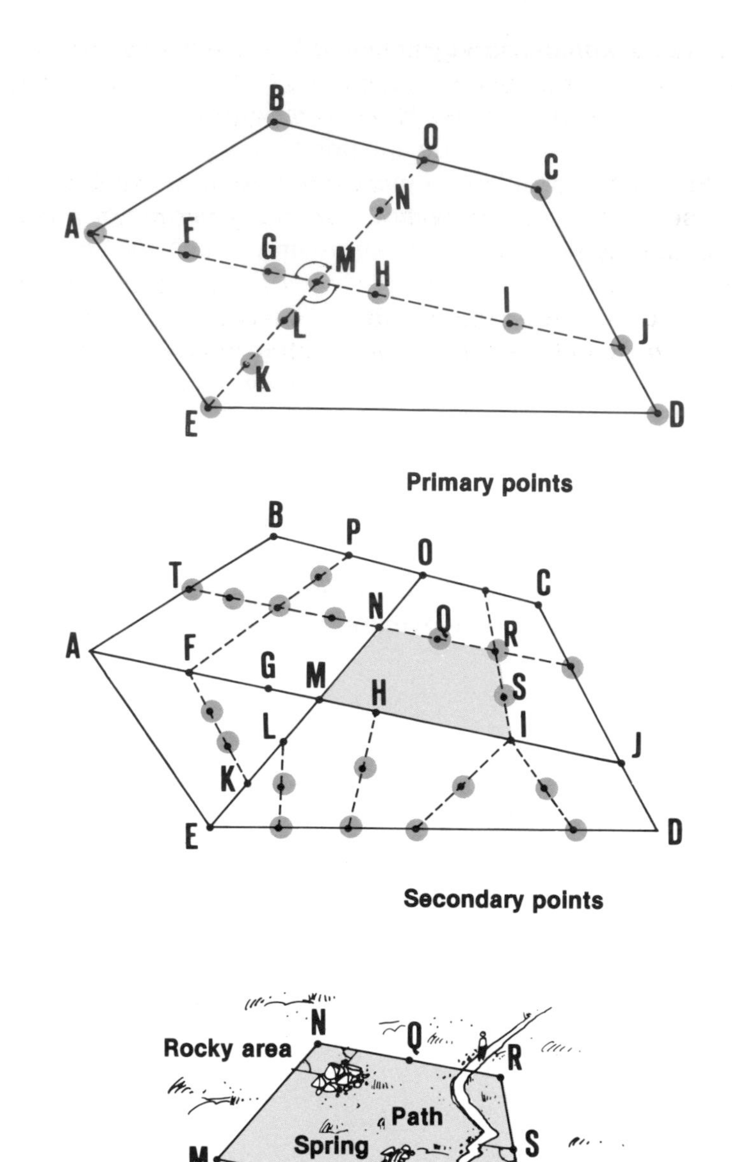

7. The way you plan a topographical survey will also depend on its **purpose**. You will use **a planning method** similar to the one described for soil surveys (see Volume 6, **Soil**, Section 24).

- First make a preliminary or **reconnaissance survey**. You can use quick methods without worrying too much about high accuracy.
- Based on the results of this survey you can plan and carry out **more detailed and accurate surveys**, such as location surveys and, last of all, construction surveys.

8. The way you plan a topographical survey will depend on the **subject you need to survey**, such as:

- a **straight line** defined by at least two points, such as the centre-lines of supply canals, pond dikes, and reservoir dams;
- a **series of lines** related to each other by horizontal angles and horizontal distances, such as the centre-lines of pond dikes in a fish-farm;
- an **area of land** such as a site chosen for the construction of a fish-farm (also see step 6 above).

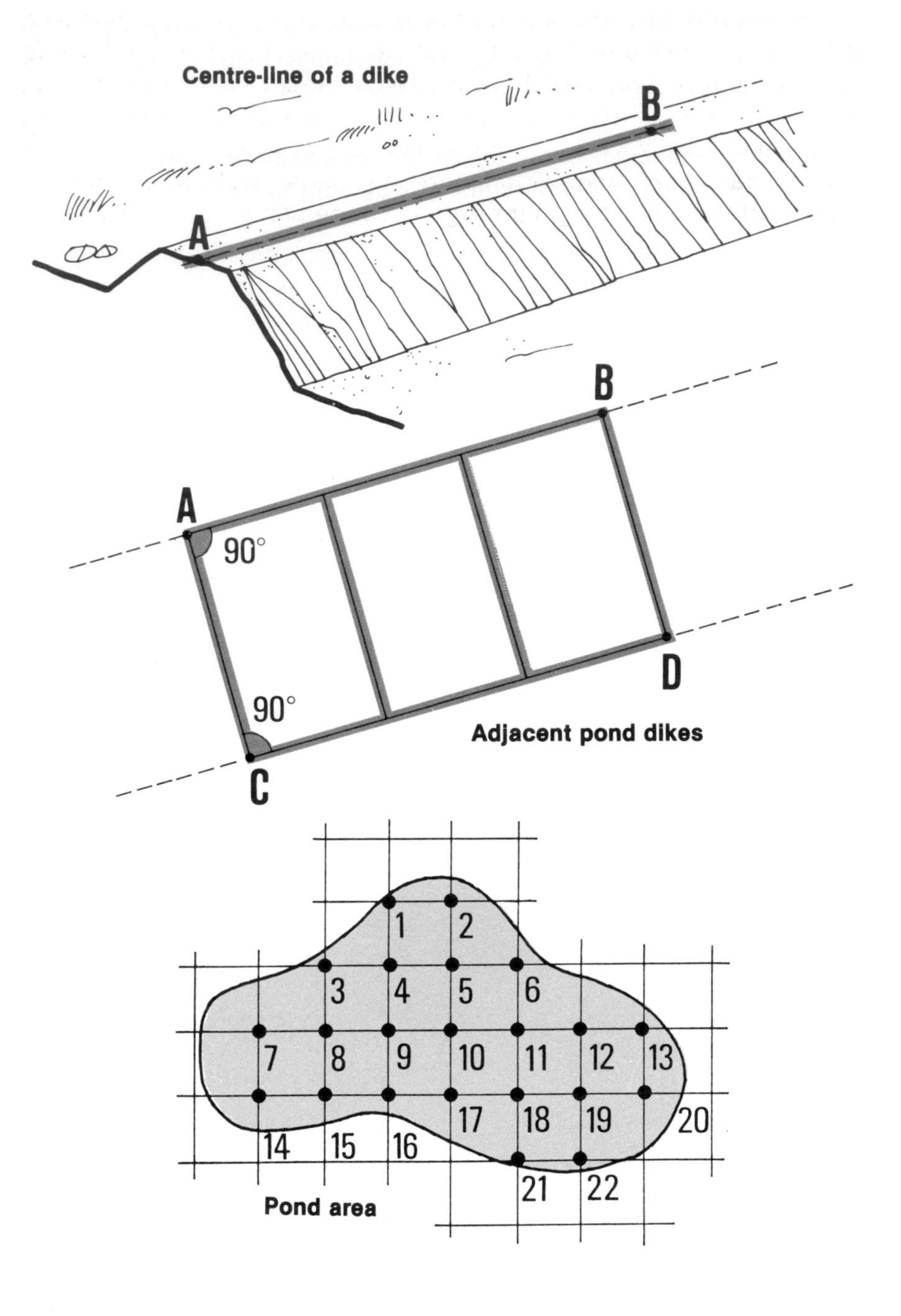

9. In **open country**, you will have no problems in plan surveying with the methods explained in the next sections. Any of the following methods should work well. In **country with thick forests**, however, you will not be able to use methods for which you need to see several points at the same time. In such areas, you will also need to rely on existing paths and roads much more than usual, and you might even need to clear lines of sight through the vegetation.

Clearing land for a survey

What are the main methods used in plan surveying?

10. There are **four main methods** used in plan surveying. You can fix the position of a point on the horizontal plane:

- from a single known point, by **traversing**, a method in which you measure horizontal distances and azimuths along a zigzag line (see Section 71);
- from a single known point, by **radiation**, a method in which you measure horizontal distances and azimuths, or horizontal angles (see Section 72);
- from a known line, by **offset**, a method in which you measure horizontal distances and set out perpendiculars (see Section 73);
- from two known points, by **triangulation** and/or **intersection**, methods in which you measure horizontal distances and azimuths, or horizontal angles (see Section 74).

Each of these methods will be explained in the next sections. When you are choosing a method, you will also need to consider which methods are suited to the measuring devices you have available. **Table 9** will help you select the most suitable plan surveying method, considering your equipment and abilities, the kind of information you need from your survey and the type of terrain you are surveying.

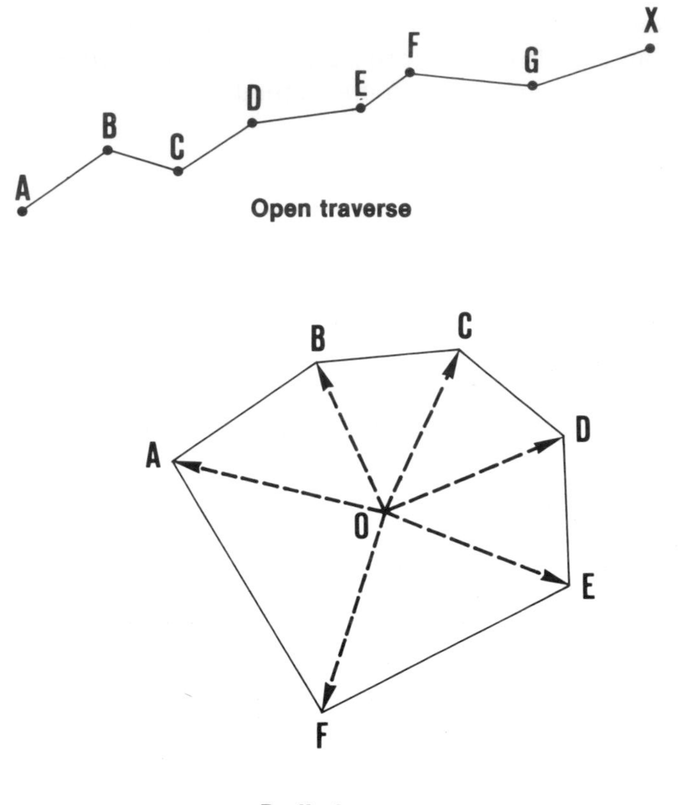

Radiation survey

TABLE 9

Plan surveying methods

Section	*Method*	*Basic elements*	*Suitability*	*Remarks*
71	Traversing, open, closed	Traverse sections and stations	Flat or wooded terrain Longitudinal or cross-section profiles Compass traverse, rapid reconnaissance and details	Traverse sections may be equal lengths, longer than 25 m and are best at 40 to 100 m Careful checks for errors needed
72	Radiating, central and lateral stations	Observation station	Small land areas For location of points only	All points should be visible and at angles greater than 15°
73	Offset	Chaining line	Details surveys next to a chaining line	Chaining line should not be more than 35 m away
74	Triangulation	Base line	Very large land areas Hilly or open terrains Inaccessible locations	Often combined with traversing and needing elaborate preliminary reconnaissance Best with angles of about 60°
74	Plane-tabling, traversing, radiating, triangulation		Reconnaissance and details surveys Open terrain and good weather Irregular lines and areas	Mapping is done in the field Rapid method after practice

71 How to survey by traversing

What is a traverse?

1. A traverse line or **traverse** is a series of straight lines connecting **traverse stations**, which are established points along the route of a survey. A traverse follows a **zigzag** course, which means it changes direction at each traverse station.

2. **Traversing** is a very common surveying method in which traverses are run for plan surveying. It is particularly suitable to use in flat or wooded terrain.

3. There are two kinds of traverses:

- if the traverse forms a closed figure, such as the boundary of a fish-farm site, it is called a **closed traverse**;
- if the traverse forms a line with a beginning and an end, such as the centre-line of a water-supply canal, it is called an **open traverse**.

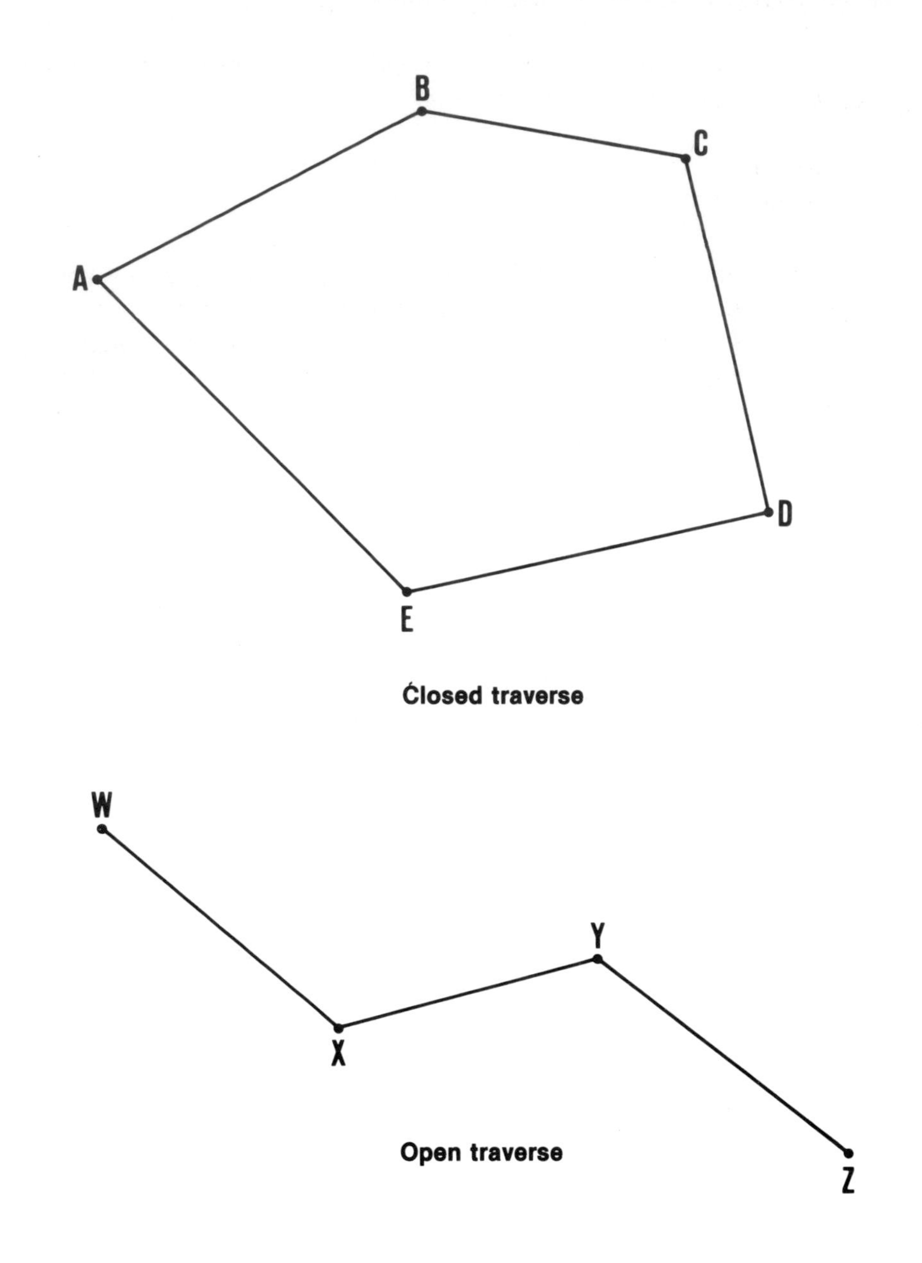

Closed traverse

Open traverse

Which method should be used for traversing?

4. When you survey by traversing, you need to make **measurements** to find information on:

- the distance between traverse stations;
- the direction of each traverse section.

5. If you have a theodolite (also called a transit), you can make a **transit traverse**. You will measure horizontal distances using the stadia method (see Section 28 in Book 1*), and you will measure horizontal angles using the method described in Section 35 for use with the theodolite. Similarly, but with much less accuracy, you could use a clisimeter (see Section 27) and a graphometer (see Section 31).

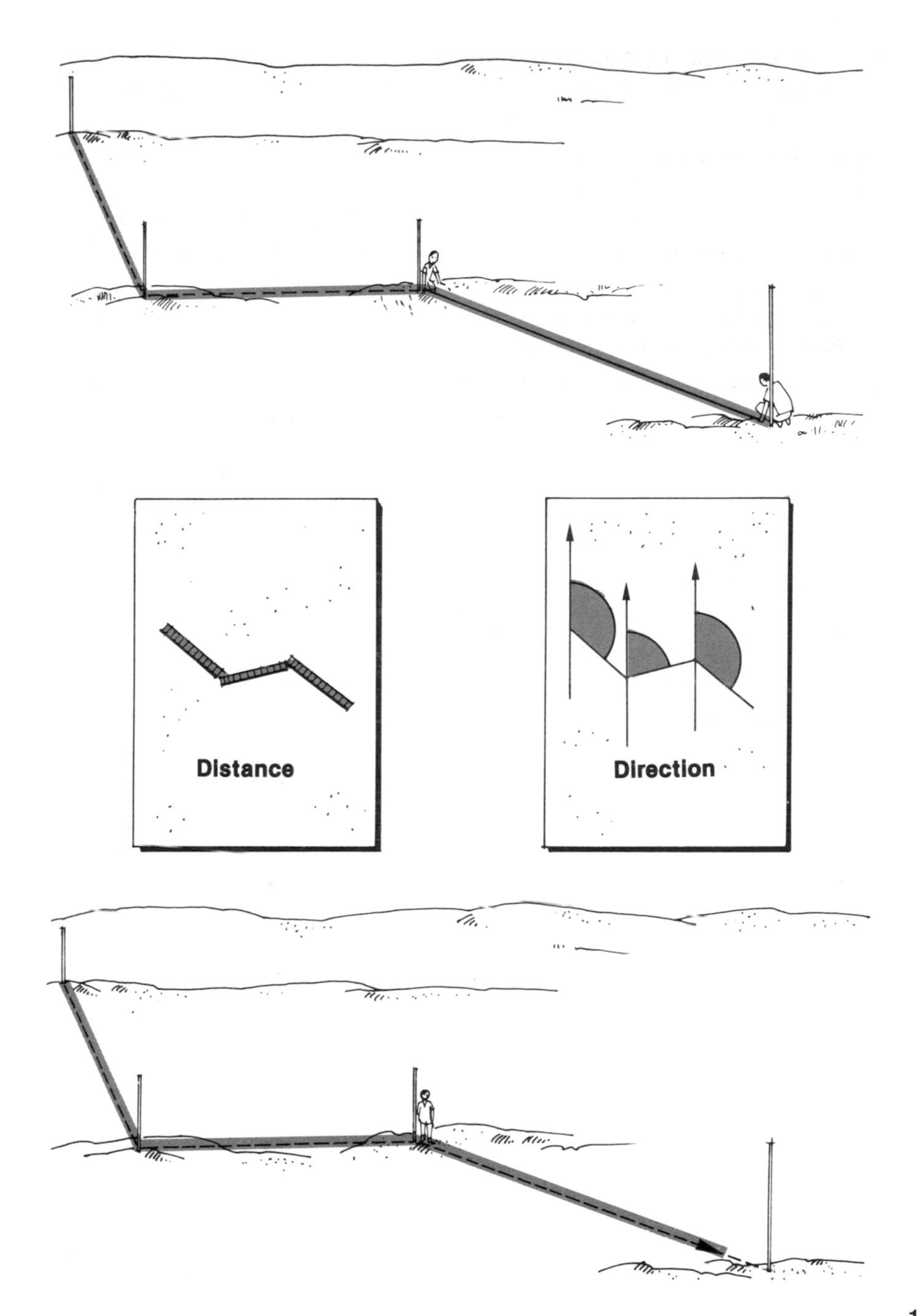

* Chapters 1 to 6, covering Sections 10 to 69, are in Book 1 of this manual.

6. If you have a magnetic compass, you can make **a compass traverse**. You will measure the horizontal distances by pacing (see Section 22) or by chaining (see Section 26), and you will measure azimuths with the magnetic compass (see Section 32). Compass traverses are very useful for getting a general picture of the terrain. They also help to fill in details on surveys that have already been done.

7. If you have a plane-table (see Section 75), you can make **a plane-table traverse**. You will measure distances either by pacing or by chaining, and you will measure horizontal angles using a graphic method (see Section 33).

8. When you need to make a quick **reconnaissance survey**, you can traverse with a simple compass (see Section 33, steps 1-9) and by pacing (see Section 22).

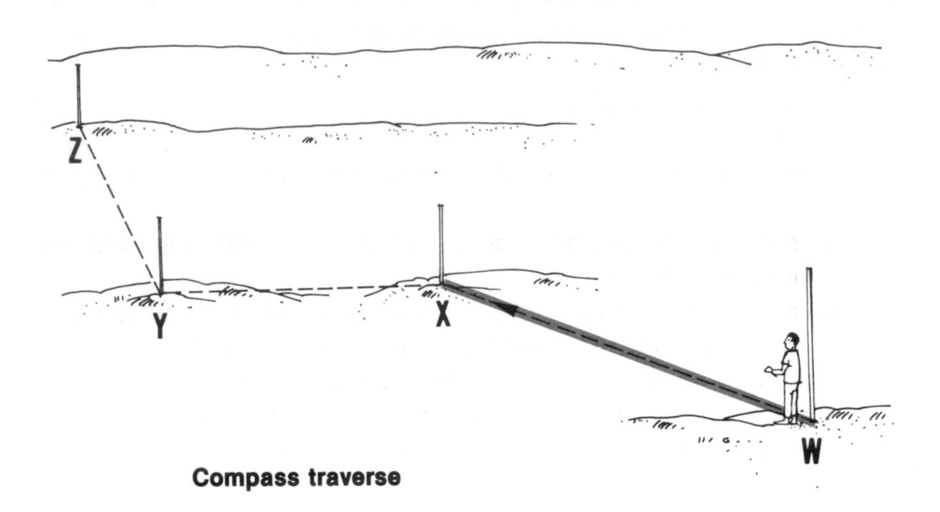

Compass traverse

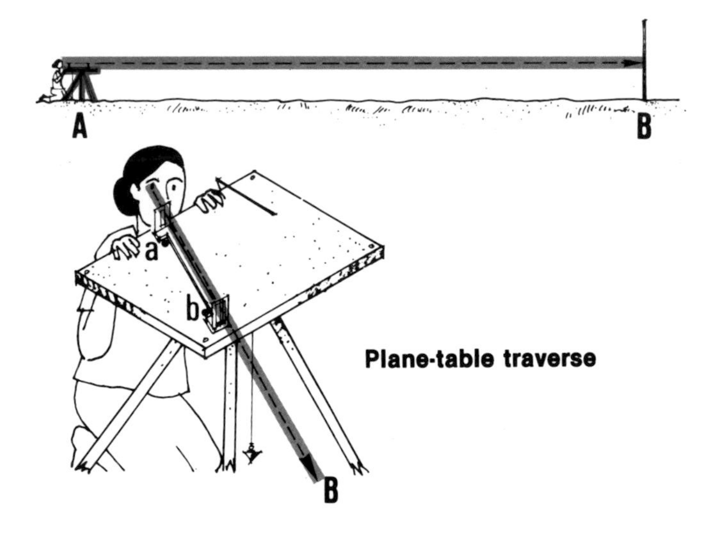

Plane-table traverse

9. In this section you will learn about compass traversing. You may use similar procedures for transit traverses. Further details on plane-table traverses will be given in Section 92.

Choosing the route of a traverse

10. When selecting the route a traverse will follow, you should try to:

- make each straight section of the traverse as **long** as possible (40-100 m);
- make the traverse sections as **equal in length** as possible;
- avoid **very short** traverse sections – under 25 m long;
- choose lines which can be **measured easily**;
- choose lines along routes which **avoid obstacles** such as heavy vegetation, rocks, standing crops and property.

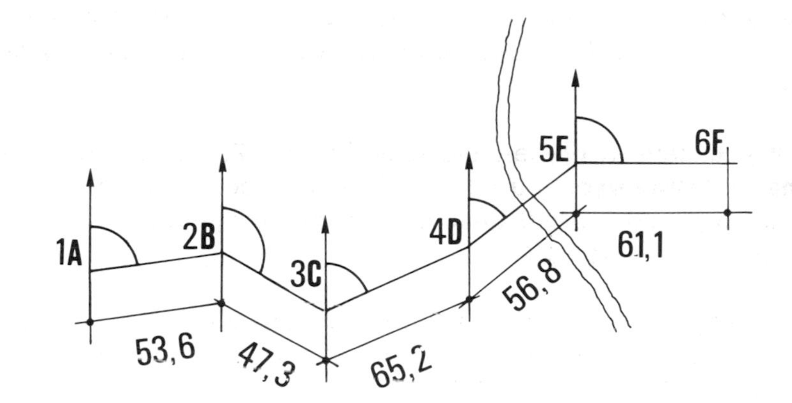

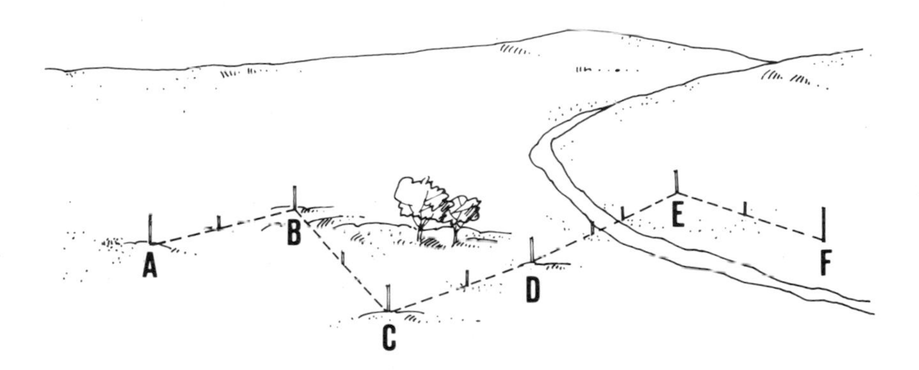

Surveying an open traverse with a magnetic compass

11. You need to survey traverse AF for a future water supply canal. First, walk along the traverse. Mark its course by placing high stakes about every 50 m. If necessary, place additional stakes at important traverse stations, such as where the traverse changes direction, where hills or other changes in elevation reduce visibility between traverse stations, or where there are particular landscape features such as a road, a river, or rocks.

12. If necessary, clear any tall vegetation from the path of the traverse, so that you will be able to see each marked point from the one before it.

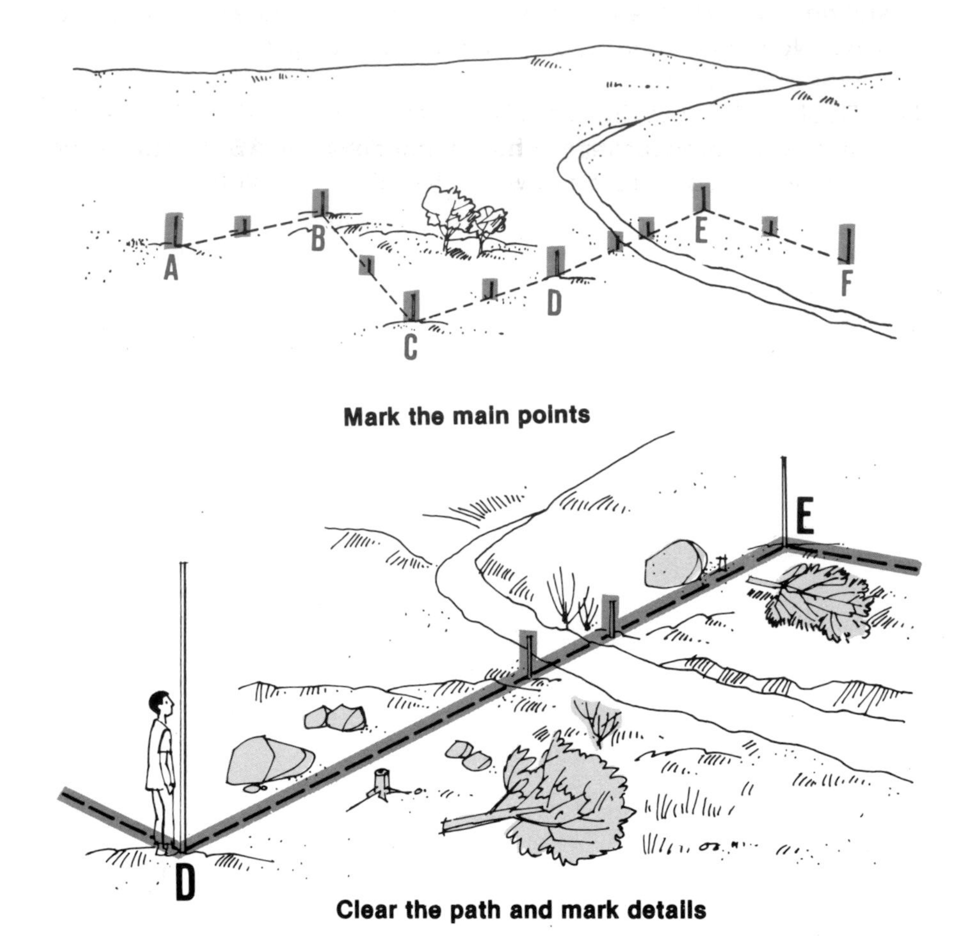

Mark the main points

Clear the path and mark details

13. Start traversing at the first point A. Remove the ranging pole and stand at point A. With the magnetic compass, **measure the azimuth*** of the line joining point A to point B, the next visible point. Point A becomes station 1. The direction you measure from there to point B, or station 2, is called a **foresight* (FS)** because you are measuring **forward**. Note down this value in a table (see step 17).

14. Replace the ranging pole at station 1 (point A) and move to station 2, while **measuring the horizontal distance AB** by pacing or chaining. Note this distance down in the table (see step 17).

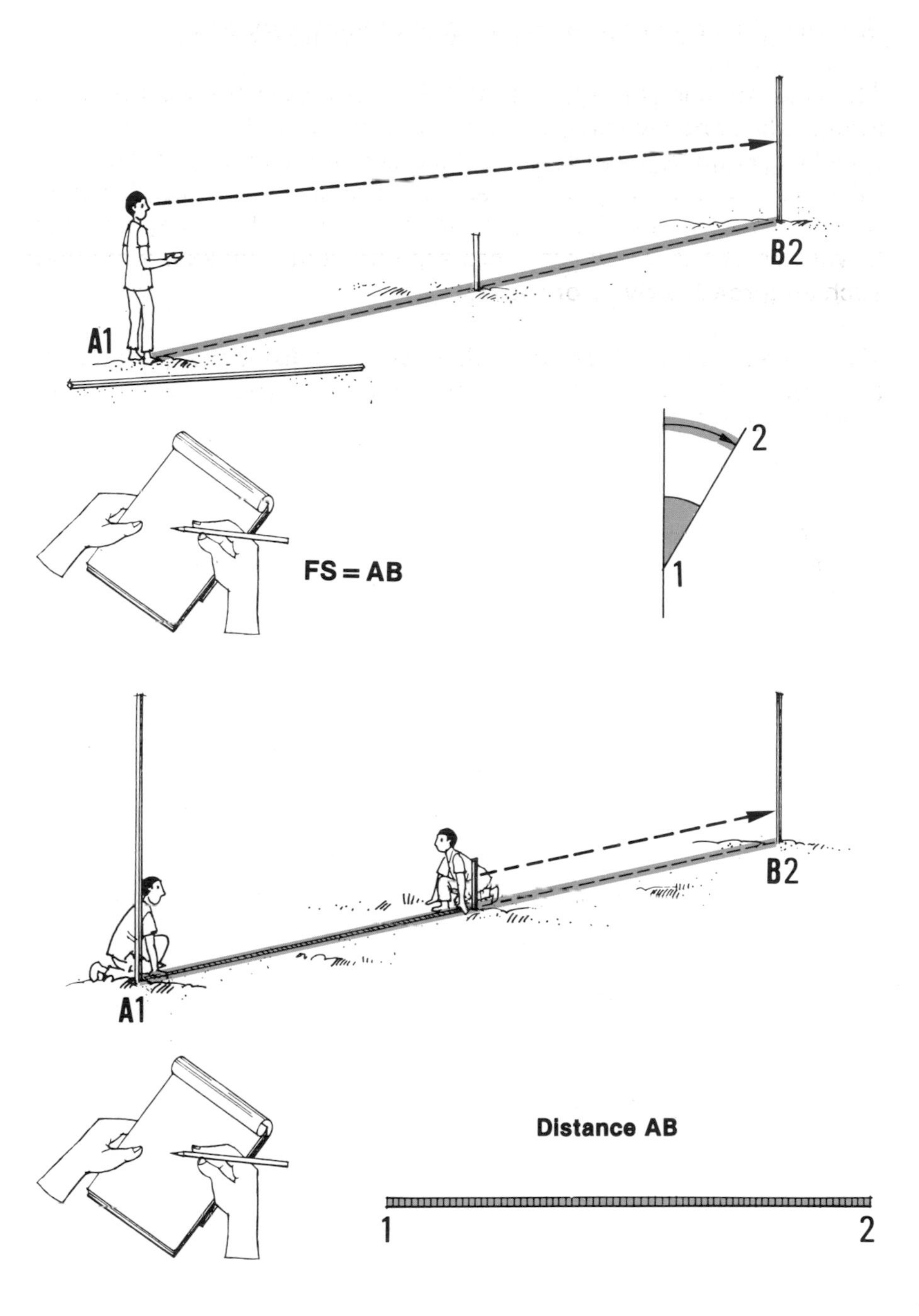

15. At station 2 (point B), remove the ranging pole and stand over the point holding the compass. Look back at station 1 and measure the azimuth of line BA, which is called a **backsight (BS)**. Then look forward at the next point C, or station 3, and measure the azimuth of line BC, **a foresight (FS)**. Measure distance BC while moving forward along the traverse. Note these values down in the table (see step 17).

Note: the difference between the foresight and backsight should be 180º. A difference of only 1 or 2 degrees between the FS and BS is acceptable and may be corrected later (see step 19). If the error is greater, you should make the measurement again before moving on to the next station.

16. Repeat this procedure, measuring horizontal distances from station to station and measuring **two azimuths** (a BS and a FS) **for each point**. However, from the last station at the end of an open traverse, you will only have a BS measurement, just as you had only an FS from station 1.

Note: if the land slopes and you need to use a more accurate method, you can use a special method to measure or calculate **horizontal distances** (see Sections 26 and 40).

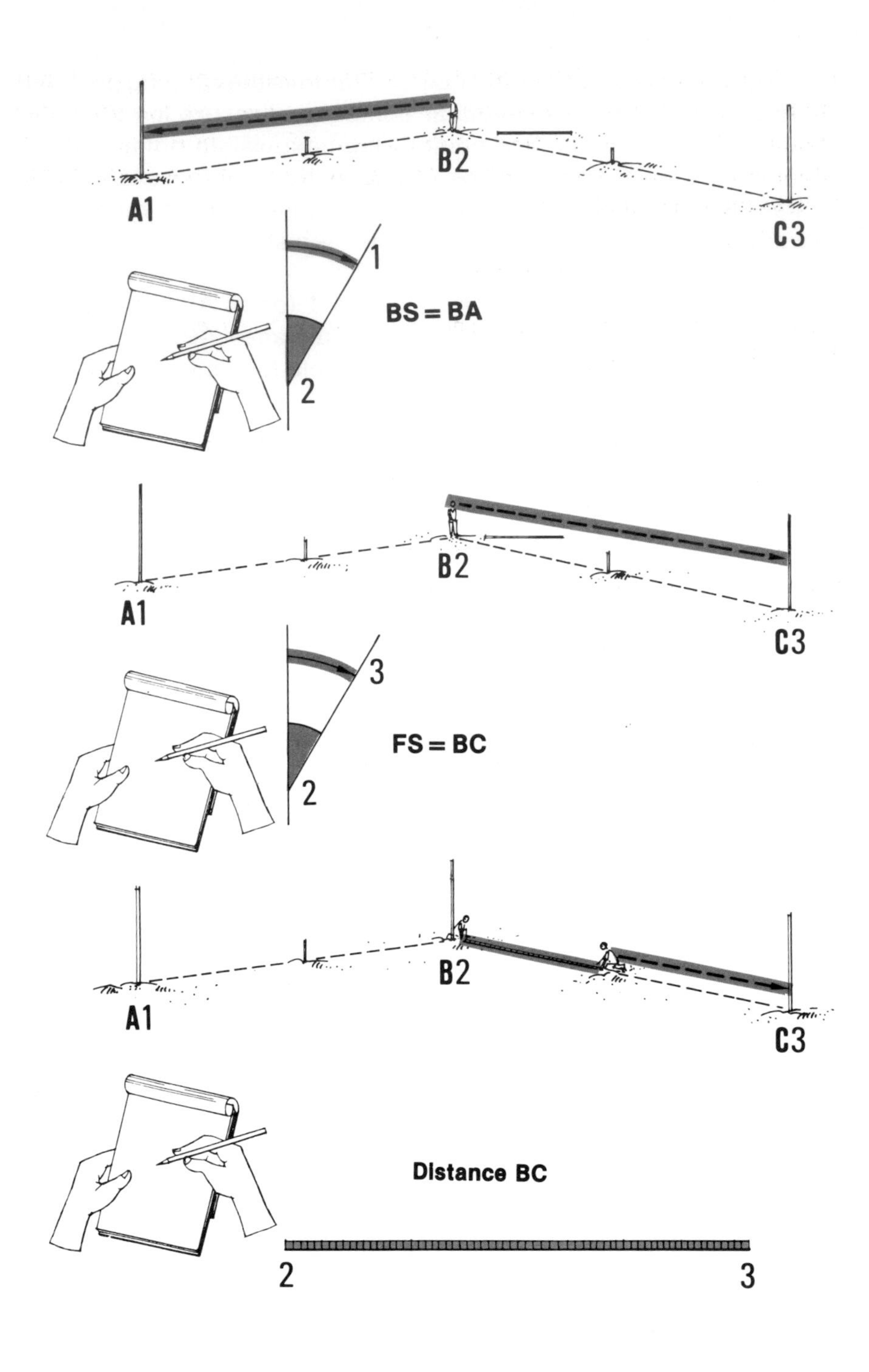

17. You should carefully note down all the measurements you have made in **a field book**. You can use **a table** like the one shown in the example or you can make **a rough sketch** of the open traverse on square-ruled millimetric paper, noting down your measurements next to the correct stations in it.

Example

Measurements observed for the beginning of compass traverse AX made of 12 stations:

Stations		*Distance (m)*		*Azimuths (degrees)*		*Calculated difference*
From	*To*	*Individual*	*Cumulative*	*FS*	*BS*	*FS/BS (degrees)*
1	2	53.6	53.6	82	261	179
2	3	47.3	100.9	120	301	181
3	4	65.2	166.1	66	248	182
4	5	56.8	222.9	51	229	178
5	6	61.1	284.0	91	270	179
...	...	...	...	...	...	...

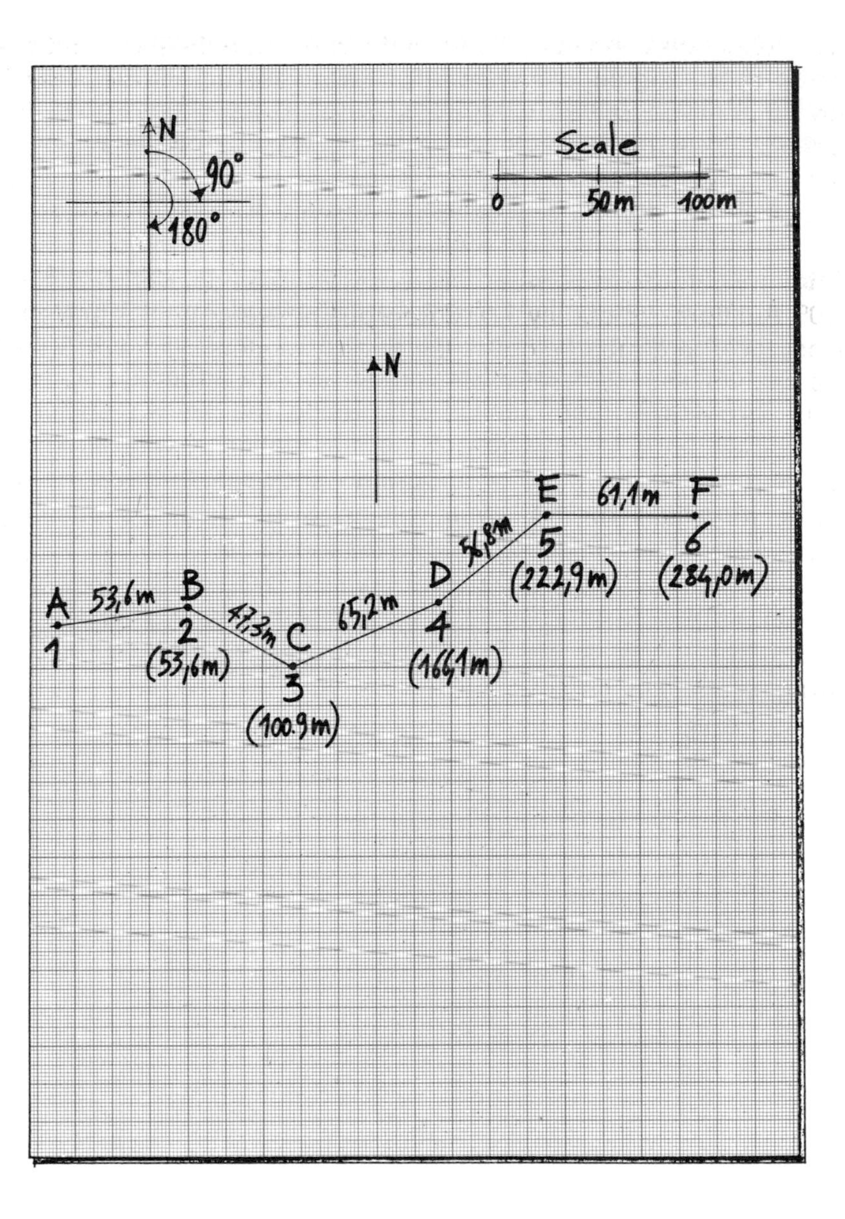

18. You must always **check on such a compass traverse**, particularly if you do not know the exact position of its starting and ending stations beforehand from studying previous surveys or existing maps. To check on your compass traverse, do the following:

- if the starting and ending traverse stations A and X are unknown, check on your first traverse by making a second compass traverse in the opposite direction, from X to A;
- if these two stations A and X are known, draw the traverse on paper as you have measured it. To do this, use a protractor for the angles (see Section 33) and an adequate scale for distances (see Section 91). Using the known station **A**, compare the position of the last station X with its known position X′. If this comparison shows a large error (**the closing error XX′**), you will need to adjust the observed traverse AX. To do this, see the next step.

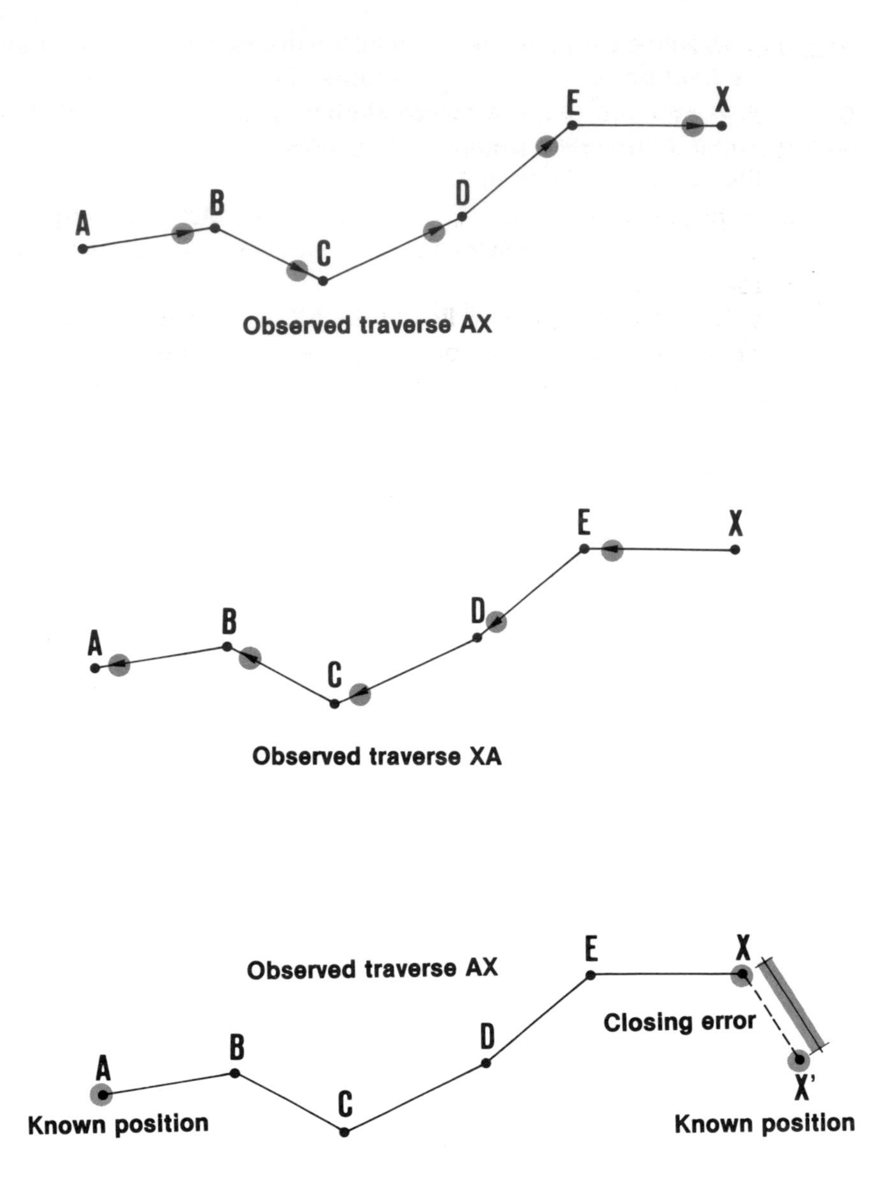

19. To adjust the observed traverse AX for the closing error XX′, it is easiest to use **the graphic method**, as follows:

- on paper, draw a straight horizontal line AX equalling the total measured length of the observed traverse, drawn at an adequate scale;
- at X, draw XX′ **perpendicular** to AX and in proportion, in length to the closing error, using the same scale as above;
- join A to X′ with a straight line;

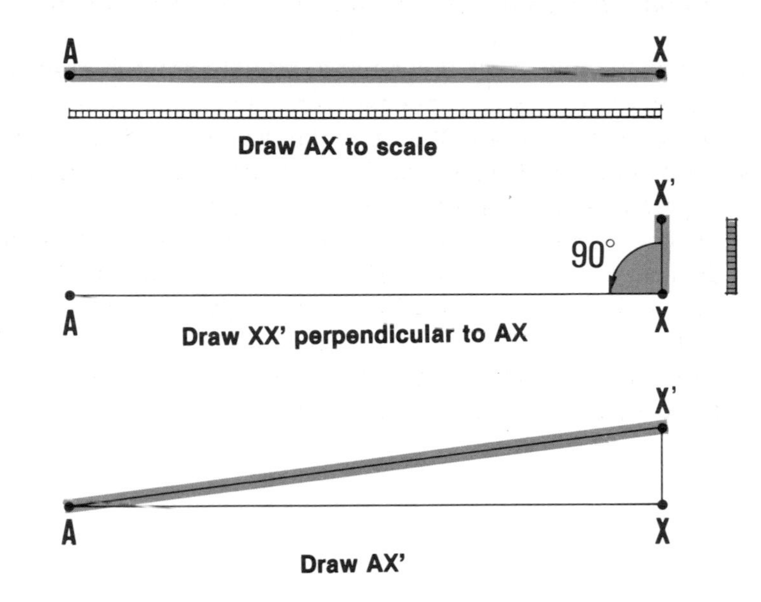

- on AX, find lengths AB, BC, CD, DE, and EX in proportion to the field measurements, using the same scale as above;
- at points B, C, D, and E, draw lines BB′, CC′, DD′ and EE′ **perpendicular** to AX;
- measure the lengths of lines BB′, CC′, DD′ and EE′, which show by how much you need to adjust each traverse station;

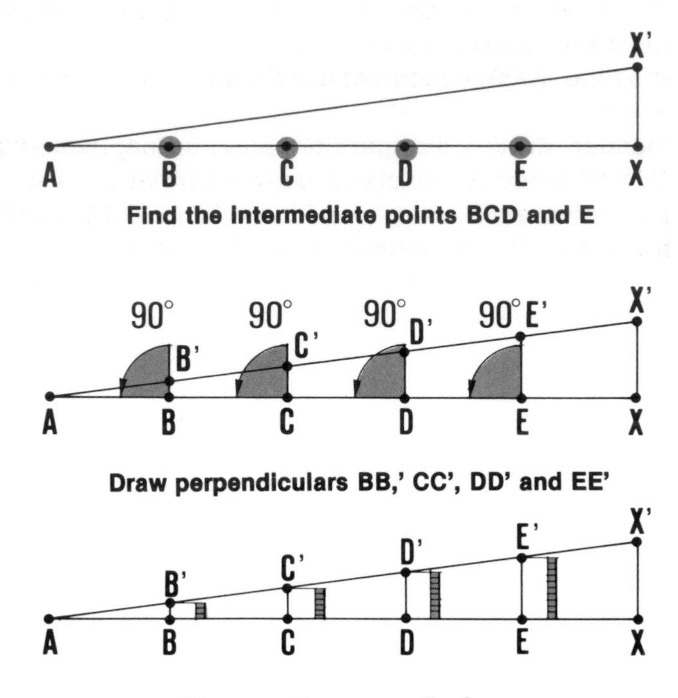

- adjust your **drawing of the traverse** by:
  - joining the observed position X of the last traverse station to its known position X′;
  - drawing short **lines parallel to XX′** through stations B, C, D and E;
  - marking on these lines the calculated adjustments BB′, CC′, DD′ and EE′, using the same scale as above;
  - joining points A, B′, C′, D′, E′ and X′ to find **the adjusted traverse**.

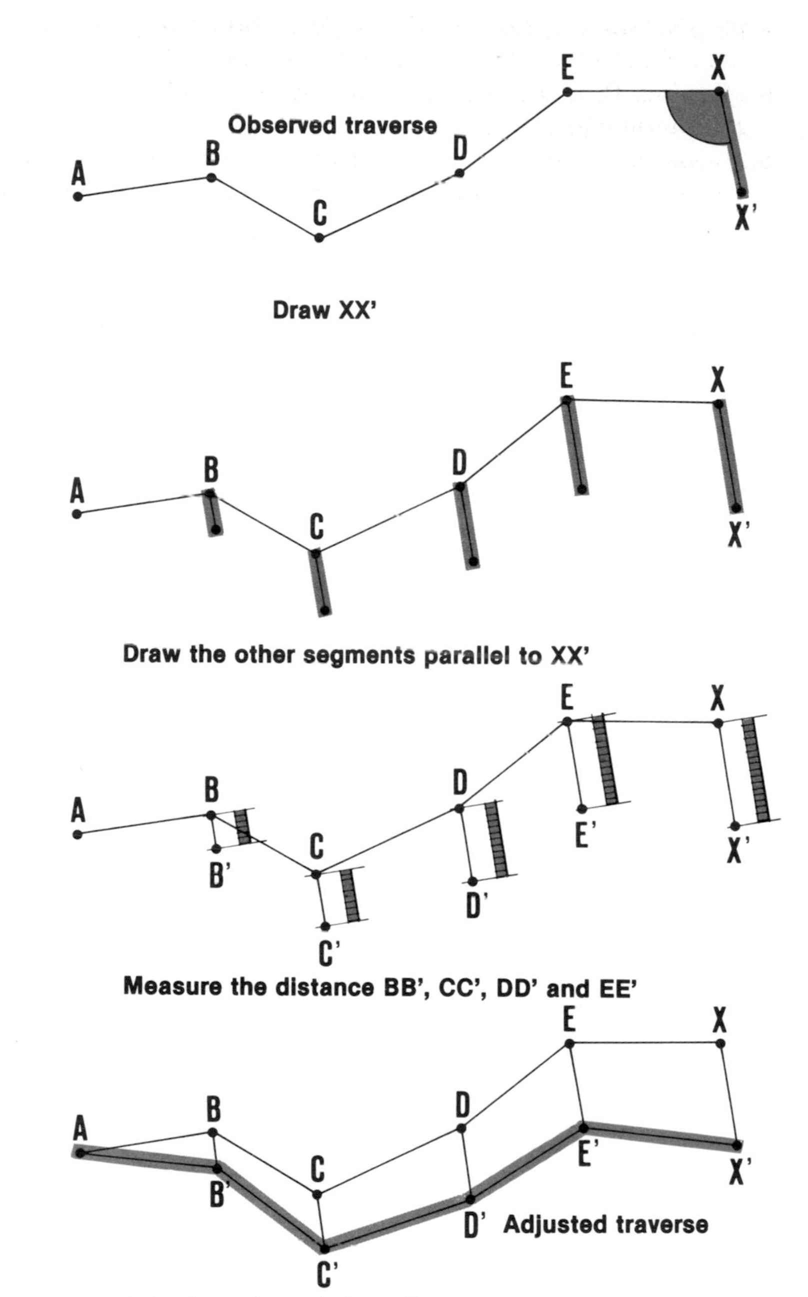

Join the points of the adjusted traverse

Surveying a closed traverse with a magnetic compass

20. You can lay out a **closed traverse** ABCDEA in exactly the same way as an open traverse, except that you will connect the last point to the initial point A.

21. To survey an irregular enclosed area of land ABCDEA (such as a site for a fish-farm) by compass traversing, proceed as follows:

- walk over the area and locate **traverse stations** A, B, C, D and E;
- mark them with ranging poles or stakes;
- if necessary, clear away any vegetation so that you can see stations A and B, B and C, C and D, etc. from each other;

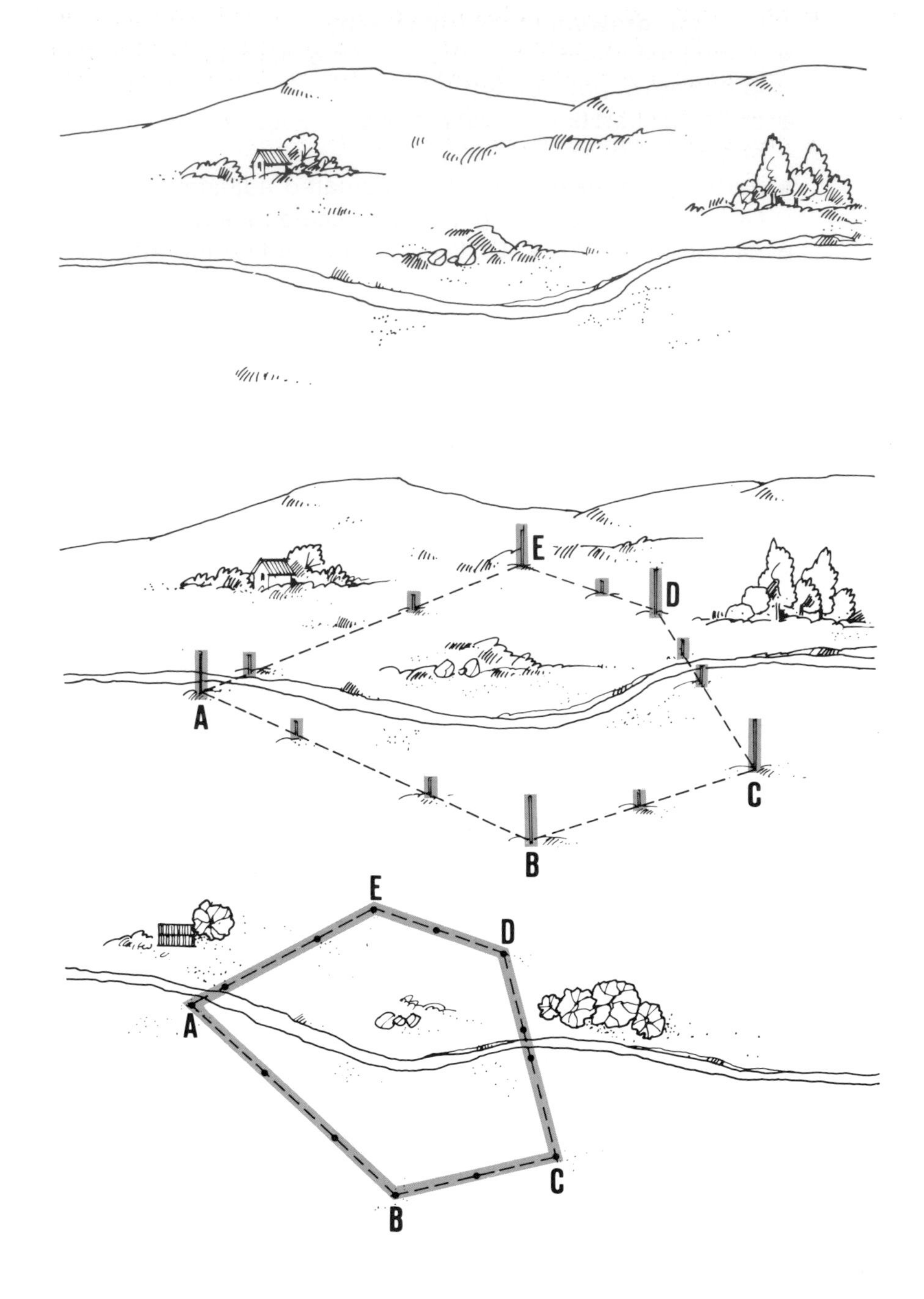

- remove the ranging pole from point A (station 1) and stand at this station. Find azimuth AB – a **foresight** – from the centre of this station with the compass. Replace the ranging pole exactly at station 1;
- measure distance AB with a measuring line;

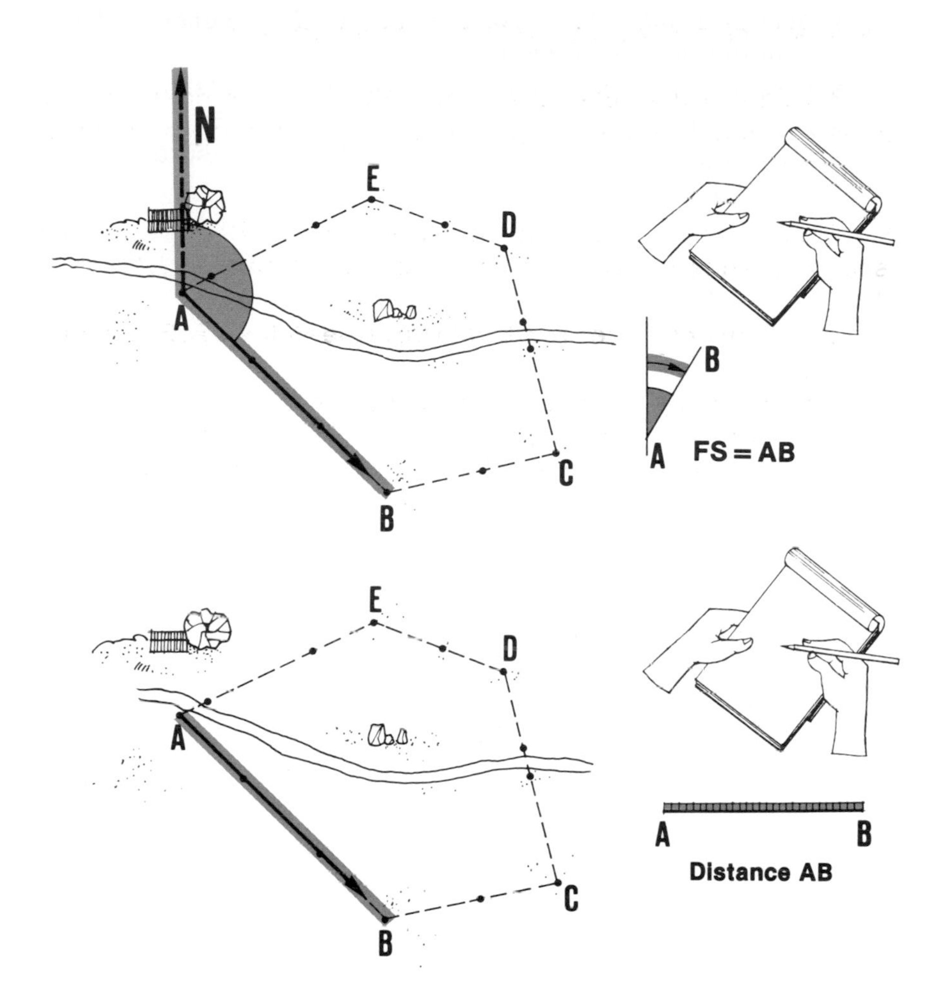

- at point B (station 2), measure azimuth BA – **a backsight** and azimuth BC – **a foresight**;
- measure distance BC as you move to point C (station 3);
- proceed in the same way at stations 3, 4 and 5;
- when you reach point A again (station 1), measure azimuth AE – a **backsight.**

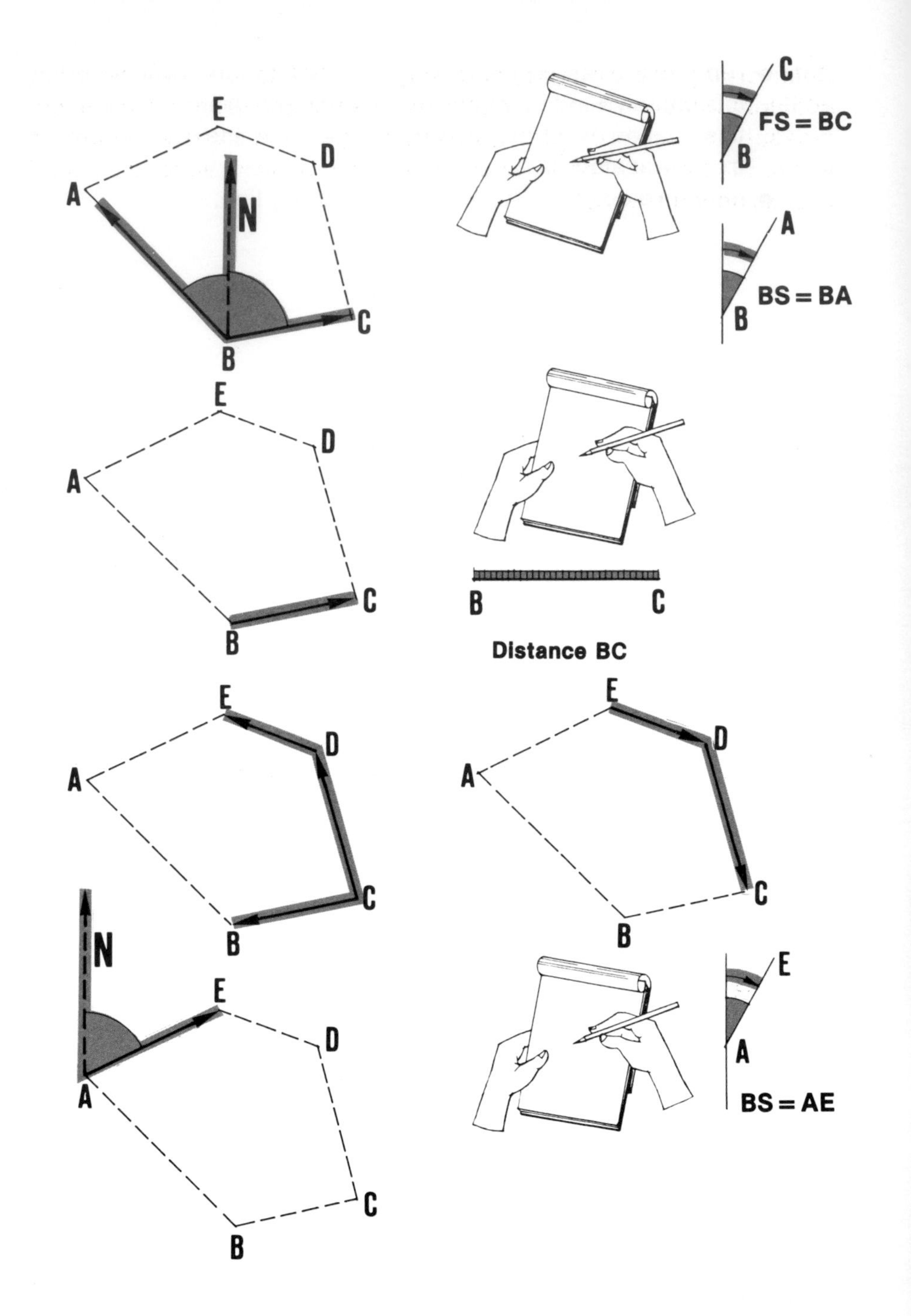

Note: during the traverse, you may be able to see one or more **additional stations** from the station where you are standing. If you do, measure the azimuths of the lines running toward them. An example is line BD from station B. These additional observations are useful checks on your work.

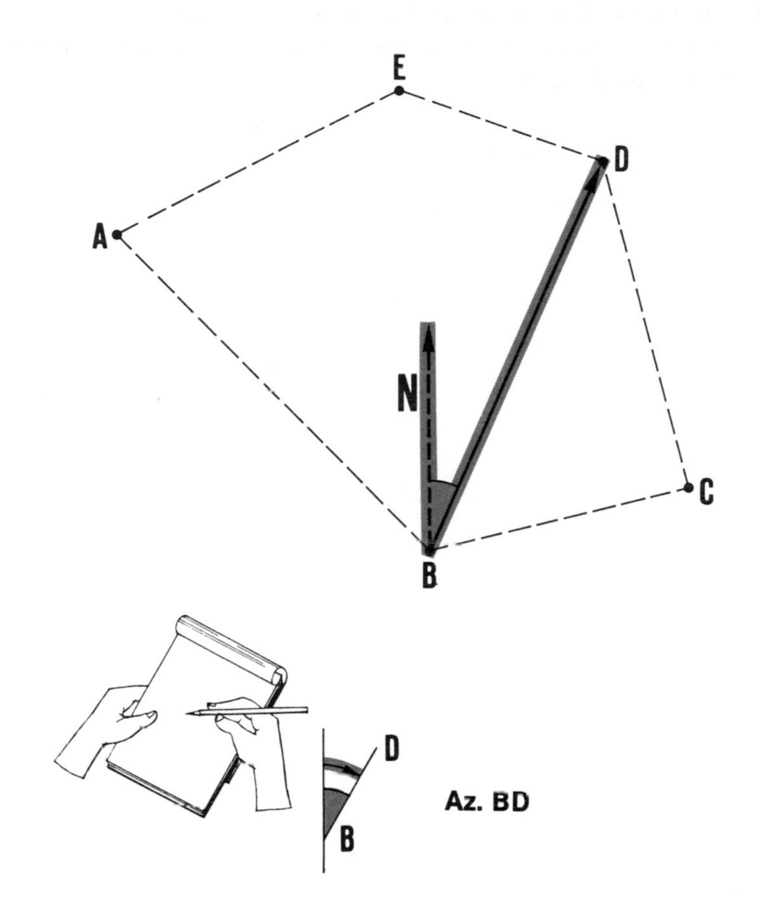

22. In a field book, carefully **note down** all your measurements. You can use a **table** similar to the one suggested for the open traverse (see step 17). You should also make a **sketch** of the traverse, on a separate square-ruled page, and write in the measurements. At the same time, check to see that the foresights and backsights differ by 180°.

Example

You have surveyed site ABCDEA with a closed traverse and your field notes are as follows:

Stations		*Distance (m)*	*Azimuths (degrees)*		*Calculated difference*
From	*To*		*FS*	*BS*	*FS/BS (degrees)*
1	2	90.8	136	315	179
2	3	53.5	78	259	179
3	4	68.7	347	168	179
4	5	44.6	292	110	182
5	1	63.7	241	63	178

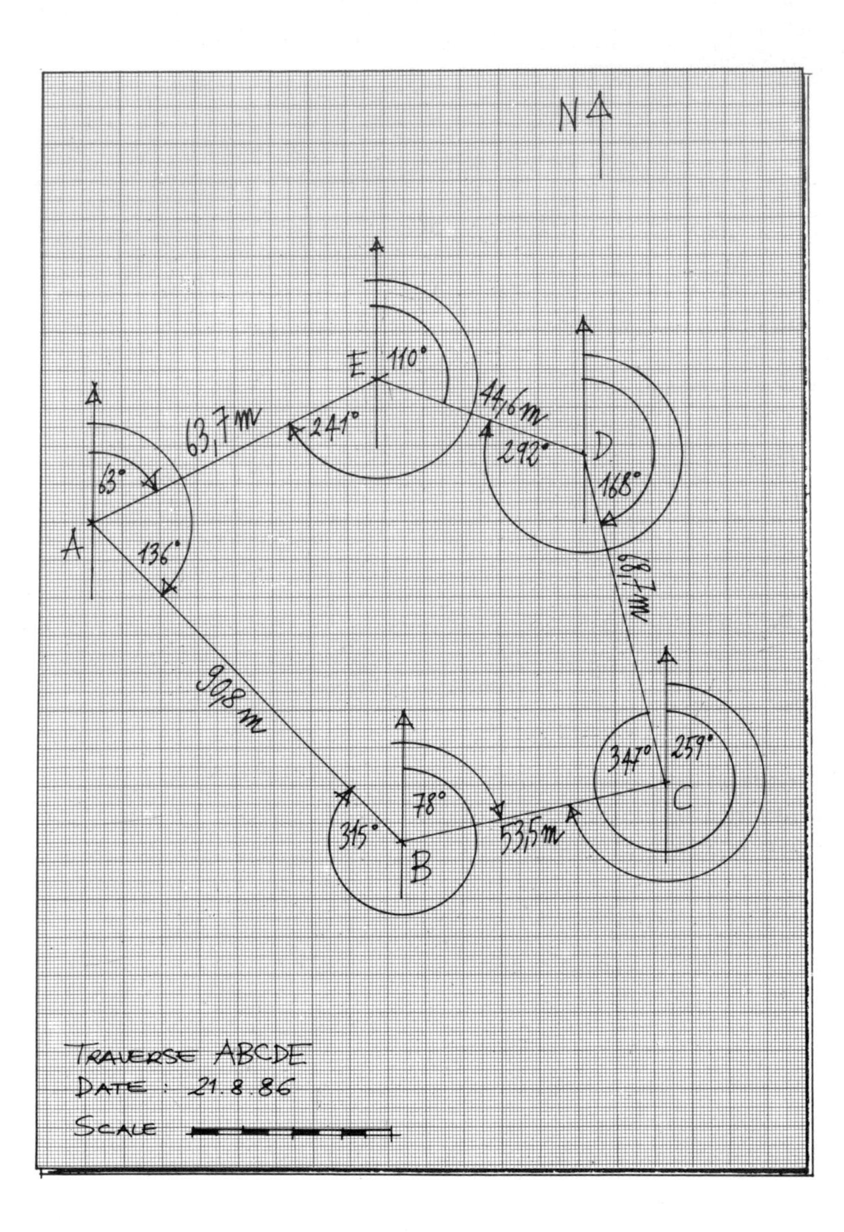

23. You have learned that in any closed polygon* of N sides, the sum of all the interior angles should be equal to (N − 2) × 180° (see Section 30). This rule will help you to check your azimuth measurements after you calculate the interior angle for each station (see Section 32, steps 10 and 11).

Example

Using the observations given in the previous example, calculate the sum of the interior angles of polygon ABCDEA as follows:

Station	Azimuth differences (degrees)	Interior angle (degrees)
1	AB − AE = 136 − 63	73
2	(BA − BC = 315 − 78 = 237)	123[1]
3	CD − CB = 347 − 259	88
4	DE − DC = 292 − 168	124
5	EA − ED = 241 − 110	131
Sum of interior angles		539[1]

[1] Since the magnetic north falls inside the angle, you must calculate it as 360° − (the azimuth difference) or 360° − 237° = 123°.

According to the general rule, the sum of the five interior angles should be equal to (5 − 2) × 180° = 3 × 180° = 540°, which closely agrees with the above result.

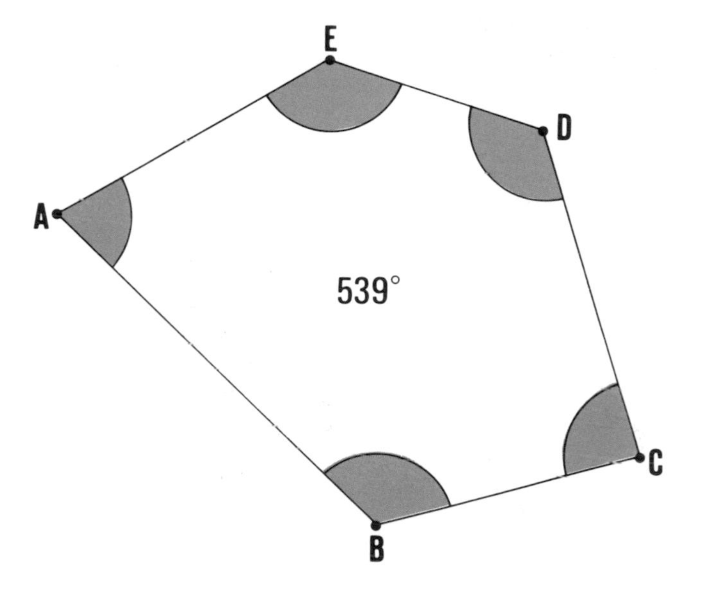

Adjusting a closed traverse

24. Starting from station 1 (A), draw the observations of your compass traverse on square-ruled paper. Use a **protractor** to measure the azimuths (see Section 33), and an adequate **scale** for the measured distances (see Section 91). If there is **a closing error**, adjust your drawing by using the **graphic method** described for an open traverse (see step 19, above).

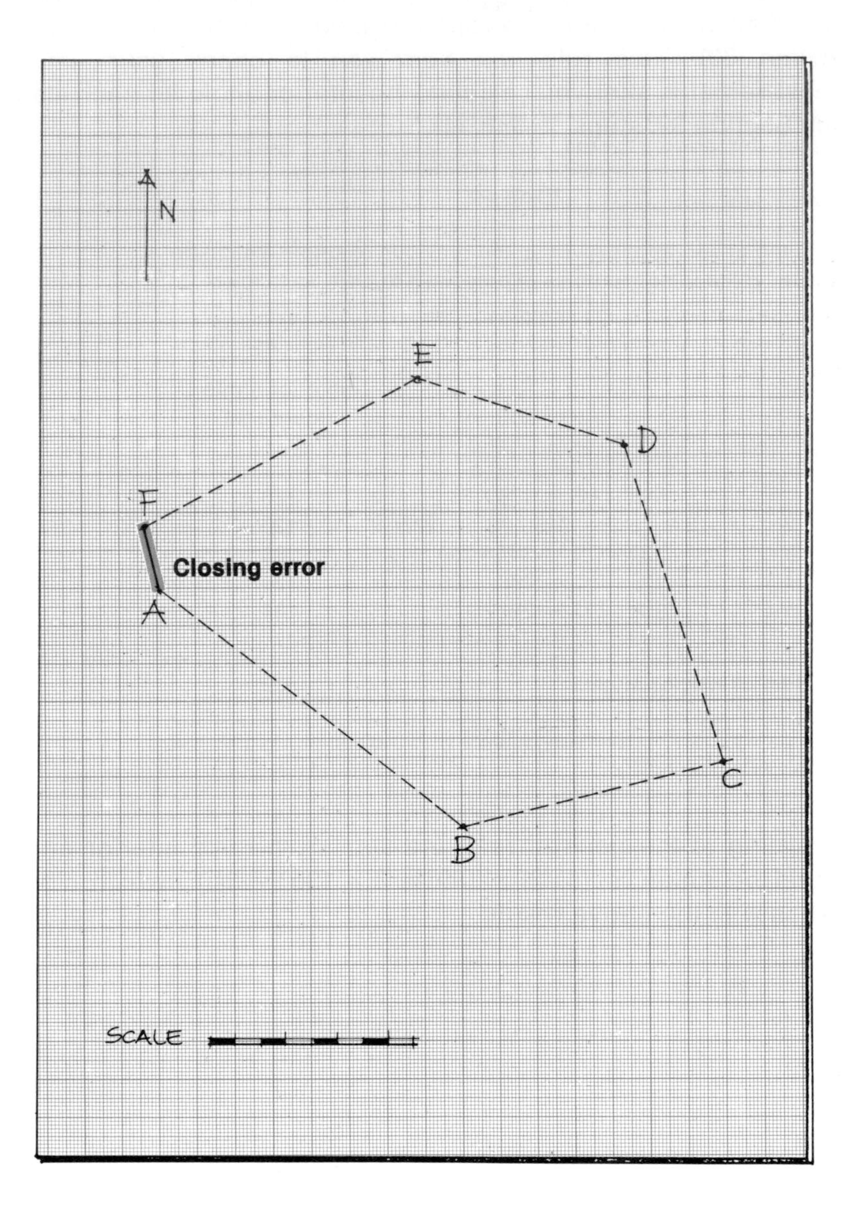

Example

For the above example, the closing error is FA. Adjust it as follows:

- using the correct scale, draw a horizontal line AF whose length equals **the total measured length** of the observed traverse;
- at F, draw FA′ perpendicular to AF, using the same scale as above. The length of FA′ should be in proportion to the closing error;
- join A to A′ with a straight line;
- on AF, draw lengths AB, BC, CD, DE and EF in proportion to the field measurements, using the same scale as above;
- at points B, C, D, and E draw lines BB′, CC′, DD′ and EE′, which show how much you must adjust each traverse station;

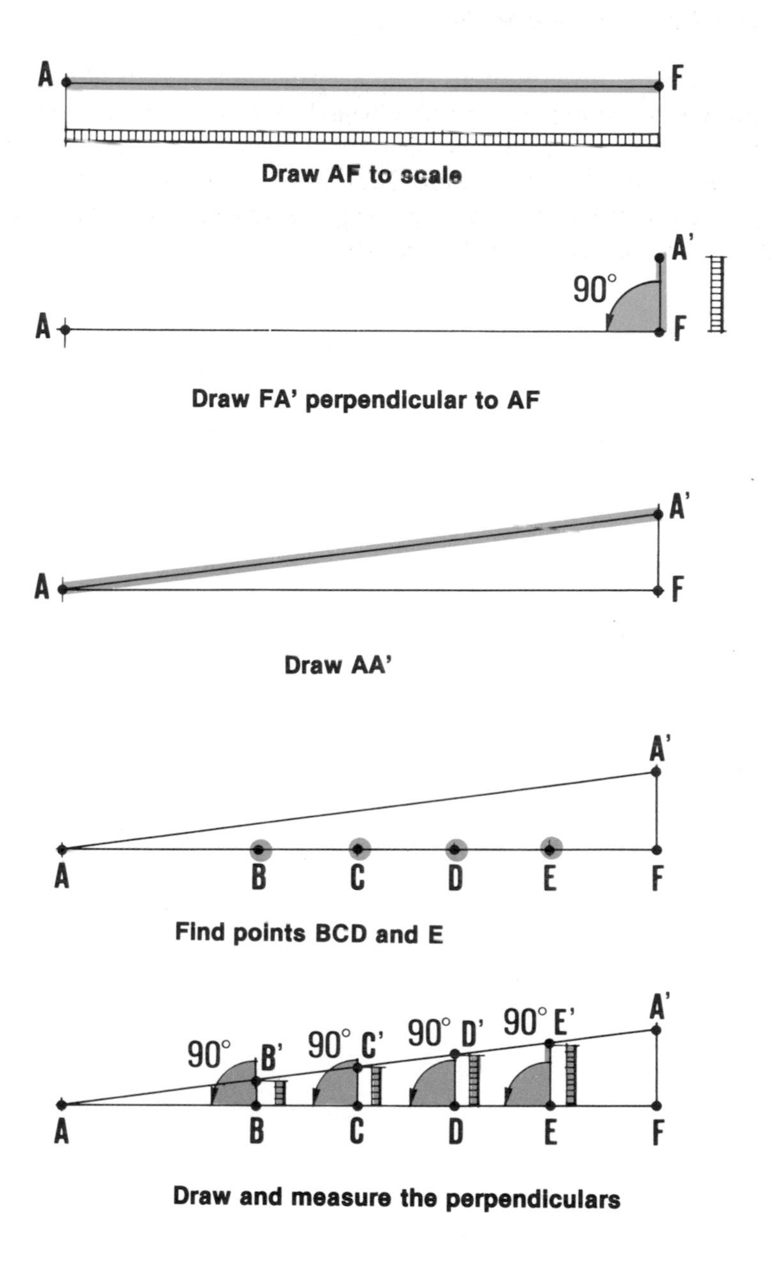

- adjust your drawing of the traverse by:

- joining the observed position F of the last station to its **known position A**;
- drawing short lines **parallel to FA** through the other stations B, C, D, and E;
- marking on these lines the **calculated adjustments** BB′, CC′, DD′ and EE′, using the same scale as above;
- joining points A, B′, C′, D′, E′ and A to determine the adjusted traverse.

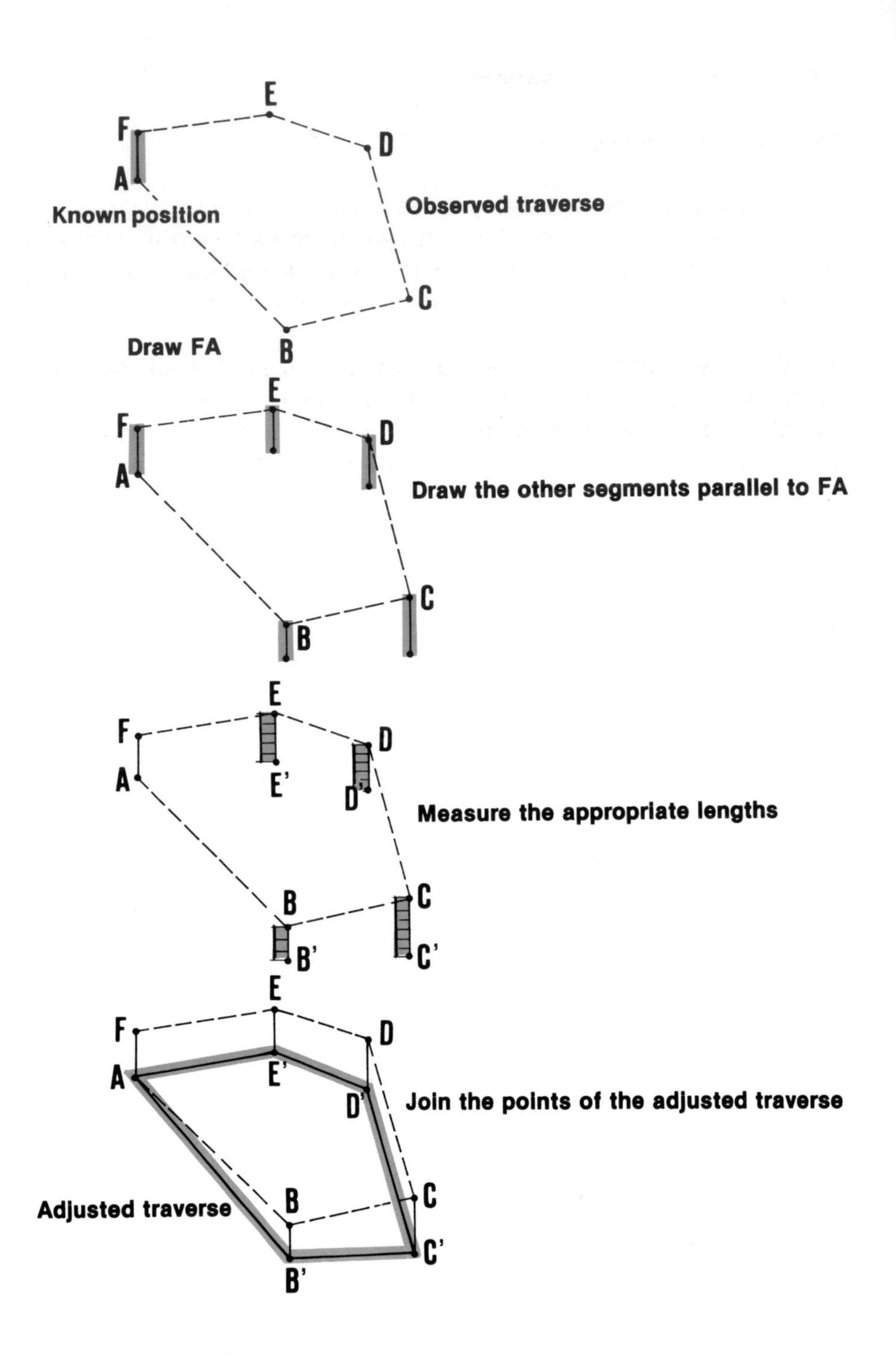

72 How to survey by radiating

What is a radiating survey?

1. When you plan a survey by radiation, you will choose one convenient observation station, from which you will be able to see all the points you need to locate. This method is excellent for surveying **small areas**, where you need **to locate only** points for mapping.

2. When you make a radiating survey of a **polygonal*** site, you connect the observation station to all **the summits of this area** by a radiating series of sighting lines. In this way, a number of **triangles** are formed. You will measure one horizontal angle and the length of two sides for each triangle.

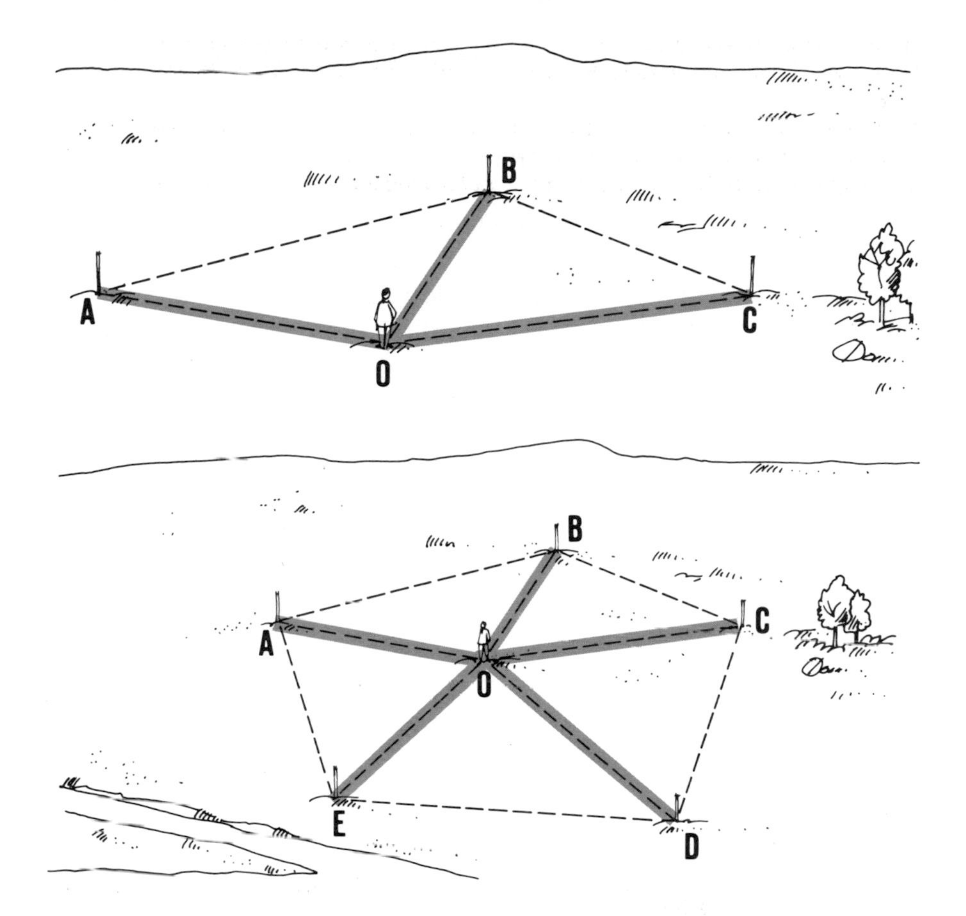

Choosing the observation station

3. You should be able to reach **the observation station** easily. This station should also be a located so that:

- you can see all the summits of the area you need to survey;
- you can measure the lines joining it to these summits;
- you can measure the angles formed by these lines.

4. When choosing the observation station, you should be particularly careful to avoid any points from which **very small radiating angles** (less than 15 degrees) might result.

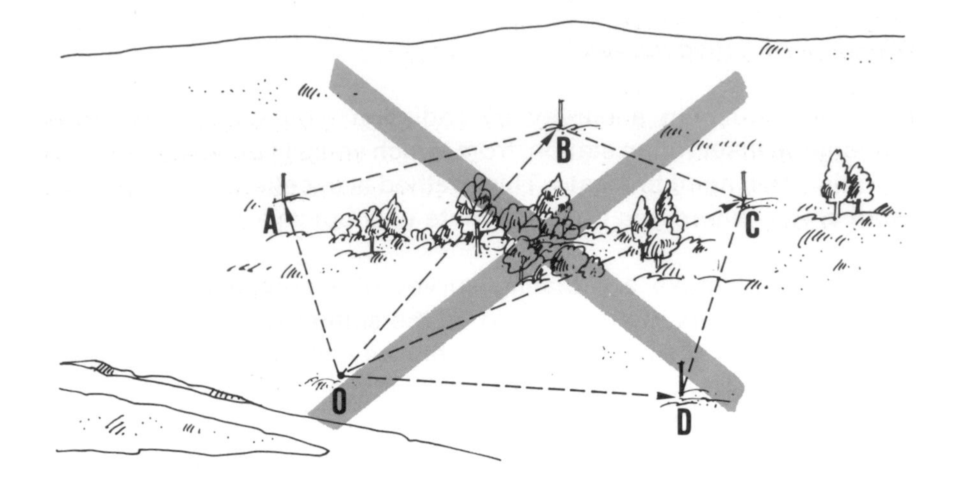

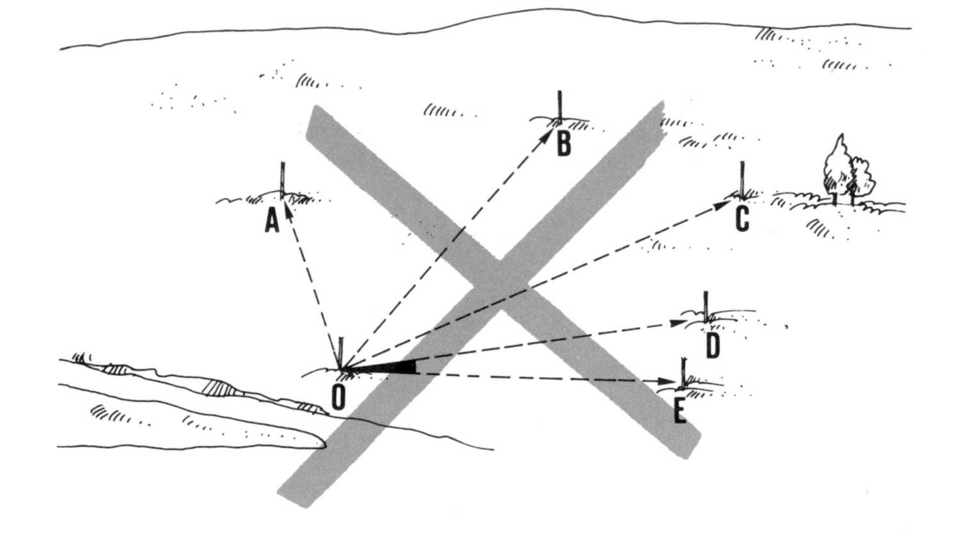

5. The observation station 0 can be **in a central position**, inside the polygon to be surveyed. In this case, you will measure as many triangles as there are sides of the polygon.

6. The observation station 0 can also be **in a lateral position** (off to the side). In this case, 0 will be one of the summits of the **polygon***. The number of triangles you need to measure will be the number of sides to the polygon, minus 2.

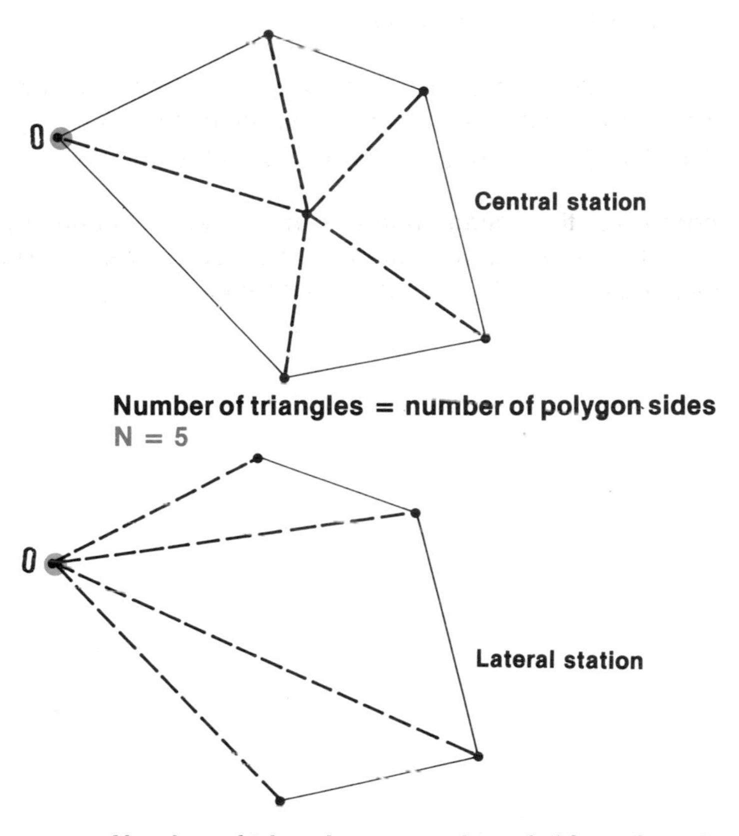

Choosing a method for radiating surveys

7. If you have a **transit** (a theodolite), you can measure horizontal angles more precisely than with the other instruments (see Section 35). A transit equipped with stadia hairs can also be used to measure distances rapidly (see Section 28).

8. If you have **a magnetic compass**, you can use it to measure the azimuths of the horizontal angles at the observation station (see Section 32). You will usually measure horizontal distances by chaining (see Section 26). To learn further details of this simple method, see steps 10-14, below.

9. If you have **a plane-table**, you can use it for mapping the area directly from the observation point (see Section 92). You will then usually measure the horizontal distances by chaining.

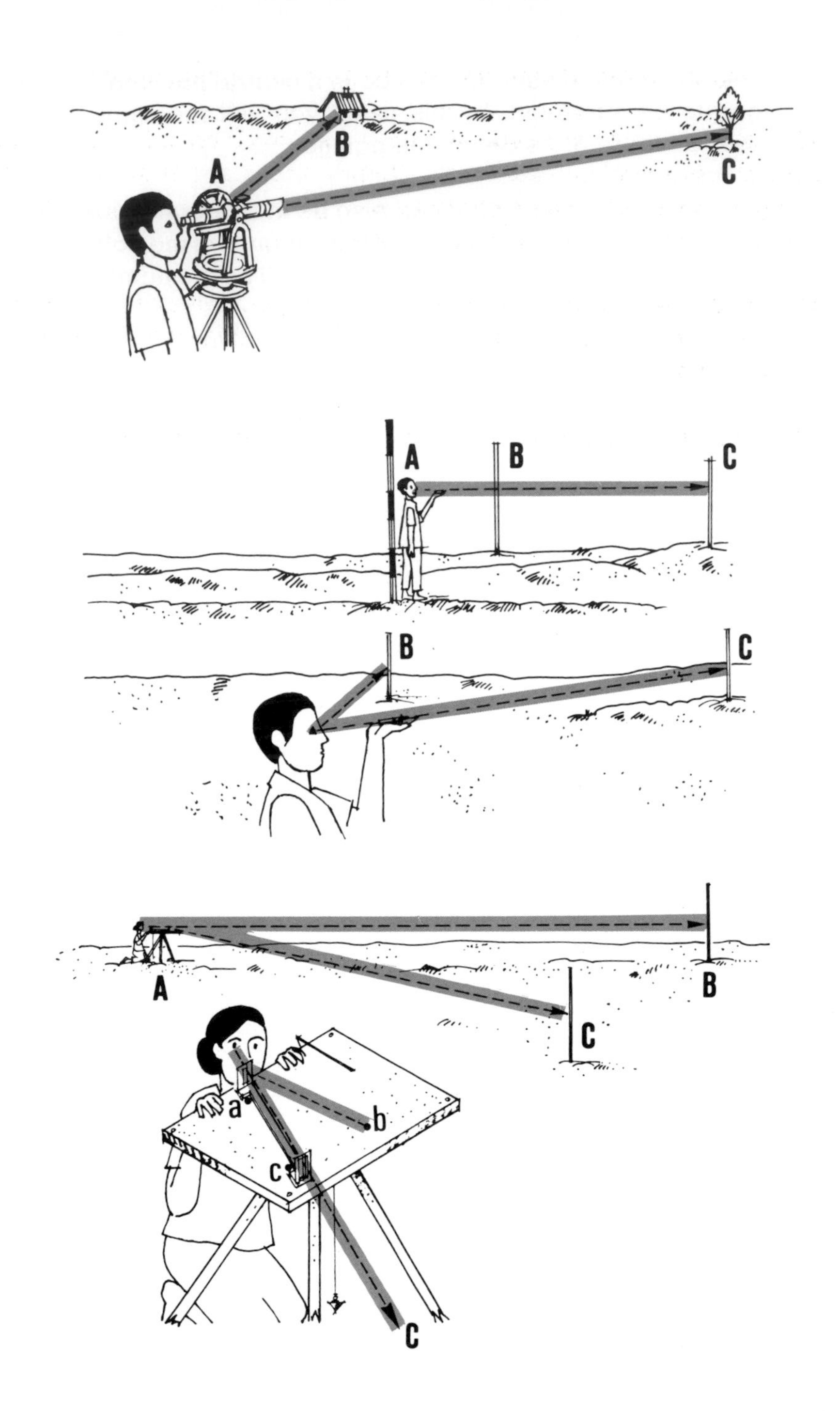

Carrying out a radiating plan survey with a magnetic compass

10. Walk over the area you need to survey and choose a convenient central **observation station 0**. Clearly mark all summits of the polygon. Clear any high vegetation along the future radiating lines of sight.

11. With your **magnetic compass**, take a position over the central station 0. Measure the **azimuths** of the six radiating lines OA, OB, OC, OD, OE and OF.

12. Measure the horizontal distance over each f these lines.

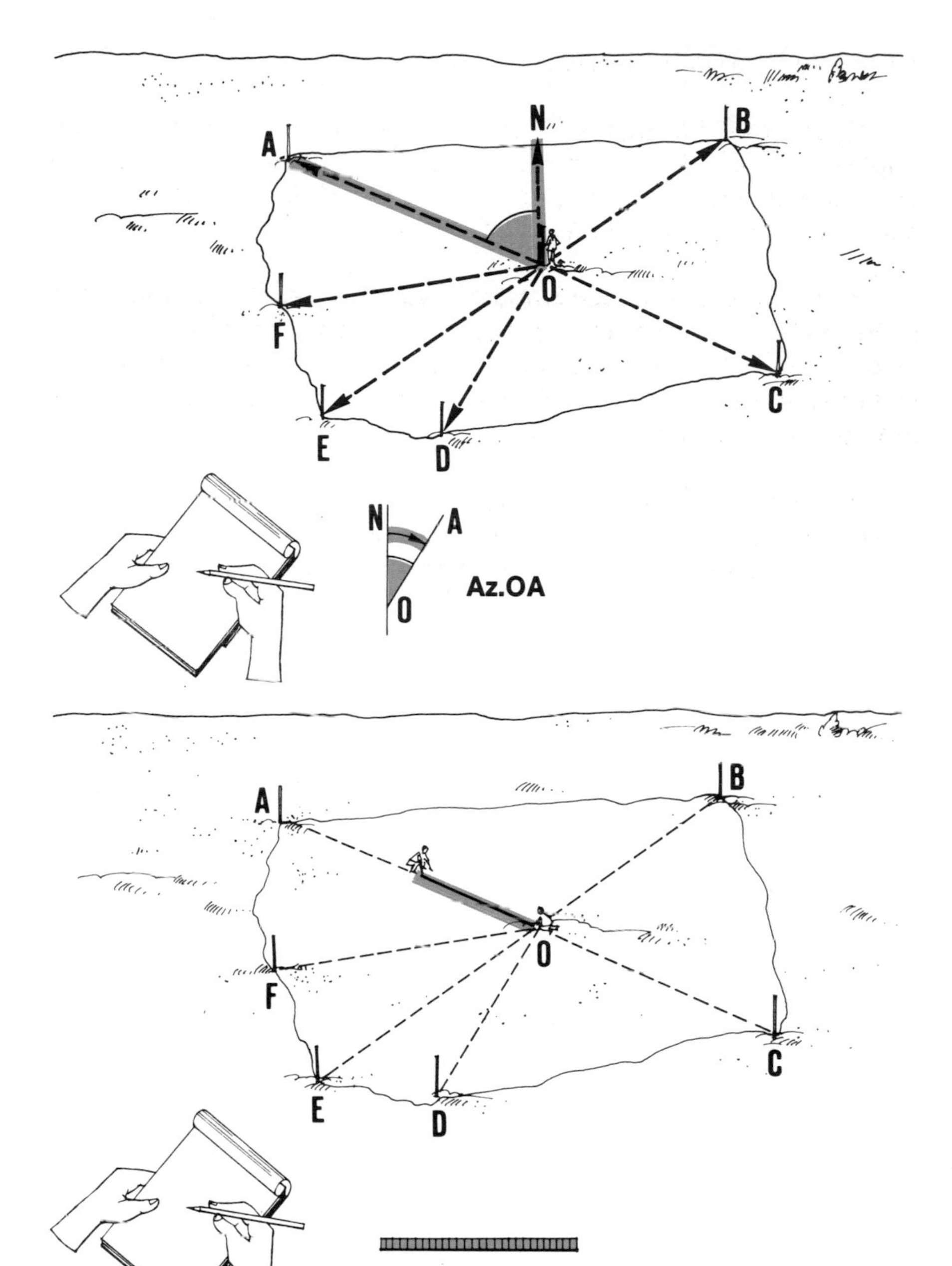

13. Carefully note down all these measurements in your **field-book**. You can use the first three columns of the **table** given in the example. Then make a **sketch** of the area, with the lines and angles and their measurements, on square-ruled paper.

14. Calculate the value of the angles between successive points (see 4th column of the table and Section 32). Check this by **adding all the values**: if you find **360°** or a figure close to that, the calculation is correct.

Example

Table for field observations from a radiating survey.

Line From	Line To	Distances (m)	Azimuths (degrees)	Angles (degrees)
O	A	65.4	265	137 [1]
O	B	58.7	42	88
O	C	51.5	130	70
O	D	89.8	200	23
O	E	41.3	223	11
O	F	43.8	234	31
	A	—	265	
Sum of the interior angles:				360

[1] Since magnetic north falls inside angle AOB, it is calculated as 360° minus the difference of the azimuths.

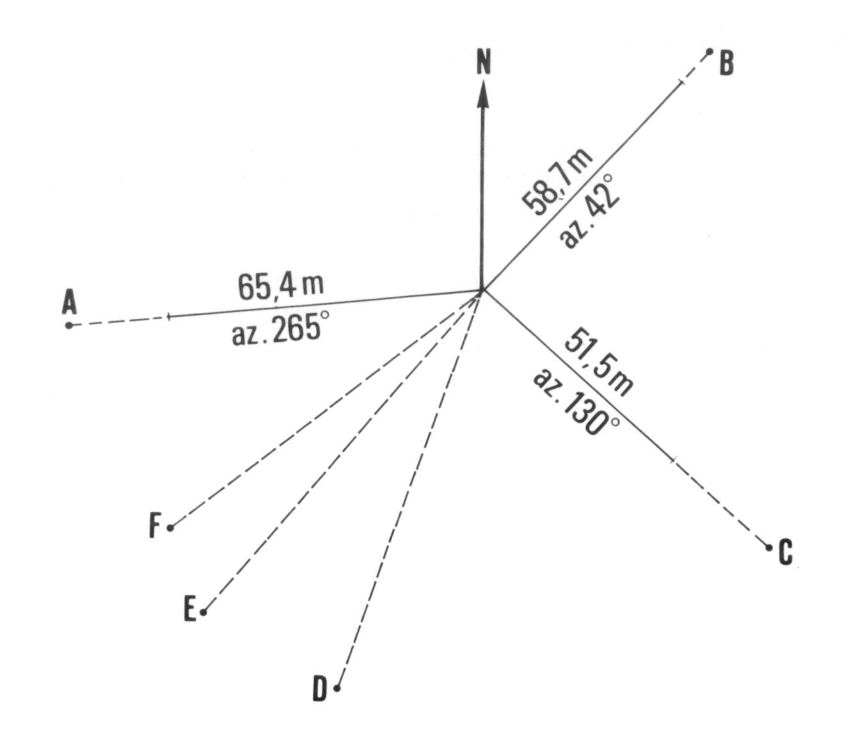

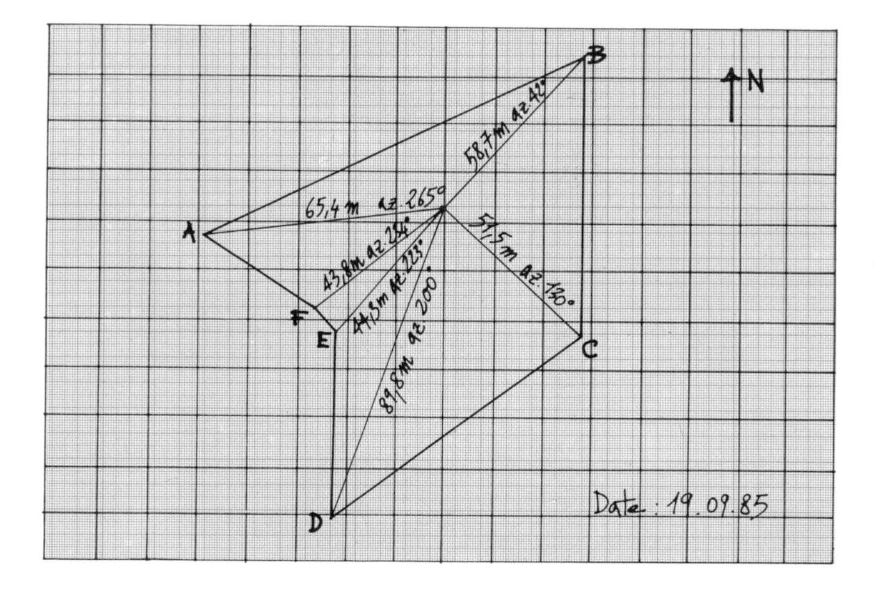

73 How to survey by offset

What is an offset?

1. In plan surveying, an **offset** is a straight line which is laid out **perpendicularly** to a line you are **chaining**.

2. Offsets are mainly used to **survey details** of the terrain (such as wells, rocks or trees) which are located close to a chaining line. Generally, offsets are less than **35 m** long.

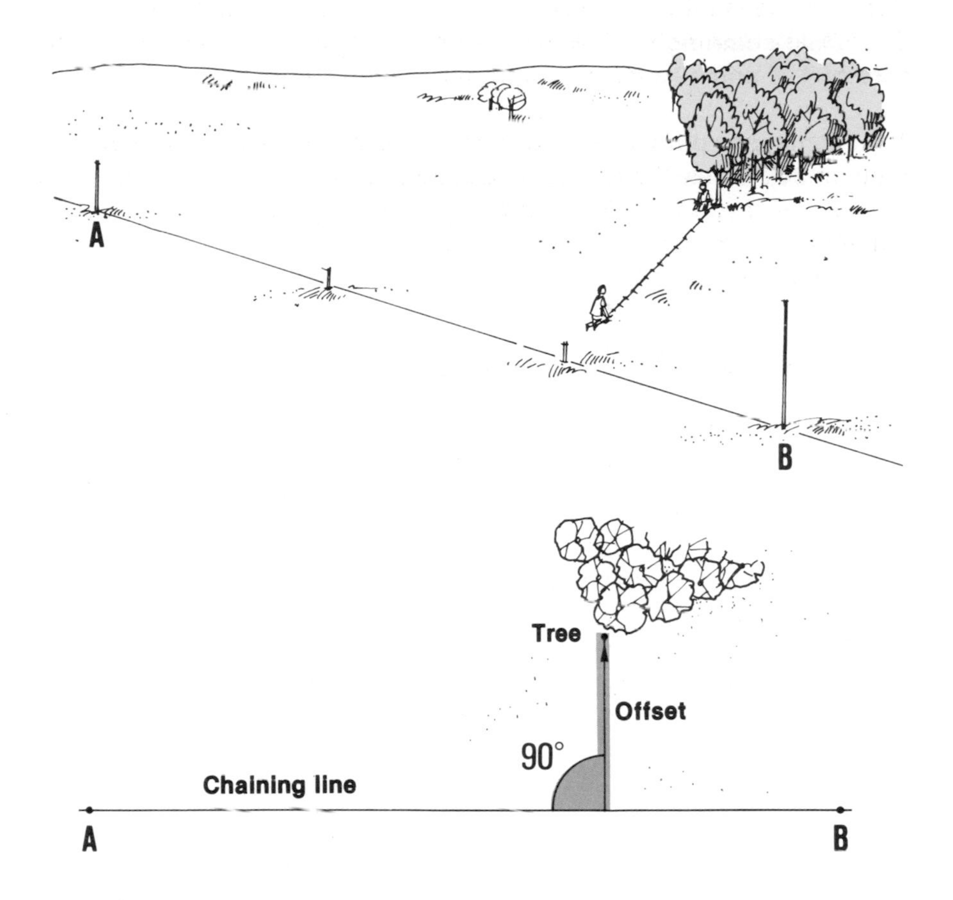

Surveying by offset

3. While chaining line AB, you see two points of interest on either side of it, X and Y, whose exact positions you want to record.

4. From these points, drop XC and YD perpendicular to line AB (see Section 36). **Lines XC and YD are offsets.**

5. Measure horizontal distances AC and CD on line AB. Measure horizontal distances CX and DY along the offsets.

6. From these measurements you can plot the exact positions of points X and Y on paper, if line AB is known (see Chapter 9).

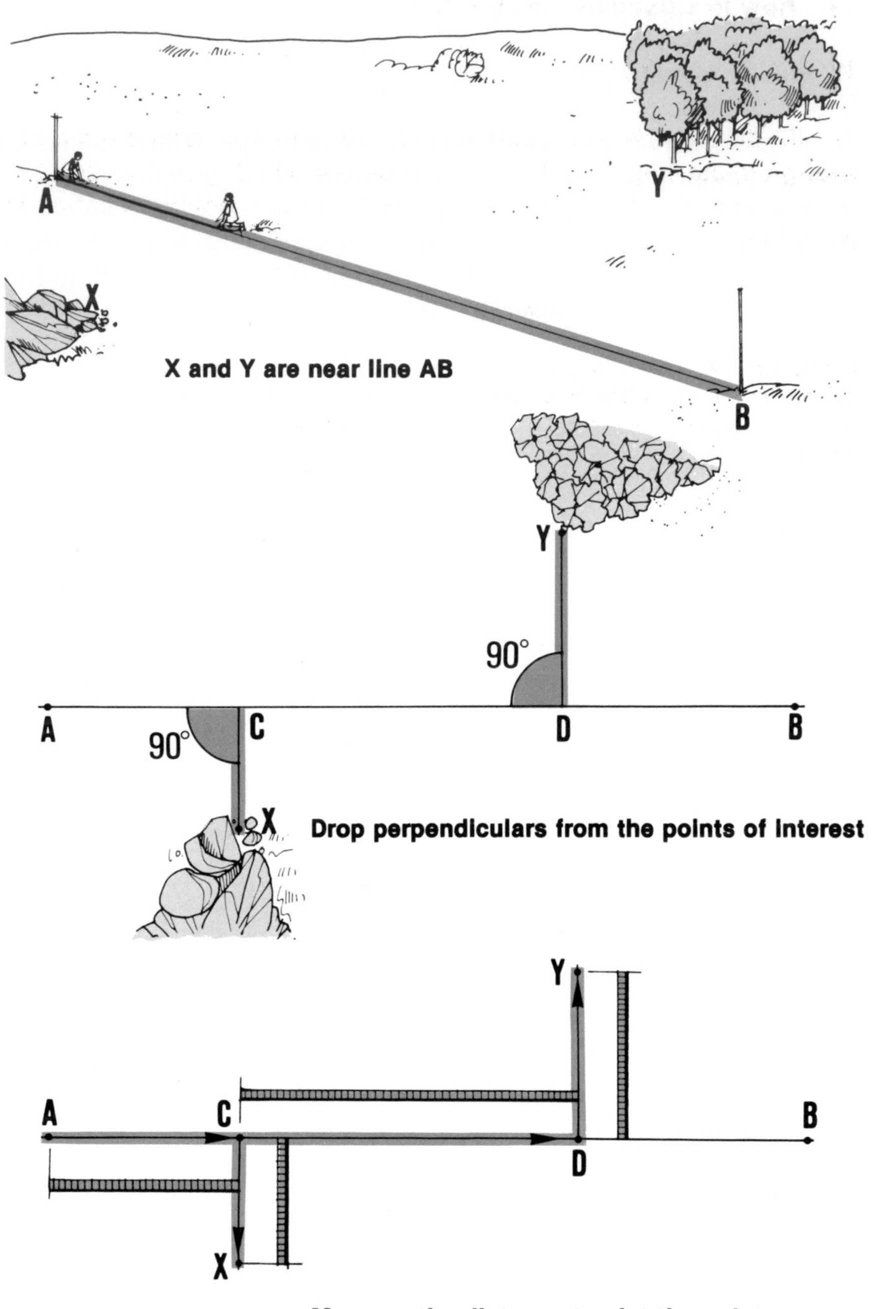

What is triangulation?

1. If you use the triangulation method, you will form **consecutive triangles**, starting from two known points which you can see from each other. The straight line joining these two points is called **the base line**.

Example

A and B are two points whose positions you know. Therefore, you can easily survey the baseline AB to find the measurements of the horizontal distance and magnetic azimuth. AB is 123 m long and azimuth AB = 150°.

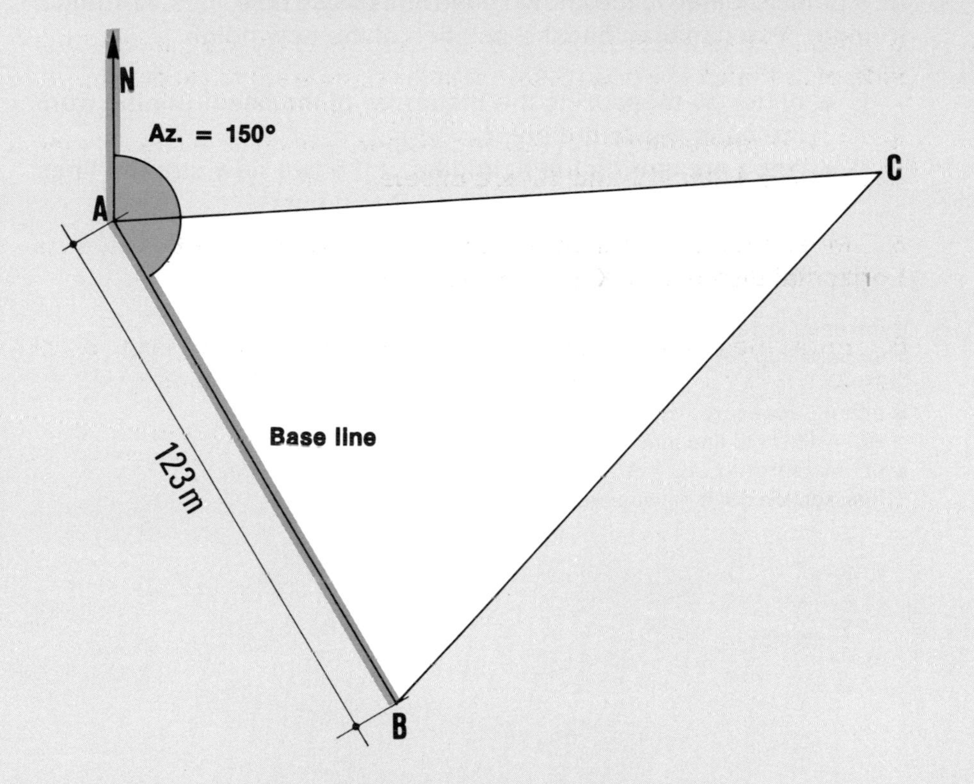

2. To determine the position of a new point C by triangulation, this new point is joined to the known base line by two new lines, forming a triangle. You can then find the position of the new point:

- either by measuring the distances of the lines running from the base line to the point;
- or by measuring the azimuths of the two new straight lines running from the points A and B to point C.

Example

If you need to determine the position of C, lay out lines AC and BC from base line AB. Then you can:

- either measure horizontal distances AC = 166 m and BC = 156 m to find intersection point C;
- or measure Az AC = 87° and Az BC = 43° to find C at the intersection point of two lines drawn with these azimuths.

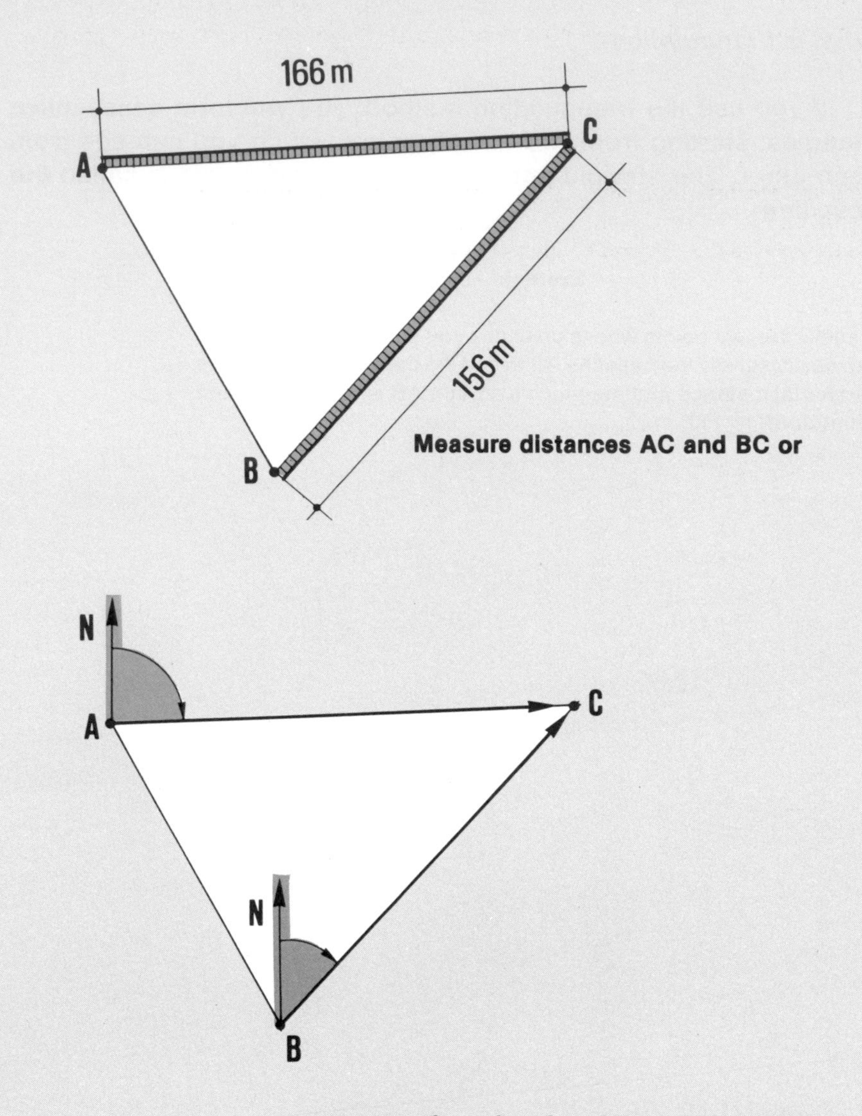

3. To find the positions of other new points, use the same procedure. As you find the positions of new points, use the most convenient existing line as the **new base line** and form **new triangles** as you work.

Example

If you need to determine the position of D, lay out triangle BCD and use BC as the base line. Similarly, to determine points E, F and G, use base lines CD, DE and EF successively.

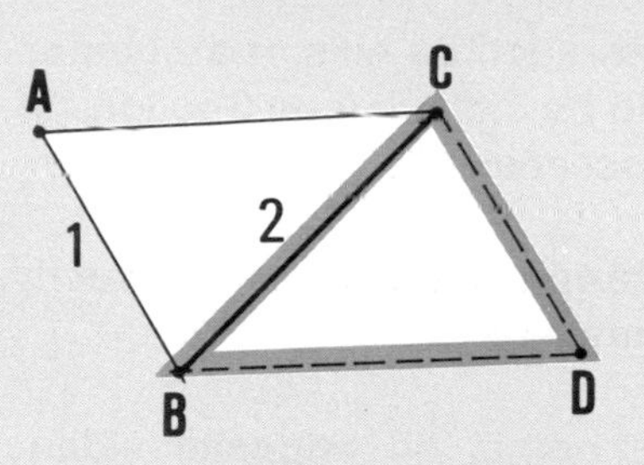

Use BC as the base line for new triangle BCD

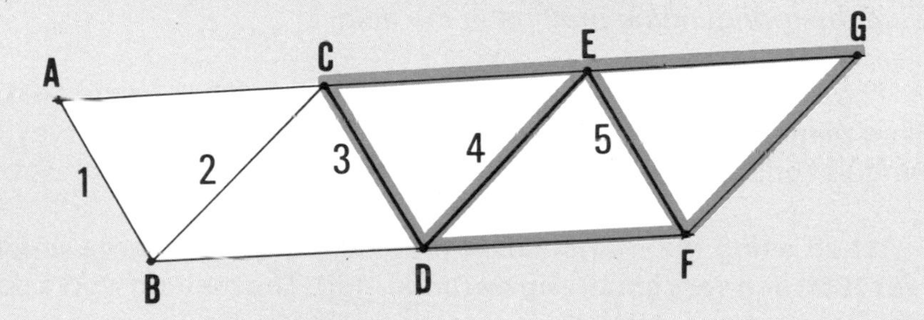

Continue making triangles until you have surveyed the whole site

Using the triangulation method

4. On terrain with many obstacles such as hills, marshes or high vegetation, where traversing would be difficult (see Section 71), you can use the triangulation method successfully.

5. When you are **traversing**, and cannot measure a line directly, you can use the triangulation method instead.

6. Triangulation makes locating points on **opposite sides** of a stream or a lake very easy.

Using the triangulation method in the field

7. The simplest way to use the triangulation method in the field is with **a plane-table** (see Section 75). You will learn how to survey by triangulation, using a plane-table, in Section 92.

8. When using the triangulation method, **avoid very large angles** (over 165°) and **very small angles** (under 15°). The method works best with angles of about 60°.

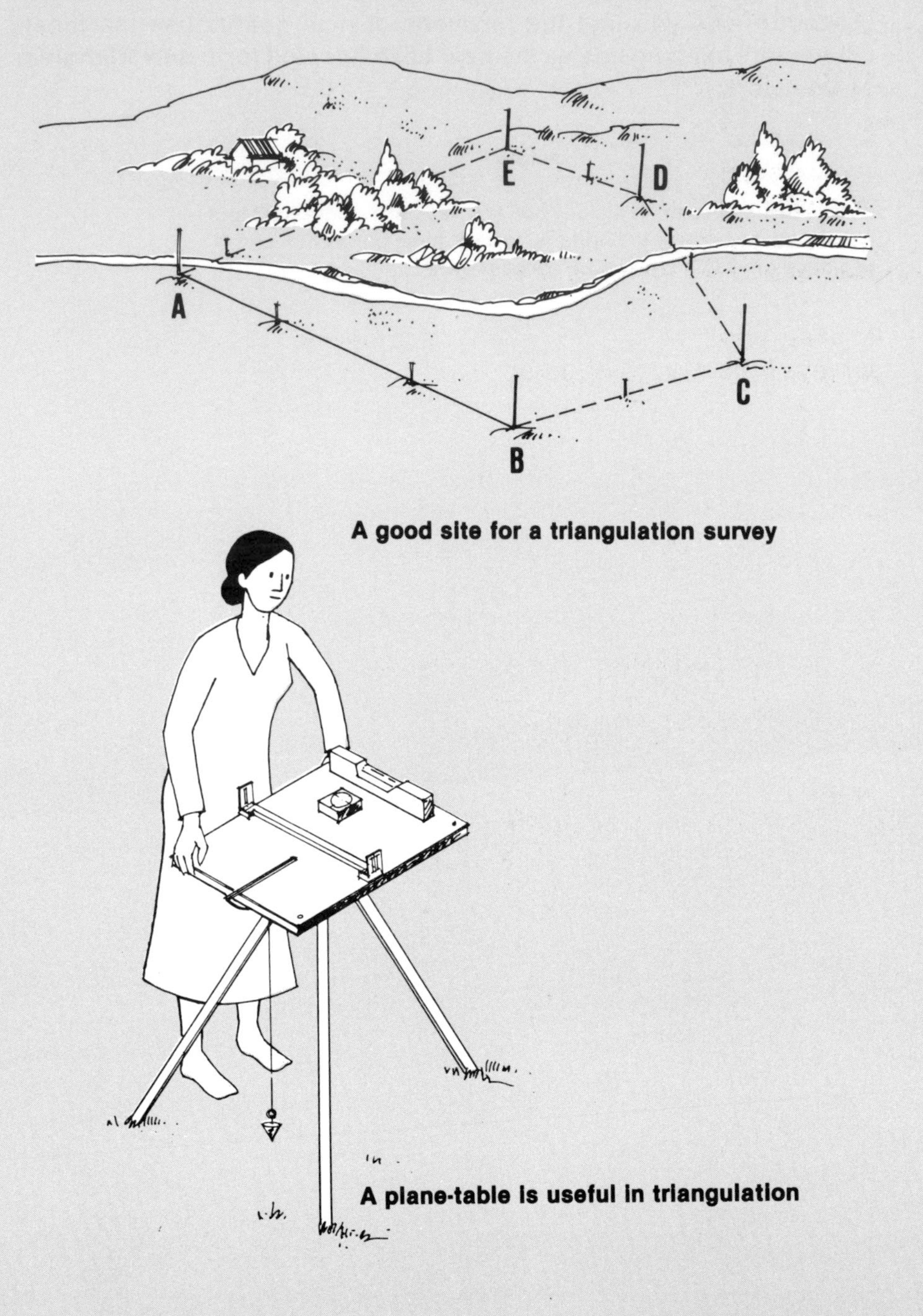

A good site for a triangulation survey

A plane-table is useful in triangulation

75 How to use the plane-table

What is a plane-table?

1. A plane-table is a horizontal **drawing-board** mounted on top of a vertical support. You use it with a **sighting device**, a **spirit level** and a **magnetic compass**.

Making a very simple plane-table

2. You can make a very simple plane-table for reconnaissance surveys from a **wooden board** and a strong **pole**.

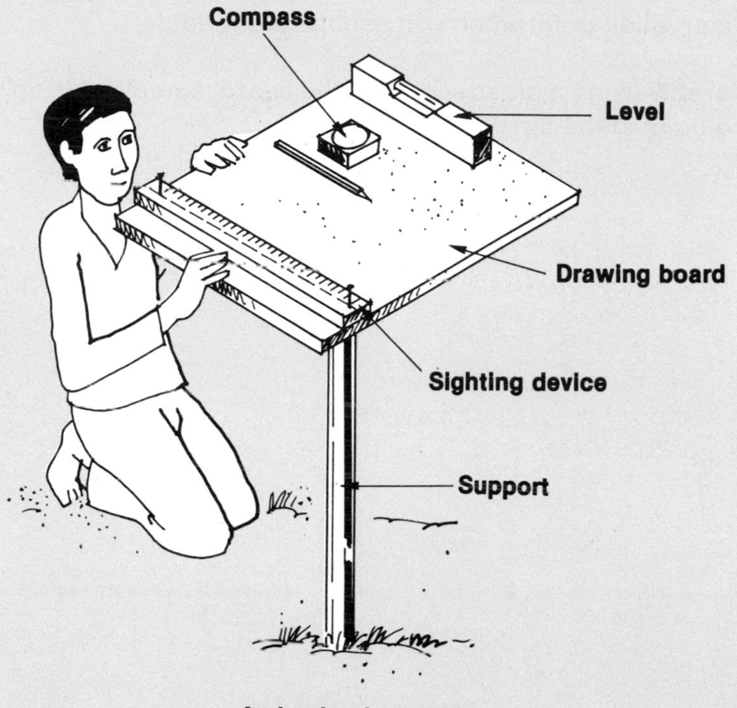

A simple plane-table

3. Get a 50 × 60 cm board of soft wood, about 2 cm thick. With sandpaper, polish one of its surfaces well until it is very smooth. Draw two diagonal lines lightly across this surface to find the centre of the board.

4. Get a straight wooden pole about 5 cm in diameter and 1 m long. Shape one end into a point. This will be firmly driven into the ground at the observation point when you use the plane-table.

5. Preferably using a brass screw, fix the board, smooth side up, by its centre-point to the top of the pole.

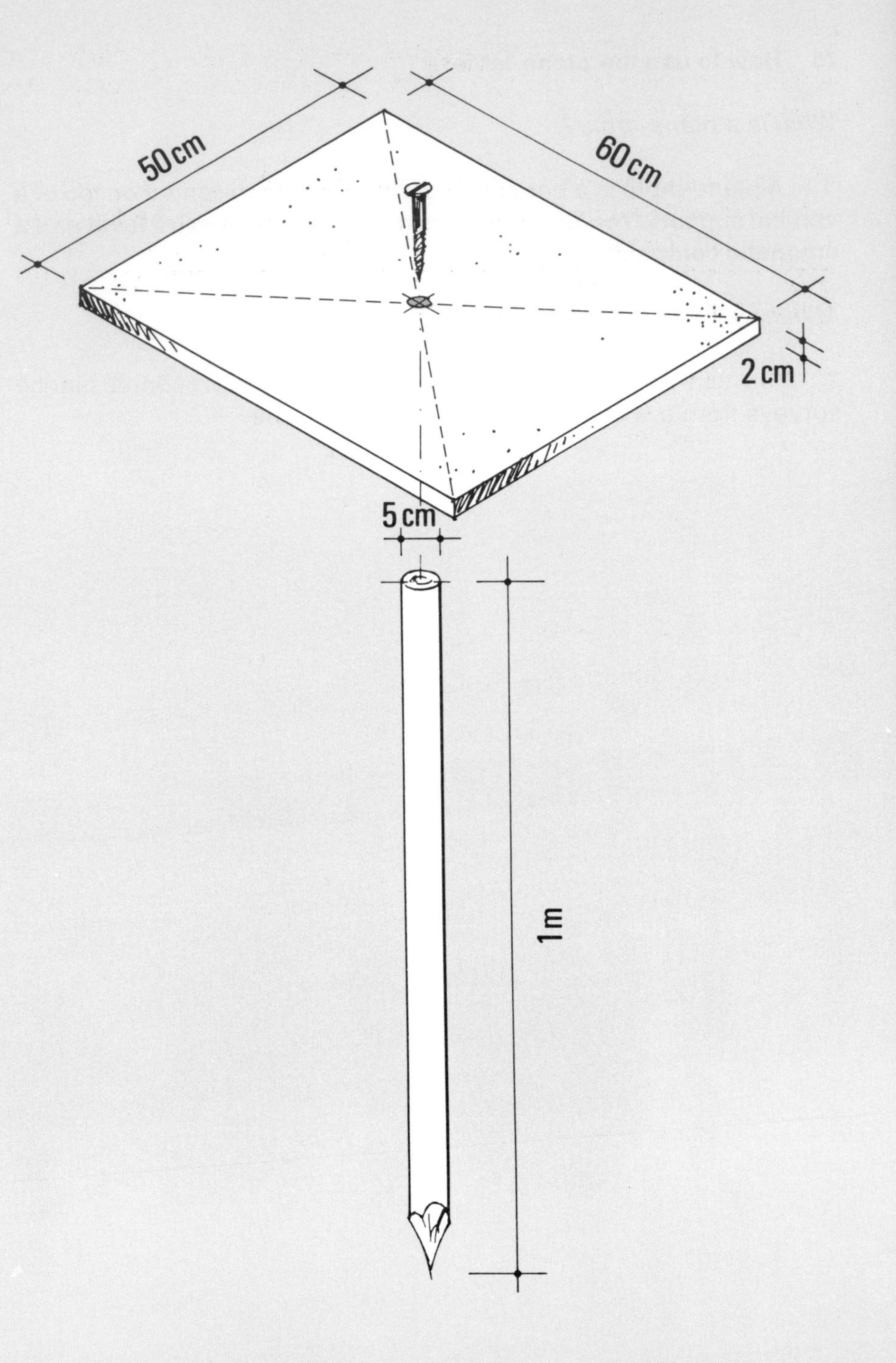

6. You can make a simple **sighting device** from an ordinary ruler about 50 cm long by driving two thin nails vertically into it along the centre-line for sighting.

7. You will also need a simple **magnetic compass** to use with the plane-table. If you have a **spirit level**, use it to set up the top board horizontally. Or simply lay a rounded object such as a small ball, a glass marble or a pencil on the board's top surface. When the object remains still, the board is horizontal.

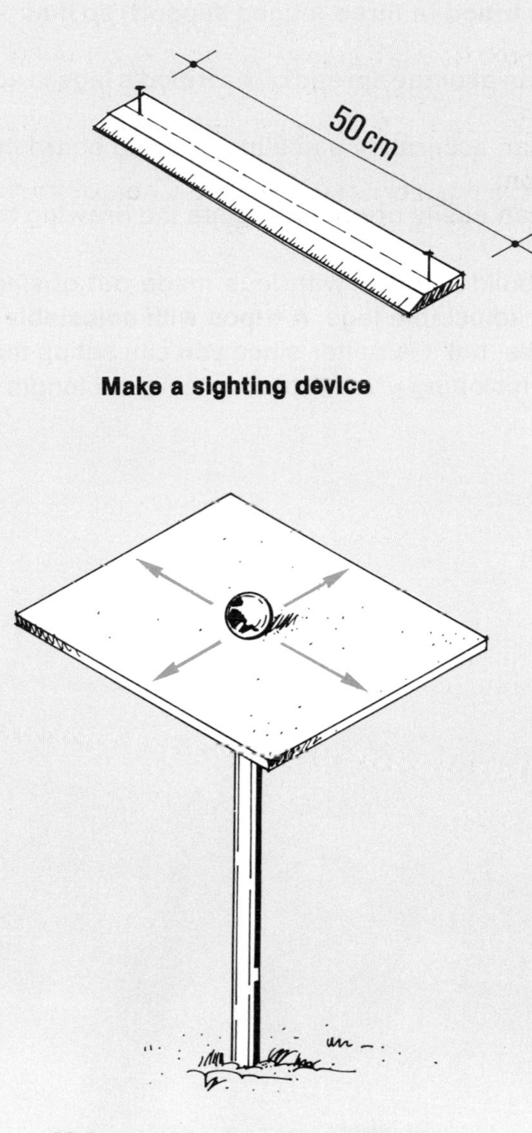

Make a sighting device

Make sure the board is horizontal

Making an improved plane-table

8. To survey more precisely, you will need a more complicated plane-table than the one just described. This plane-table will be mounted on a **tripod** (a three-legged support) so that:

- you can alter the spread of the tripod's legs to adjust to rough terrain,
- you can accurately place the drawing board in a horizontal position;
- you can easily orient and rotate the drawing board.

9. You can build a tripod with legs made out of single pieces of wood, or with adjustable legs. **A tripod with adjustable legs** is more difficult to make, but it is better since you can set up the plane-table more easily on sloping ground by changing the length of the legs.

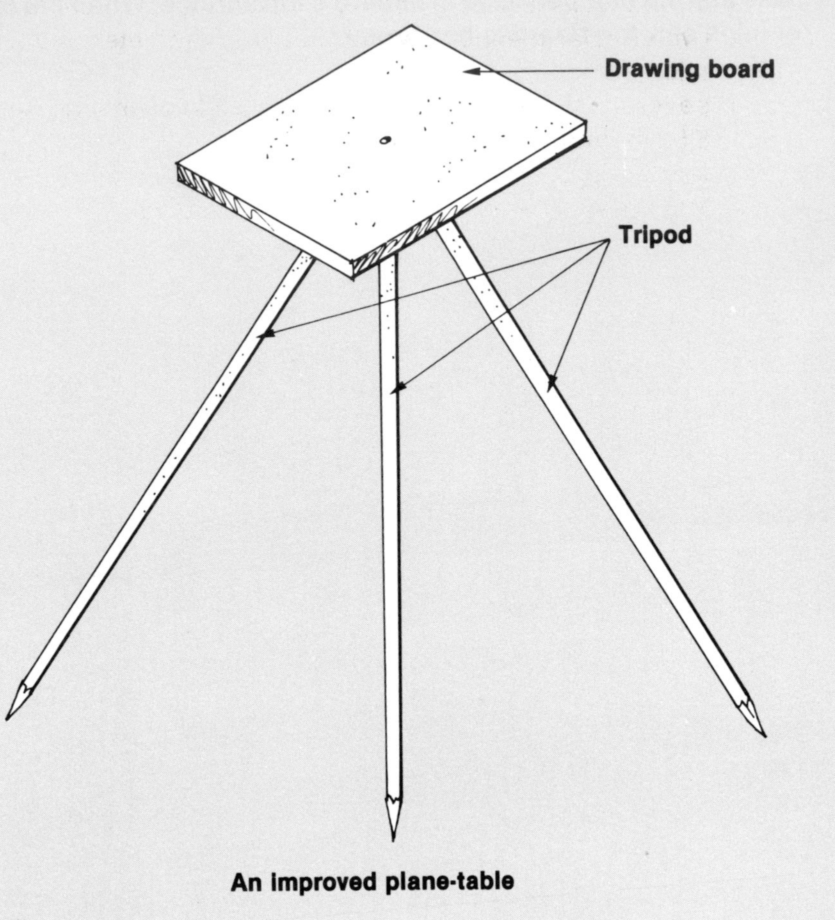

An improved plane-table

10. A plane-table with a normal tripod is adequate for surveying horizontal areas and areas with small slope gradients, which you must often survey in aquaculture. To make this type of plane-table, you will need the following materials[1]:

- one board of soft wood, about 40 × 55 cm and 2 cm thick;
- three pieces of wood, about 2.5 × 4.5 cm, and 1.4 m long;
- three blocks of wood, about 2.5 × 4.5 cm, and 7 cm long;
- two circular pieces of wood, 15 cm in diameter and 2.5 cm thick;
- several nails or wood screws, both 3.5 to 4 cm long and 6 to 6.5 cm long;
- four bolts, 6 mm in diameter and about 6 cm long;
- four washers and four wing nuts for the bolts.

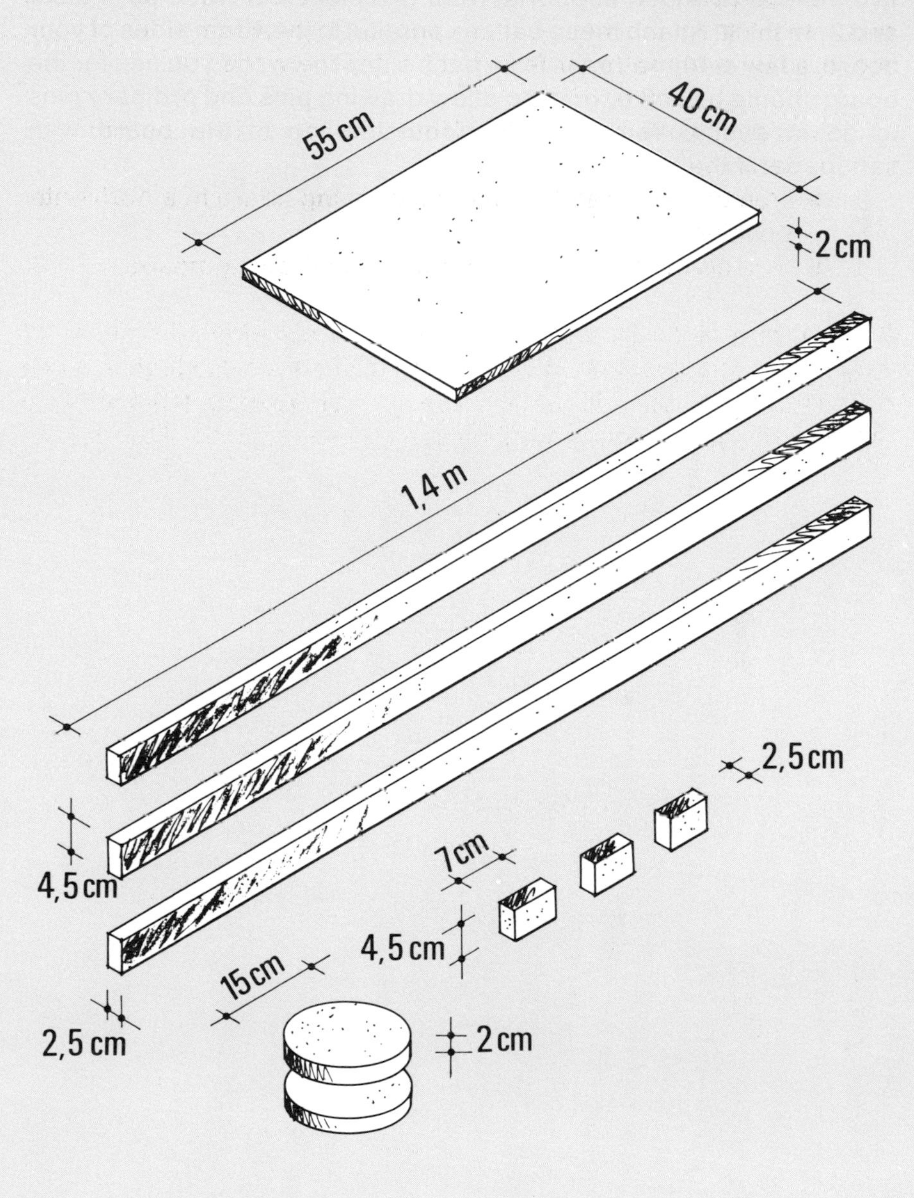

[1]Adapted from *Using Water Resources*, Maryland, USA, VITA Publications, 1977, pp. 137-140.

11. Get a piece of 40 × 55 cm plywood 2 cm thick to use for the drawing board. If the plywood you have is thinner than 2 cm, make two battens (wooden supports) from two pieces of wood 30 × 8 cm and 2 cm thick. Attach these battens parallel to the 40 cm sides of your board, a few centimetres in from each side. The wood you use for the board should be soft enough to allow drawing pins and ordinary pins to go in easily. You should smooth the top of the board with sandpaper if the surface is irregular.

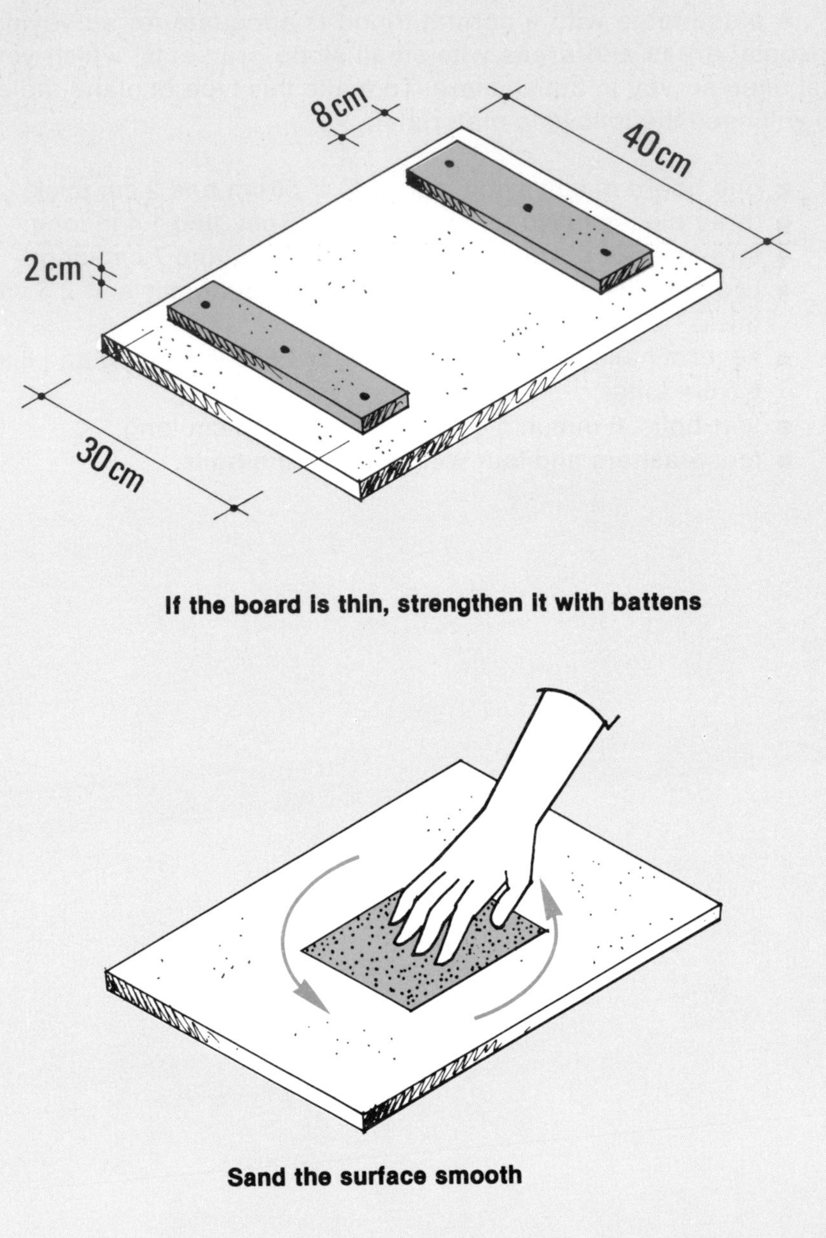

If the board is thin, strengthen it with battens

Sand the surface smooth

12. Make the **three legs** from the 1.4 m pieces of wood. Shape each into a point at one end. On the other end-face of each leg, mark a centre-line parallel to the 2.5 cm sides. Continue this line 5 cm down either side of the leg. At these two points, mark a centred perpendicular line 2.5 cm long; connect the end-points of this 2.5 cm line up the sides of the leg and over the top. Cut out this block you have marked, which will measure 2.5 × 2.5 × 5 cm, and discard it. Round off the edges of the two remaining "prongs" of wood which face toward the 2.5 cm side of the leg, using a knife and sandpaper, for example.

13. On these prongs, drill a 6 mm hole at a point 1.3 cm from the top of the leg.

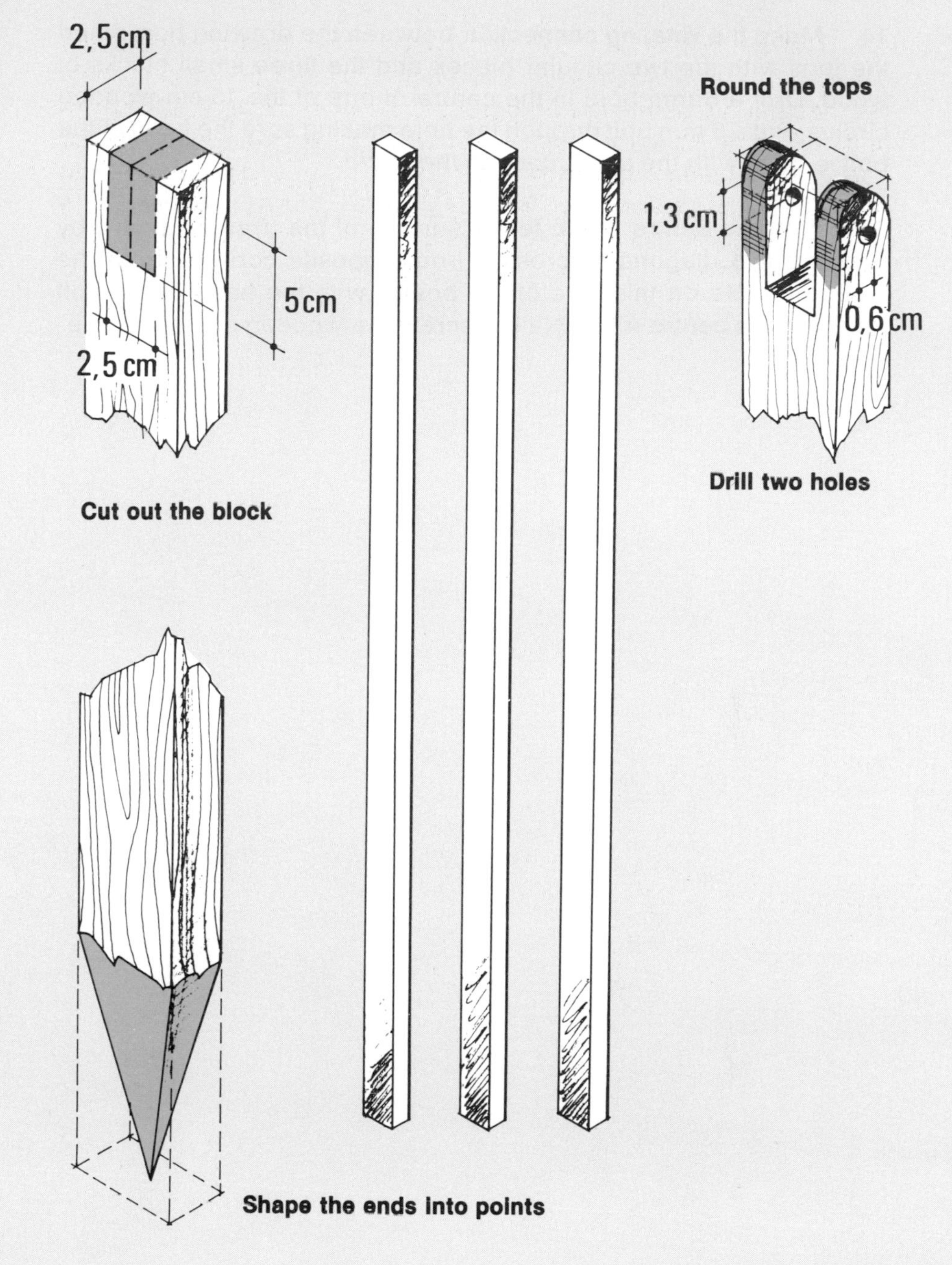

14. Make the **rotating connection** between the drawing board and the legs with the two circular pieces and the three small blocks of wood. Drill a 6 mm hole in the centre of one of the 15 cm wooden circles. Put a 6 mm bolt through the hole making sure the head of the bolt is even with the top surface of the circle.

15. Find the centre of the lower surface of the drawing board by drawing two diagonals across it from opposite corners. Hold the wooden circle on this side of the board, with the head of the bolt touching the centre mark. Nail or screw the wooden circle in place.

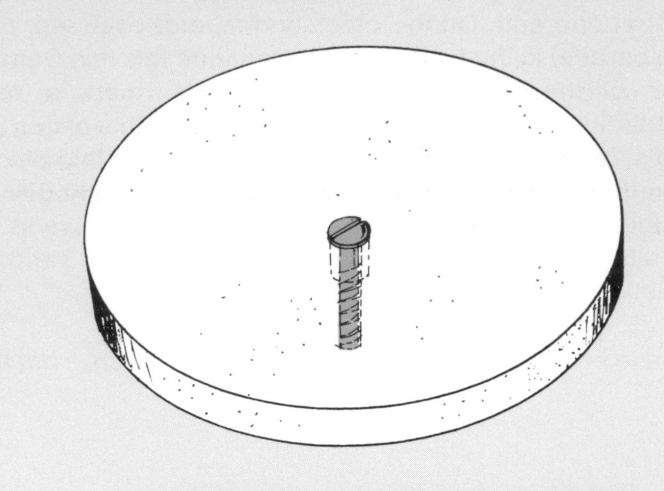

Put the bolt through the centre of the disc

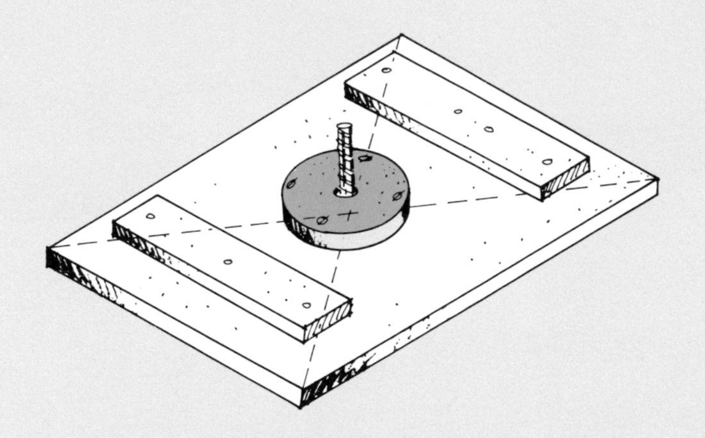

Nail the disc to the board so that the bolt sticks up

16. Take the second 15 cm circle and mark the points where you will attach the legs. To do this, first draw two perpendicular lines across the circle. They should intersect at the exact centre of the circle. Call them diameters **a** and **b**. With a protractor, using line **b** as the 0 to 180° line, draw two more lines from the centre of the circle to the edge at 45° and 135°. Call them radiuses **c** and **d**. They should divide one half of the circle into four equal, wedge-shaped sections. Then drill a 6 mm hole in the centre of the circle.

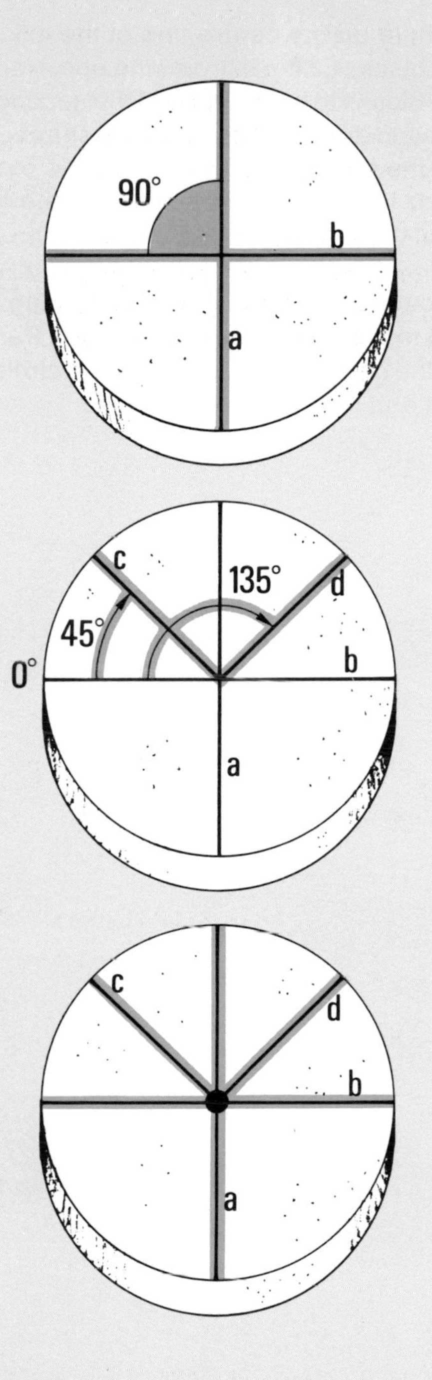

17. Drill a 6 mm hole on the centre line of the 4.5 × 7 cm face of each 7 cm wooden block, 1.3 cm in from one end. Nail or screw these three 7 cm wooden blocks to the surface of the second wooden circle, so that they join around the centre-hole in a Y-shape. To do this, align the centre-lines of the blocks' 2.5 × 7 cm faces over the lines a, c and d that you drew in step 16. The ends with the holes should be towards the edge of the circle.

18. Place this wooden circle, with the blocks facing you, against the circle already fixed to the underside of the board. Pass the bolt in the first circle through the centre-hole of the second circle. Add a washer and a wing nut to it and tighten them securely.

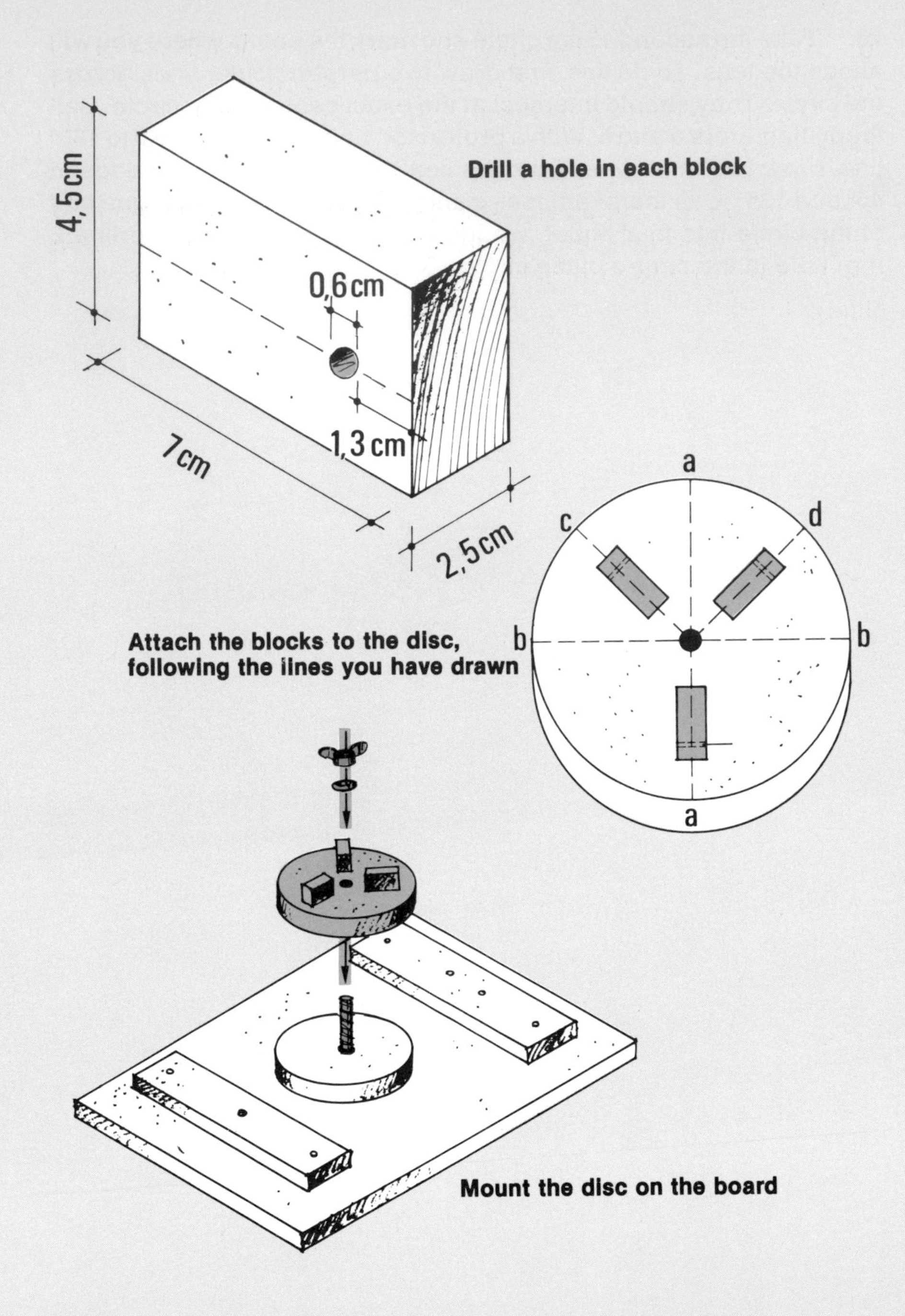

19. Align the holes in the three legs with the holes in the three blocks of wood on the underside of the board, and attach the legs with bolts, washers and wing nuts to the blocks. Your plane-table is now ready to use.

20. You will also need **a small spirit level, a magnetic compass**, and a sighting device called an **alidade**. You have already learned about one kind of alidade (see Section 31), but this one will be slightly different.

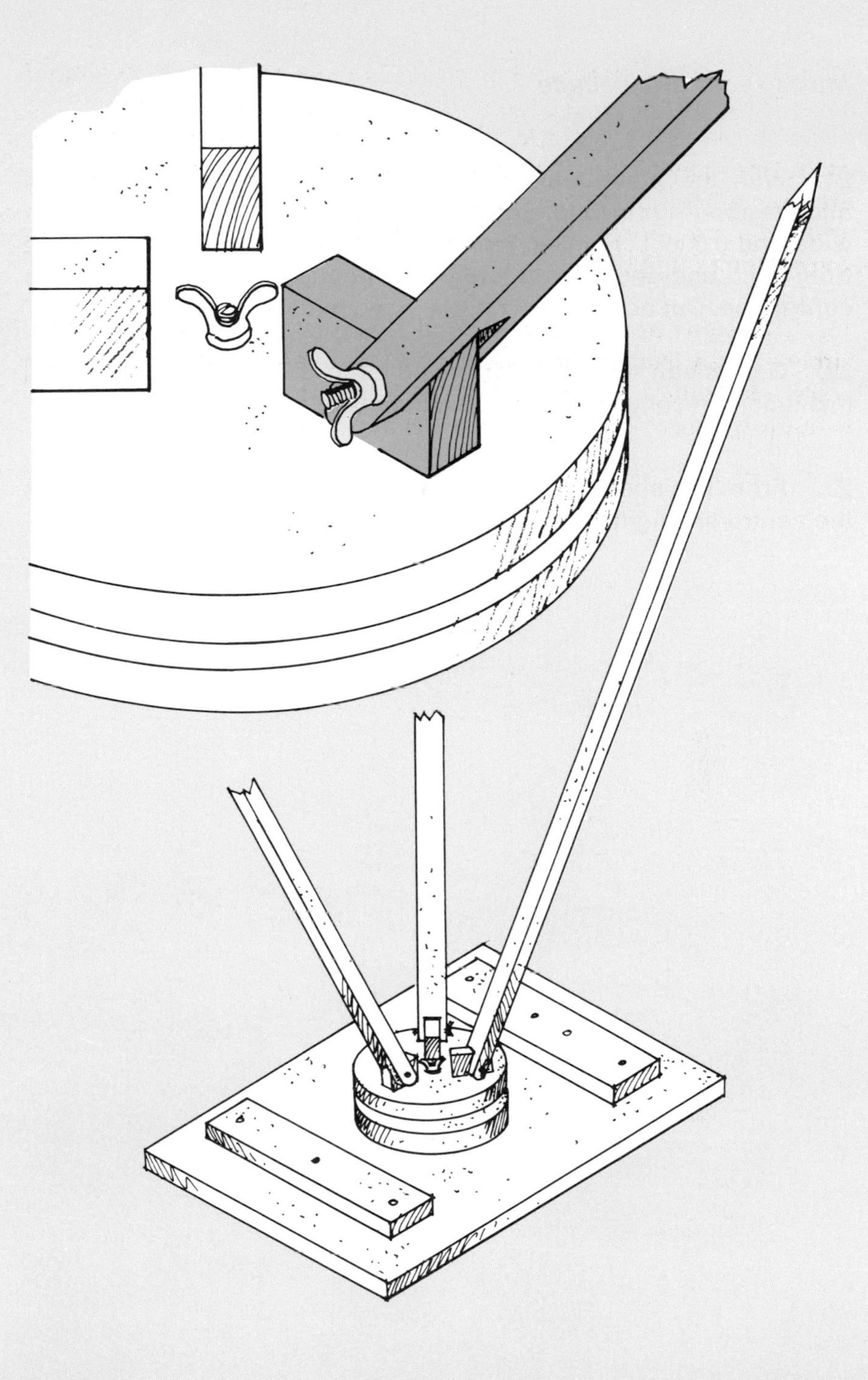

21. With the plane-table described above, you should use **an alidade** about 40 cm long. Get a straight strip of wood 40 cm long, 5 cm wide and 0.5 to 1 cm thick. Find the centre-line, then measure 5 cm from each end and draw a line from the edge of the alidade to the centre-line. Cut out the section you have marked off.

22. Get a clean, empty metal tin and remove its top and bottom. Cut this tube vertically and flatten it out to make a sheet of metal.

23. From this sheet, cut out two pieces 5 cm × 12 cm each. Mark the centre-line lightly on each, using a nail to scratch the line.

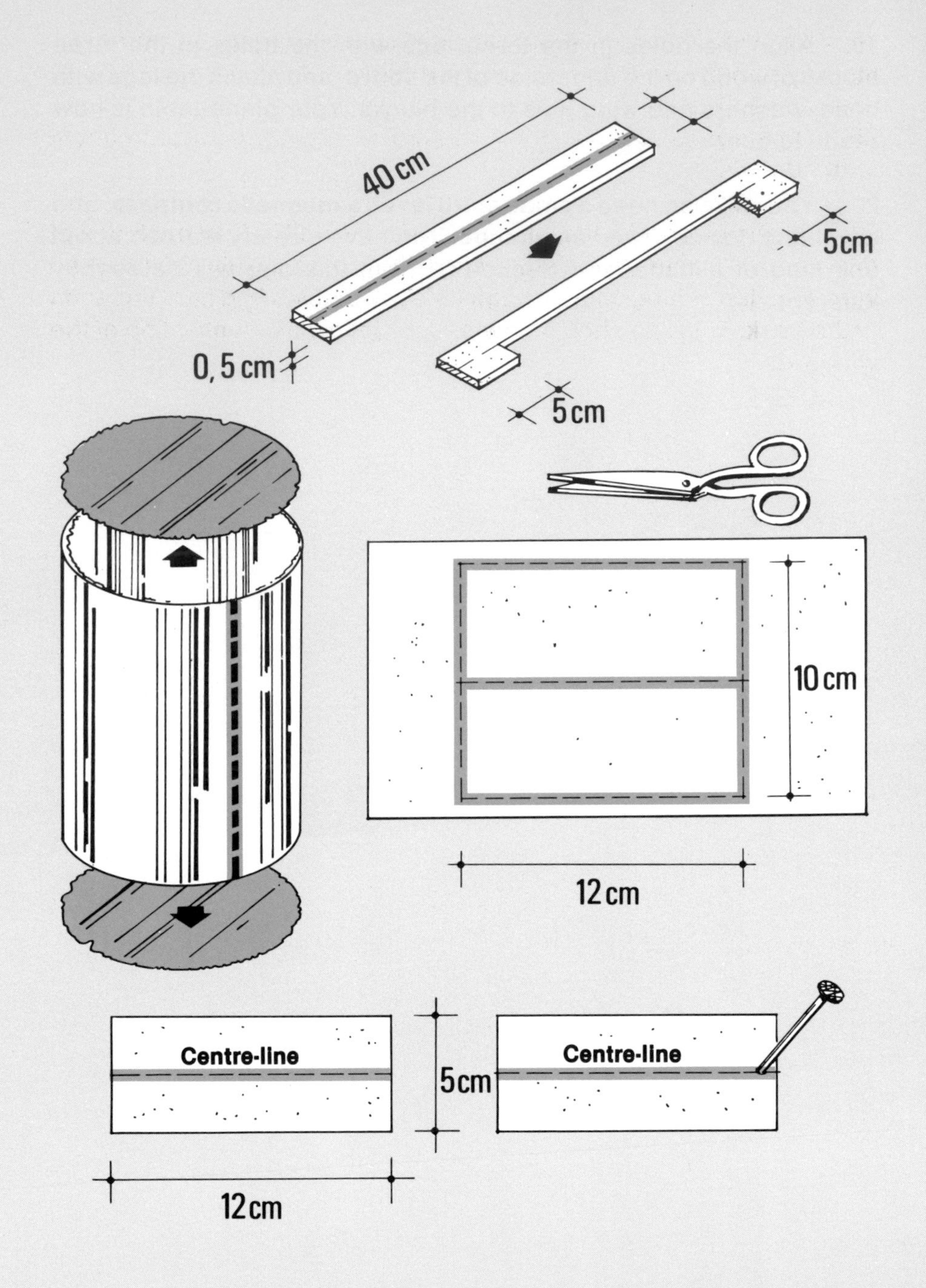

24. On one of these pieces, cut an 8 cm slit along the centre-line, starting about 1 cm in from the 5 cm edge.

25. On the second piece, cut out a 3 cm × 8 cm window, as shown in the drawing.

26. On the piece with the window, make a small hole at each end of the window "frame", along the centre-line. Thread a thin line (such as wire or nylon fishing line) through these two holes and knot the ends at the back. This line should now exactly follow the centre-line of the window.

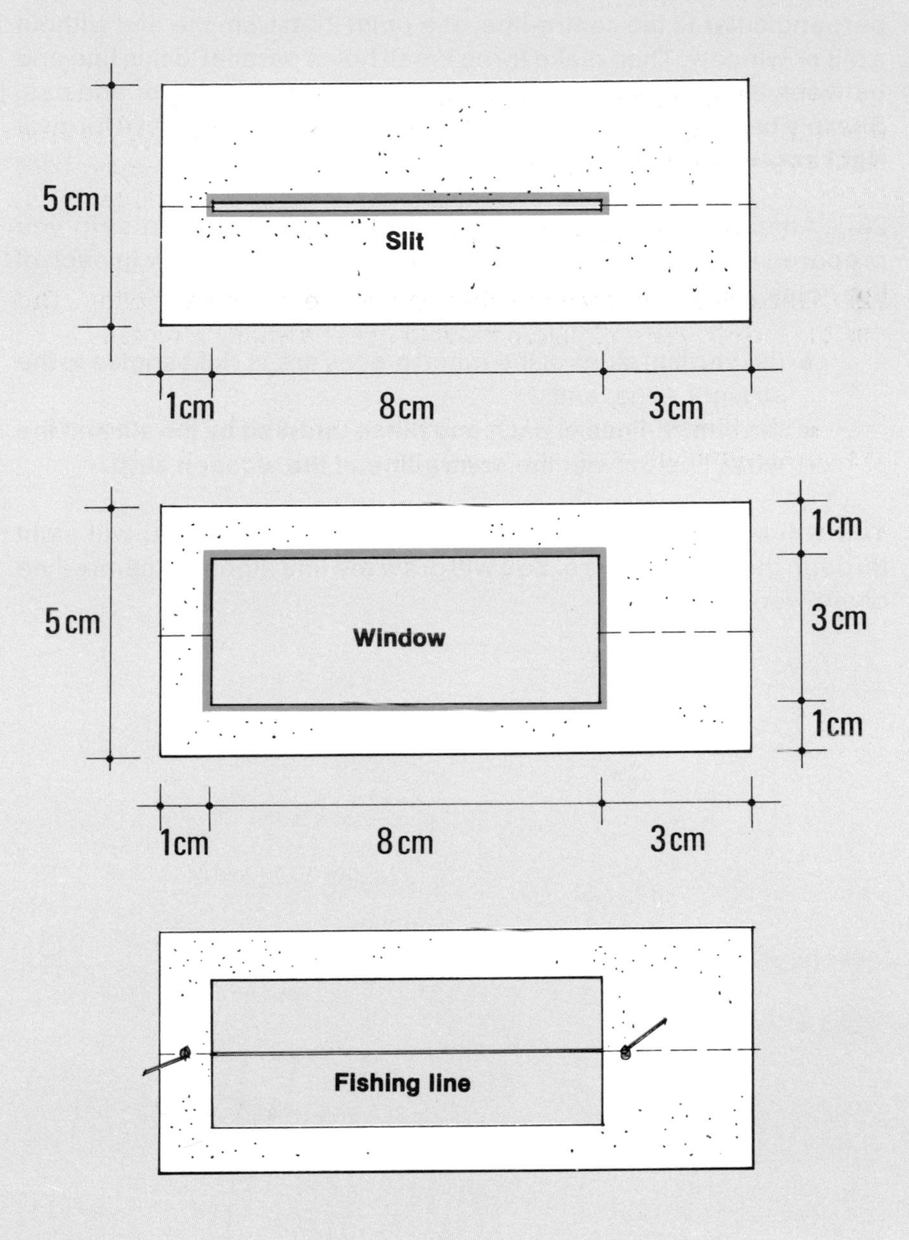

27. On each metal piece, use a nail to draw a fairly deep line perpendicular to the centre-line, at a point 2 cm from the end without a slit or window. Then make three small holes parallel to this line and between the line and the end of the piece, using a hammer and nail. Sharply bend this end of the metal along the deep line, until it forms a **right angle** with the rest of the piece.

28. Attach the metal pieces to the ends of the wooden strip you prepared in step 21. Hold them in place with a small screw in each of the holes you have made in the metal. Make sure that:

- the **vertical sides** of the metal pieces are **at right angles** to the straight edge; and
- the **centre-lines** of each end piece (**marked by the slit and the wire**) line up with the **centre-line of the wooden strip**.

You will use the alidade set flat on the plane-table. You will sight through the slit at the wire. You will draw the line along the centre-line of the wooden strip.

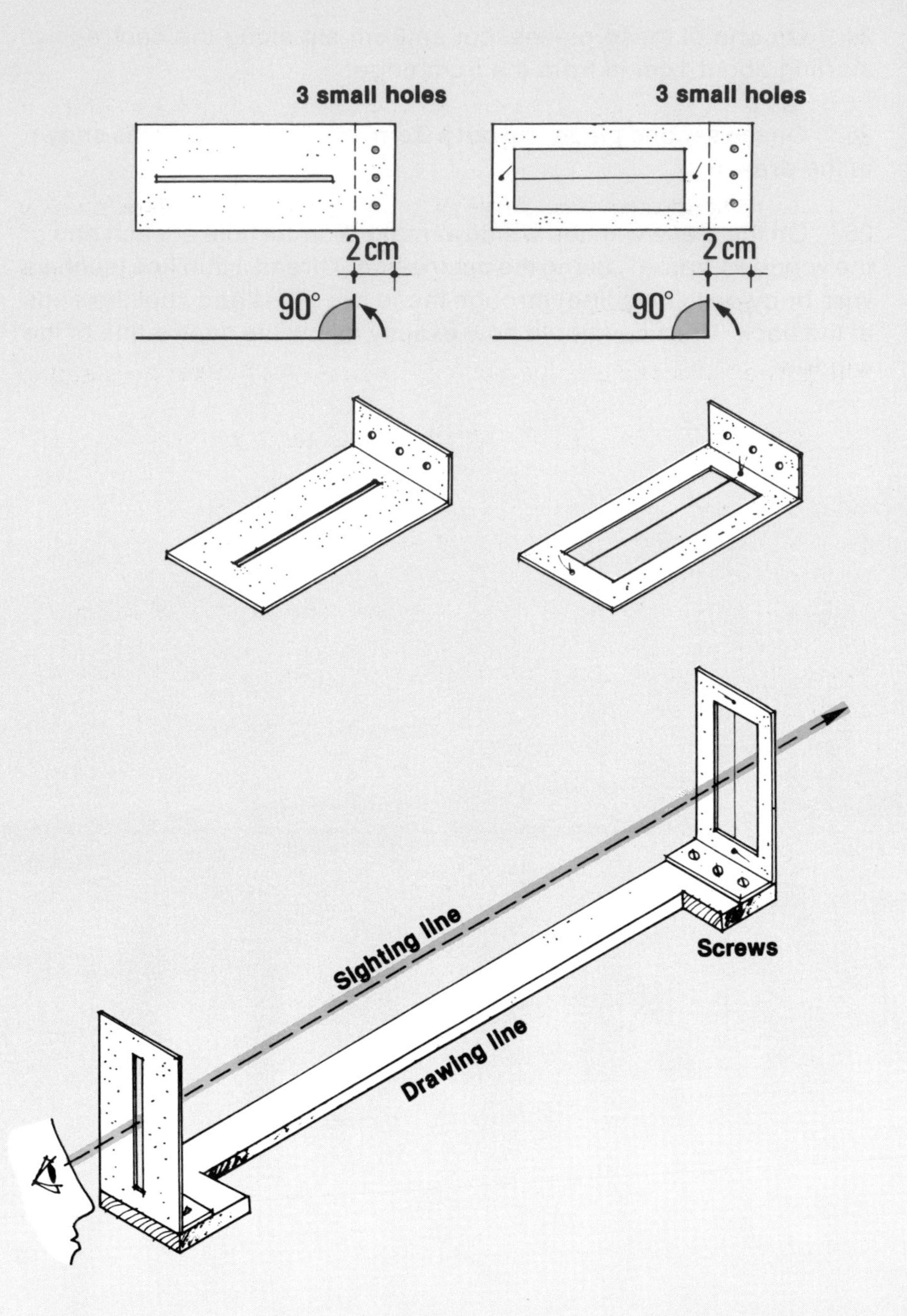

Using the plane-table

29. You can use the plane-table in two different ways, depending on the type of survey you are making:

- in **reconnaissance surveys**, to make **maps and plans** quickly in the field;
- in later surveys, to **fill in details** after you have determined the primary points.

The plane-table can also be used for measuring **horizontal angles**.

30. Before you plan survey with the plane-table, you will need to:

- fix a piece of drawing paper on the top of the board;
- set the plane-table up over the station point;
- level the drawing board, or make it horizontal;
- orient the drawing board to face the line you want to survey.

You will learn more about each of these procedures later (see steps 34-47).

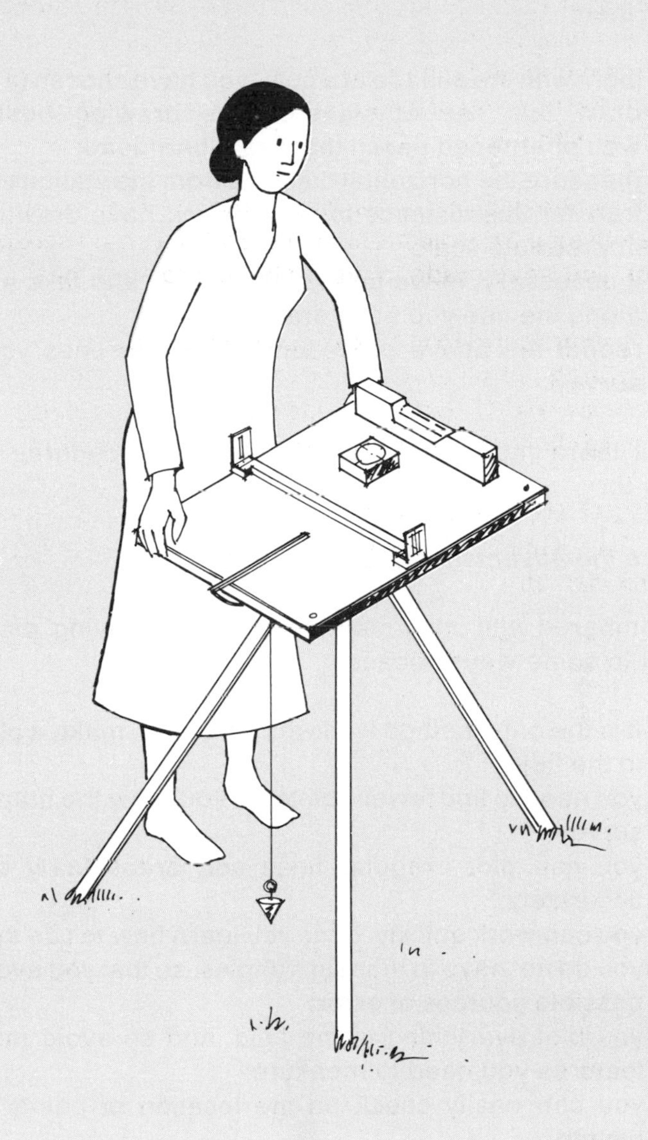

31. When you are ready to start surveying with your plane-table, you will then:

- sight with the alidade at a point you have chosen (a **foresight**);
- draw this line of sight on the drawing board with a well-sharpened pencil that has a hard lead;
- measure the horizontal distance from the station to the point;
- transfer this distance to the line you have drawn, using an appropriate scale;
- if necessary, move to another station, and take **a backsight** along the line you have drawn;
- repeat the above procedure for all the lines you need to survey.

You will learn more about each of these procedures later (see Chapter 9).

What are the advantages of plane-tabling?

32. Compared with other methods of plan surveying, plane-tabling is better in some ways because:

- it is the only method with which you can make a plan or map in the field;
- you need to find fewer points, as you draw the map while you survey;
- you can plot irregular lines and areas fairly easily and accurately;
- you can work quickly, once you learn how to use the method;
- you do not have to measure angles, so that you avoid several possible sources of error;
- you plot everything in the field, and so avoid missing any features you need to measure;
- you can easily check on the location of points you have plotted.

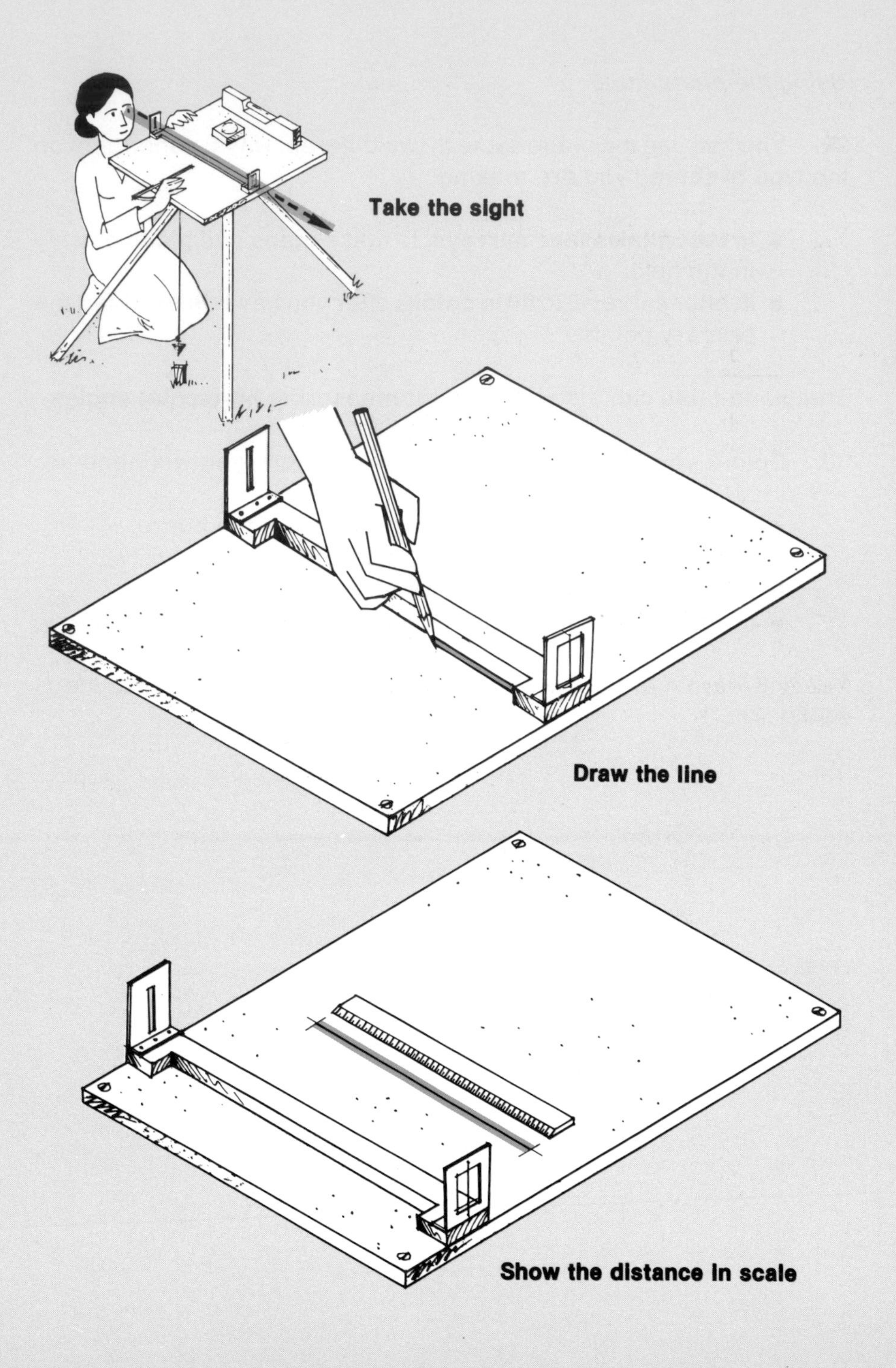

What are the disadvantages of plane-tabling?

33. Several disadvantages to plane-tabling are that:

- the plane-table and its extra equipment are heavy and fairly awkward to carry;
- learning how to use the plane-table correctly takes some time;
- you can only use the method in fairly open country, where you can see most of the points you are surveying;
- you cannot use the method in bad weather conditions, such as heavy rains or high winds.

Covering the board with drawing paper

34. You should try to find the best quality drawing paper possible to use with the plane-table. Since the paper will be exposed to outdoor conditions, you should prepare it to make it more resistant to changes in the humidity of the air. With a wet cloth, lightly dampen the paper and dry it several times before you use it. This is called **seasoning the paper.**

Note: be careful not to make the paper too wet when you season it.

35. Cut the sheet of drawing paper to a size 20 cm larger than the dimensions of your drawing board.

36. Cut the four corners of the paper off diagonally. To do this, measure 20 cm from each corner along its two sides, and mark the points. Join these points by diagonal lines, and cut along these lines.

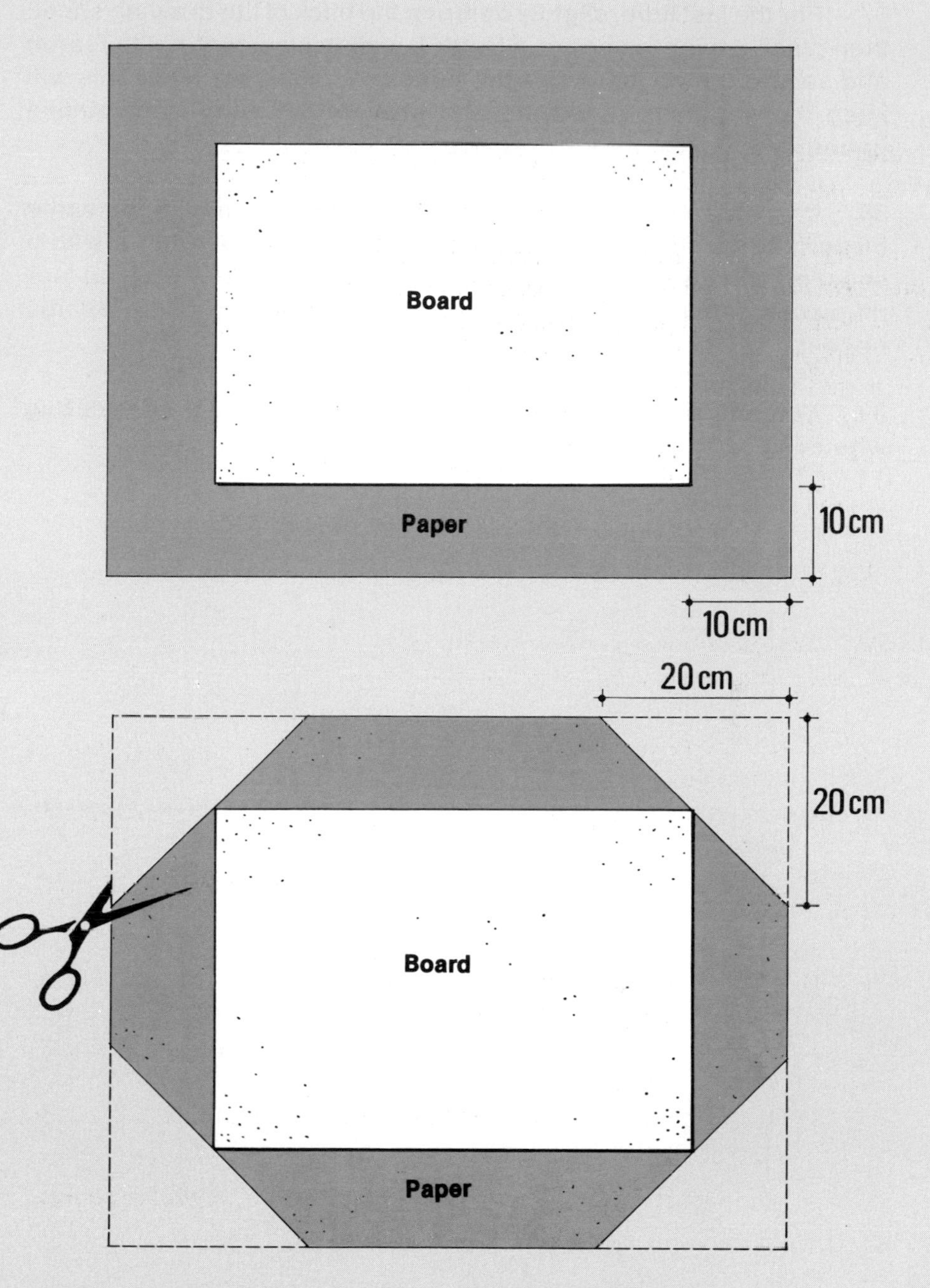

37. For the last time, **slightly dampen** the back of the drawing sheet, then place it over the board. **Stretch it well** (taking care not to tear it) and secure the edges **under the board** with **drawing pins**. This will keep the paper from moving and prevent the wind from getting underneath it.

38. If you plan to work in the field for several days with the same piece of drawing paper, you should protect it by **covering it with a sheet of smooth, heavy paper**. As you work in the field, you can tear off pieces of this cover sheet to expose the drawing paper as you need it.

39. You should keep the plane-table in a **waterproof canvas bag** when you carry it in the field.

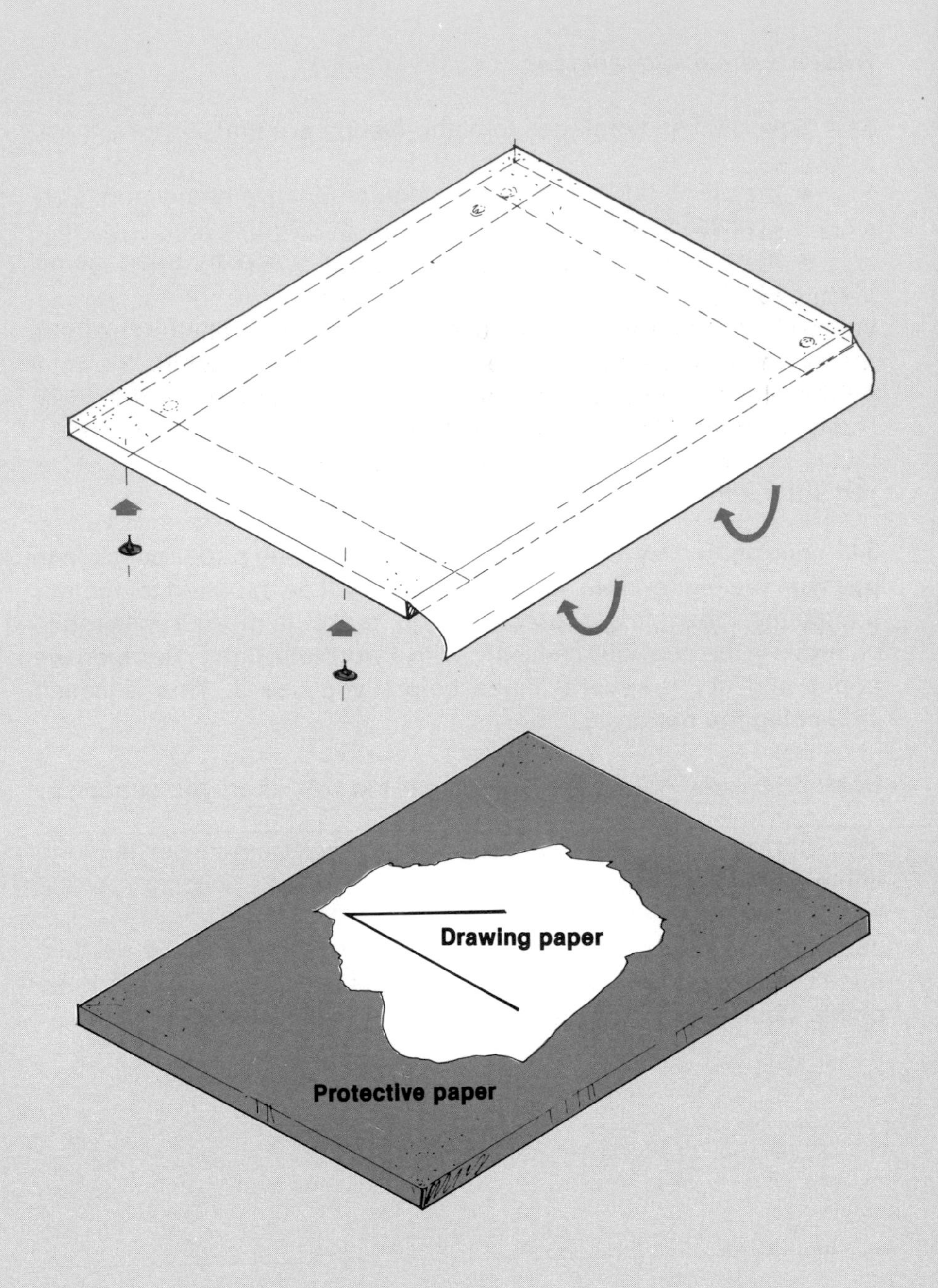

Setting up the plane-table

40. If you decide to start the survey from a selected station, first set up the plane-table over this station.

Note: you may need to set up the plane-table so that a point drawn on it is exactly over a corresponding ground point. You can use **a V-shaped metal arm and a plumb-line**, which you can easily make yourself. Otherwise, you can use calipers and a plumb-line. The metal arm or calipers should be placed with one tip touching the point on the plane-table and the other tip on the underside of the table. Hang the plumb-line from the point indicated on the underside of the table, and move the table until the plumb-line is directly over the ground point.

41. Spread the **tripod legs** well apart, and plant them firmly in the ground. The drawing board should be waist-high, so that you may bend over it **without resting against it**.

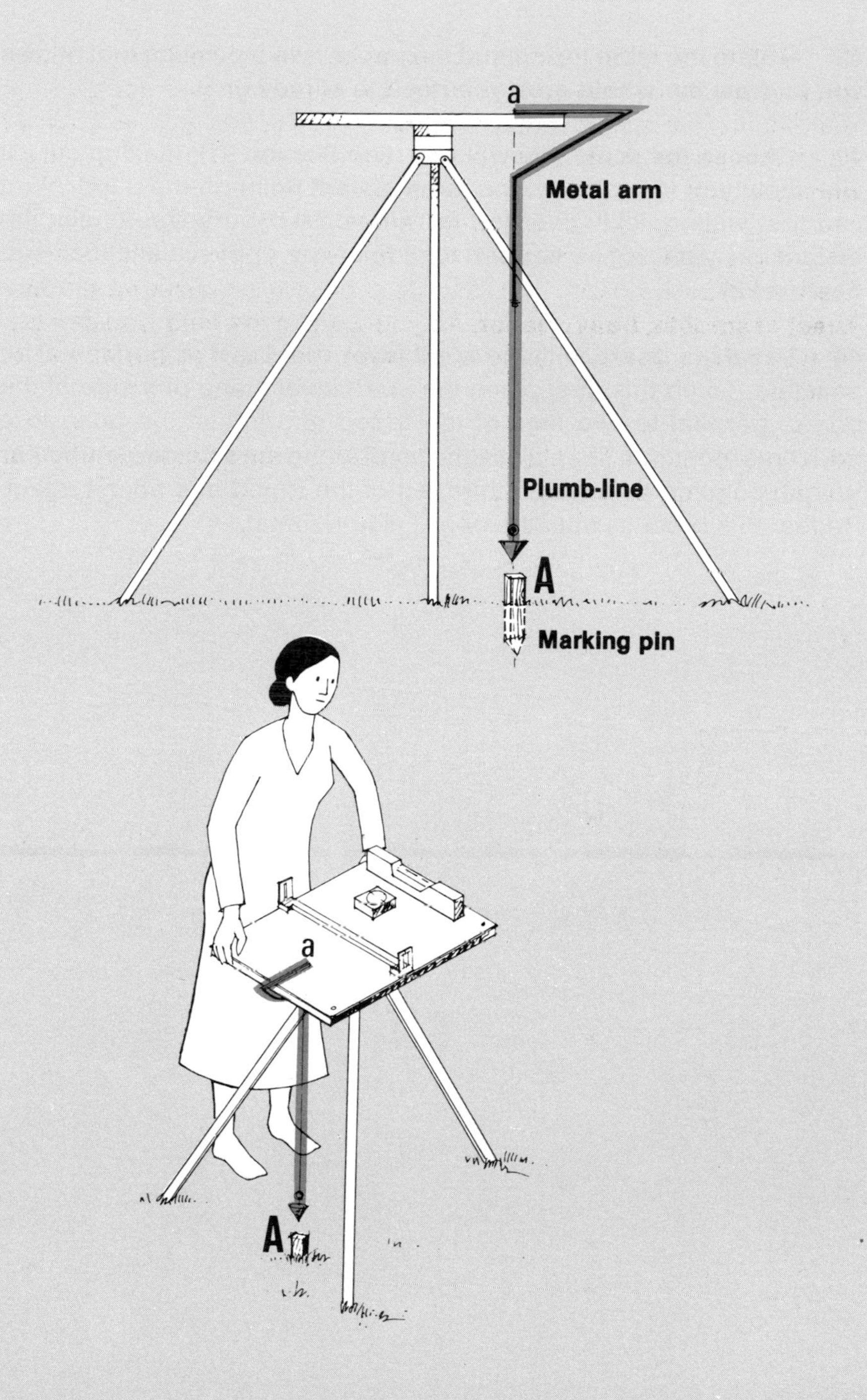

42. Rotate the table top so that the paper is in a position that allows you **to draw the whole area** you need to survey on it.

43. **Choose the scale** you will use (see Section 91), making sure it will allow you to plot even the **most distant point** on the paper. You can first walk quickly over the terrain you will survey to check the distances by pacing so you can decide on the right scale to use (see Section 22).

44. **Level** the board with the **spirit level**, making it as horizontal as possible. To do this, first place the spirit level along one side of the board, **parallel to two legs** of the tripod and adjust the table to a horizontal position. Then place the level along the **side perpendicular** to that, pointing toward the third leg of the tripod and adjust again. Repeat this process until the board is horizontal.

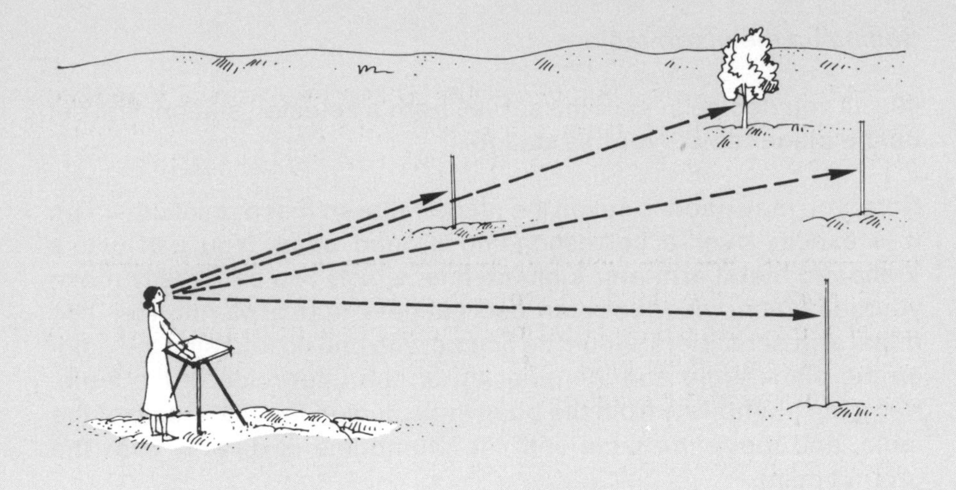

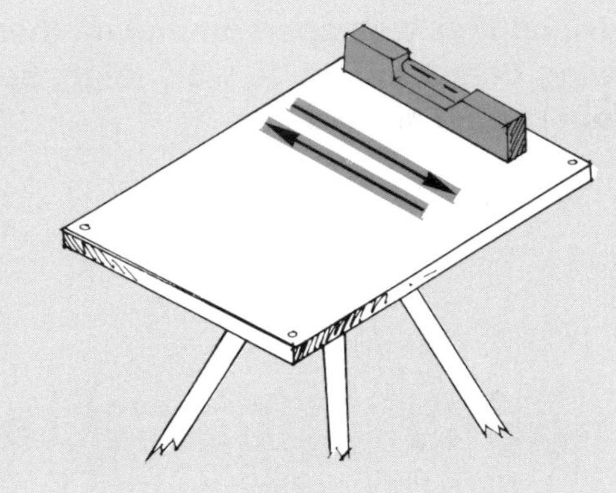

Level the table-top in both directions

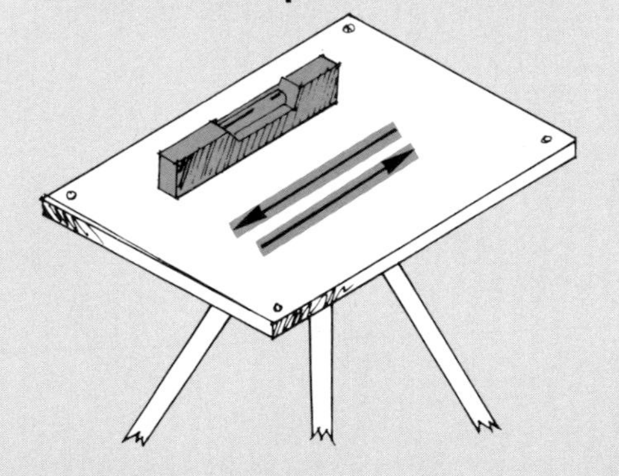

Orienting the plane-table

45. You can orient the plane-table either by using a **magnetic compass** or by **backsighting**. Usually, the board is first oriented roughly by compass, and then more precisely by backsighting.

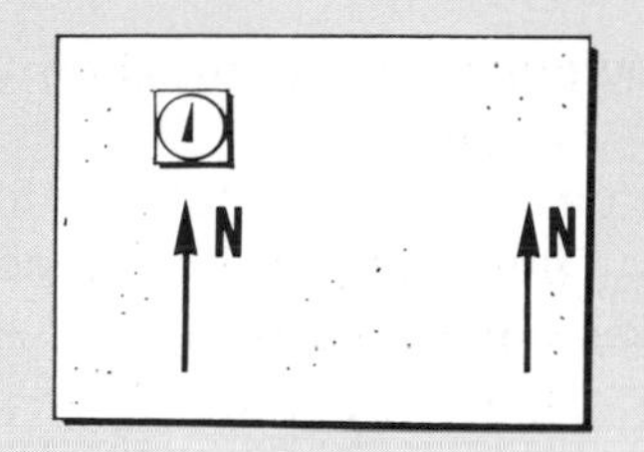

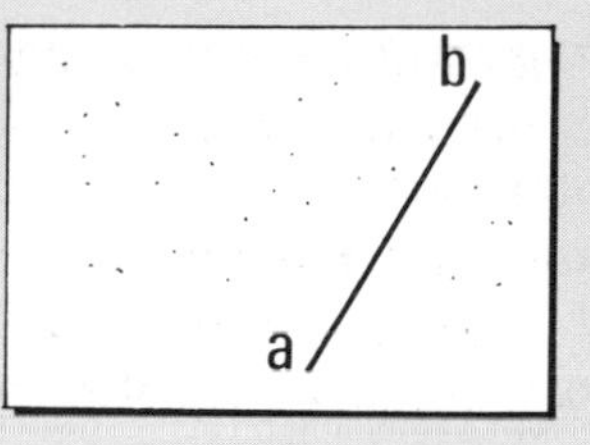

46. If you use **a magnetic compass** (see Section 32), rotate the compass until the direction of the needle lines up with the direction of south-north, or the 180º to 360º direction. Draw a line on the drawing paper showing this direction. Draw another line in the same direction on another part of the paper. Mark the north direction on these lines with an arrow and the letter N.

Note: remember to keep away from any materials which could have an effect on the magnetic needle of the compass (see Section 32, step 17).

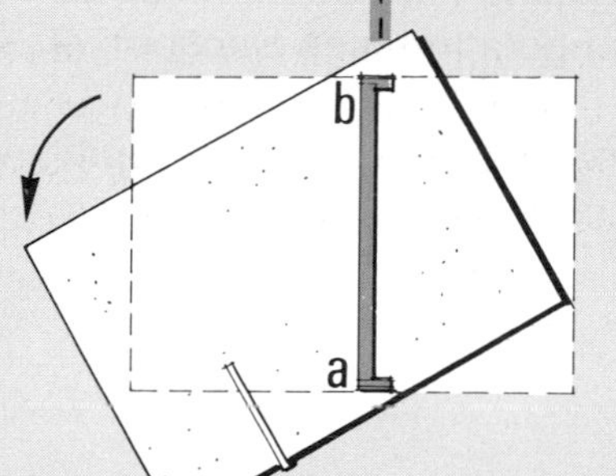

47. If at a surveying station you know the direction of a line which you have already plotted on the board, you can use that line to orient the plane-table by taking a **backsight**. It is the most precise way of orienting the plane-table and you should use it whenever possible.

Example

From station A, you have already plotted line ab. Set up the plane-table at station B. Place the centre-line of the alidade along line ba on the board. Rotate the board until the line of sight on the alidade lines up with line BA on the ground. The table is now oriented. You can proceed to survey and plot new points.

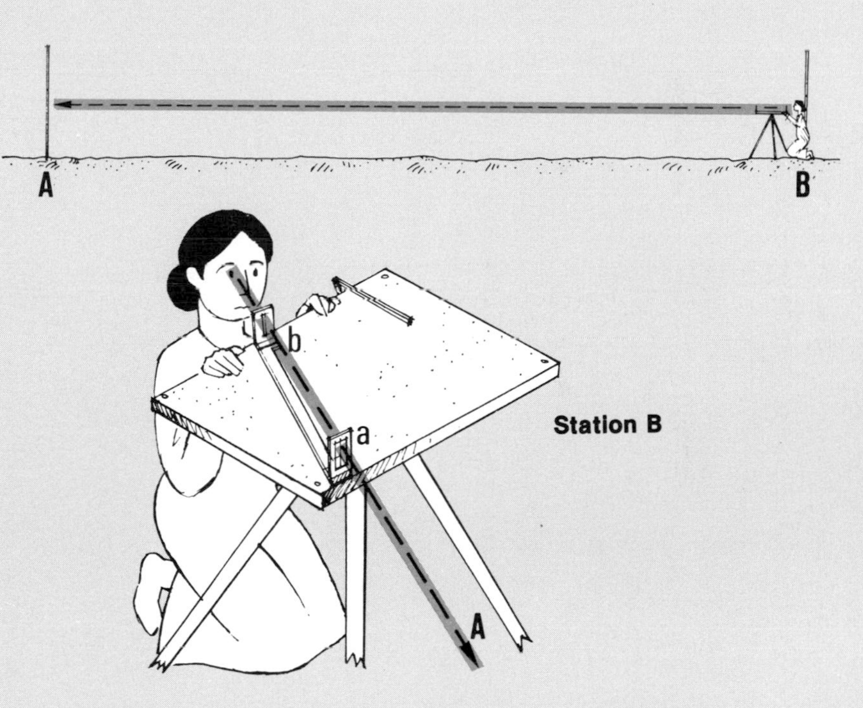

Plane-tabling methods for reconnaissance surveys

48. During reconnaissance surveys, you can use plane-tabling to quickly map out areas and open traverses. The survey will proceed by one of the methods described earlier in this chapter or a combination of them. This method may be:

- traversing (see Section 71);
- radiating (see Section 72); or
- triangulation (see Section 74).

You will learn more about mapping with a plane-table by these surveying methods in Chapter 9.

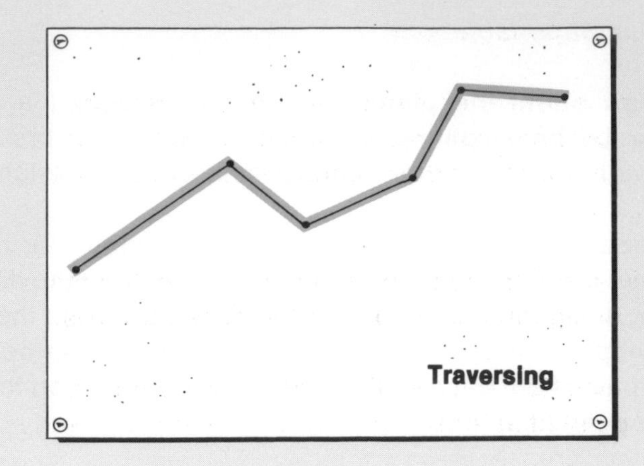

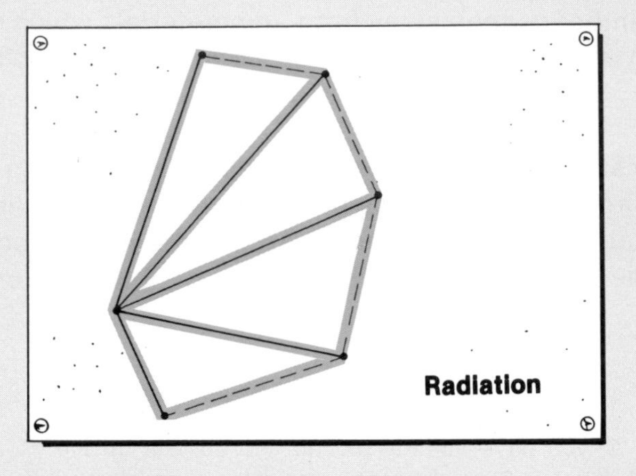

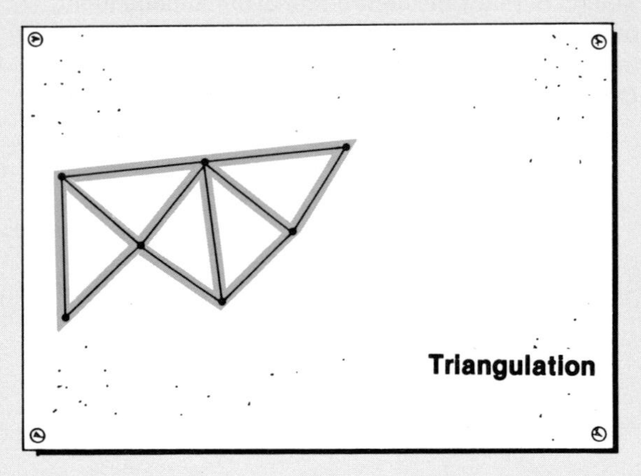

Plane-tabling for plotting details

49. When you have finished the reconnaissance survey and accurately mapped the main stations, you can further **use plane-tabling to locate details** such as rocks, buildings, a well or a group of trees.

50. To do this, set up the plane-table at each of the main **stations** in turn, and draw sighting lines to each of these features.

51. You can locate each detail on the drawing board by finding **the intersection point** of at least three sighting lines. You will not have to take any more measurements.

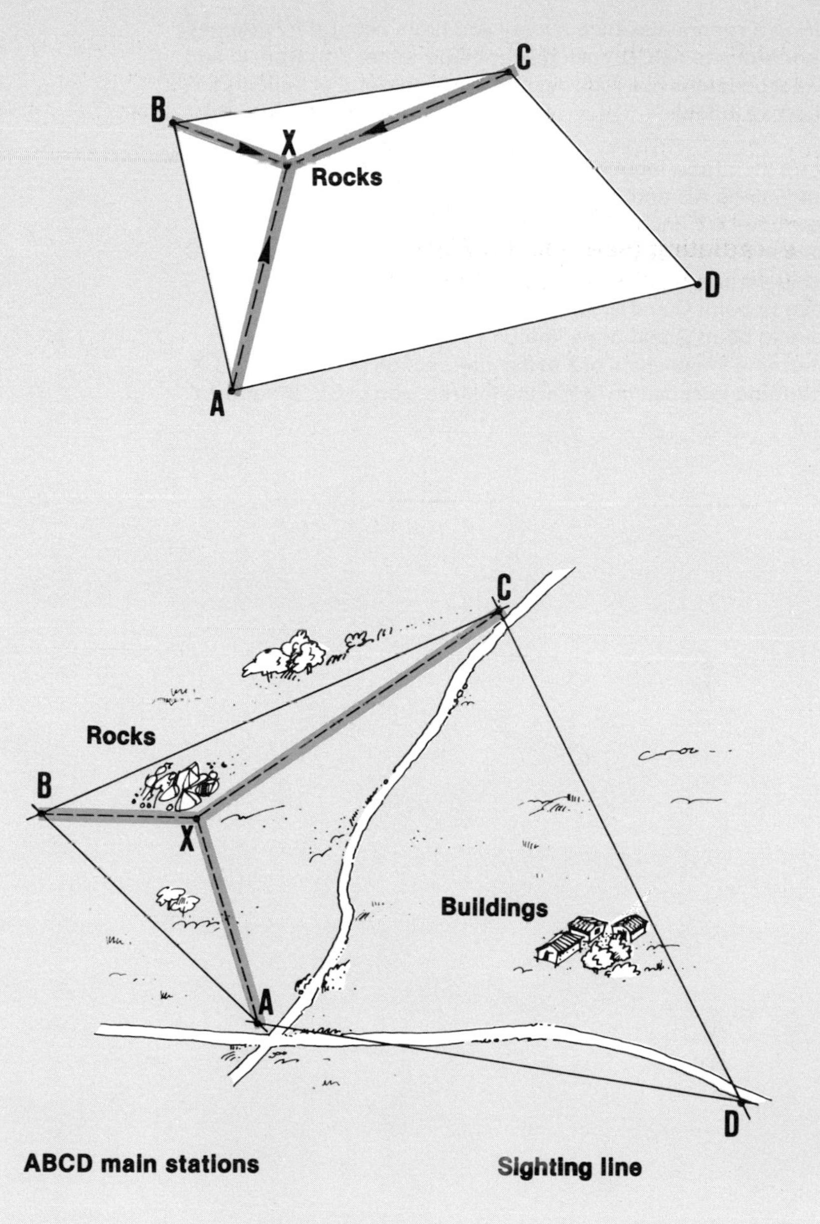

Example

During a reconnaissance survey you have accurately mapped the fish-farm site ABCDA using your plane-table. You want to add the exact positions of a rock outcrop X and a group of buildings Y. Proceed as follows:

- set up the plane-table over point A, orienting it by backsighting known lines AB and AD;
- draw lines AX and AY;
- move the plane-table to point B, orienting by lines AB and AC and draw line BX;
- move to point C and draw lines CX and CY;
- move to point D and draw line DY;
  determine the position of X at the intersection of AX, BX and CX;
- determine the position of Y at the intersection of CY, DY and AY.

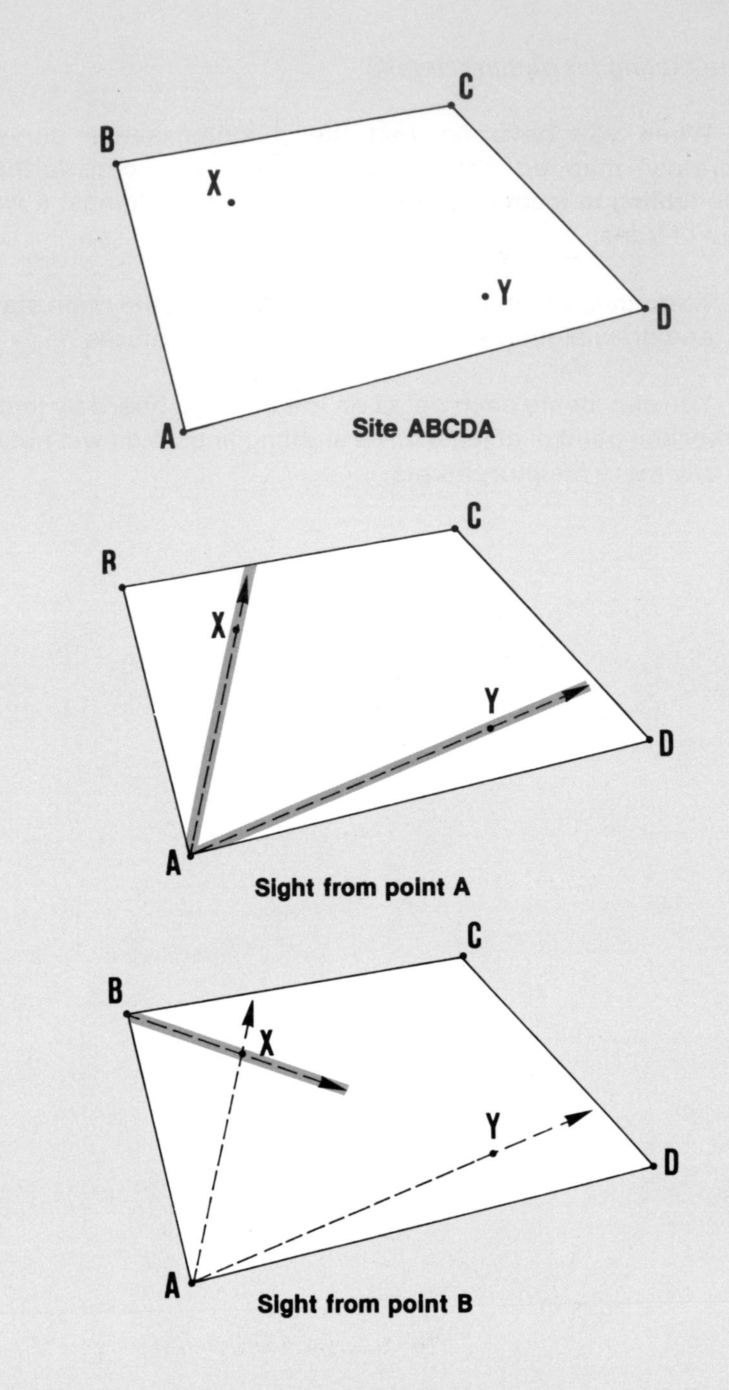

Site ABCDA

Sight from point A

Sight from point B

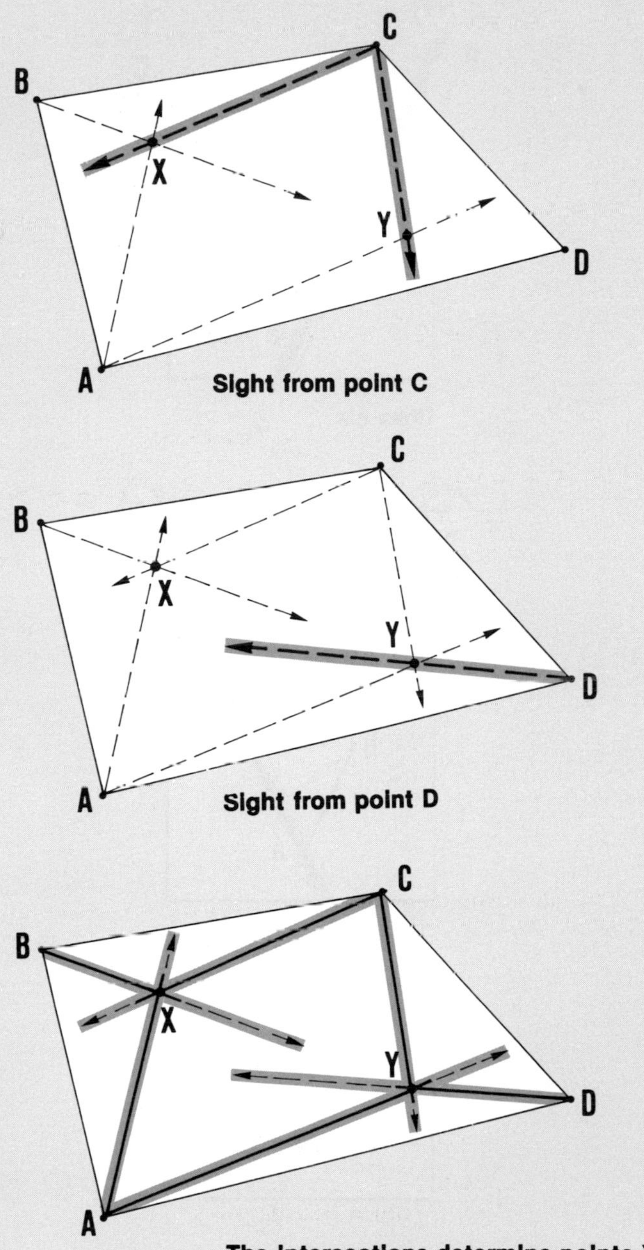

The intersections determine points X and Y

Measuring horizontal angles by plane-tabling

52. You can measure horizontal angles fairly accurately by drawing sighting lines on a plane-table and measuring this angle with a **protractor** (see Section 33).

Example

- You need **to measure angle BAC** formed by straight lines AB and AC, which have been well-marked in the field. Begin by setting up the plane-table at station A.
- Place the alidade so that it passes through point a, and sight at point B, and draw line ab.
- With the alidade passing through point a, sight at point c and draw line ac.
- Measure angle bac with a protractor.

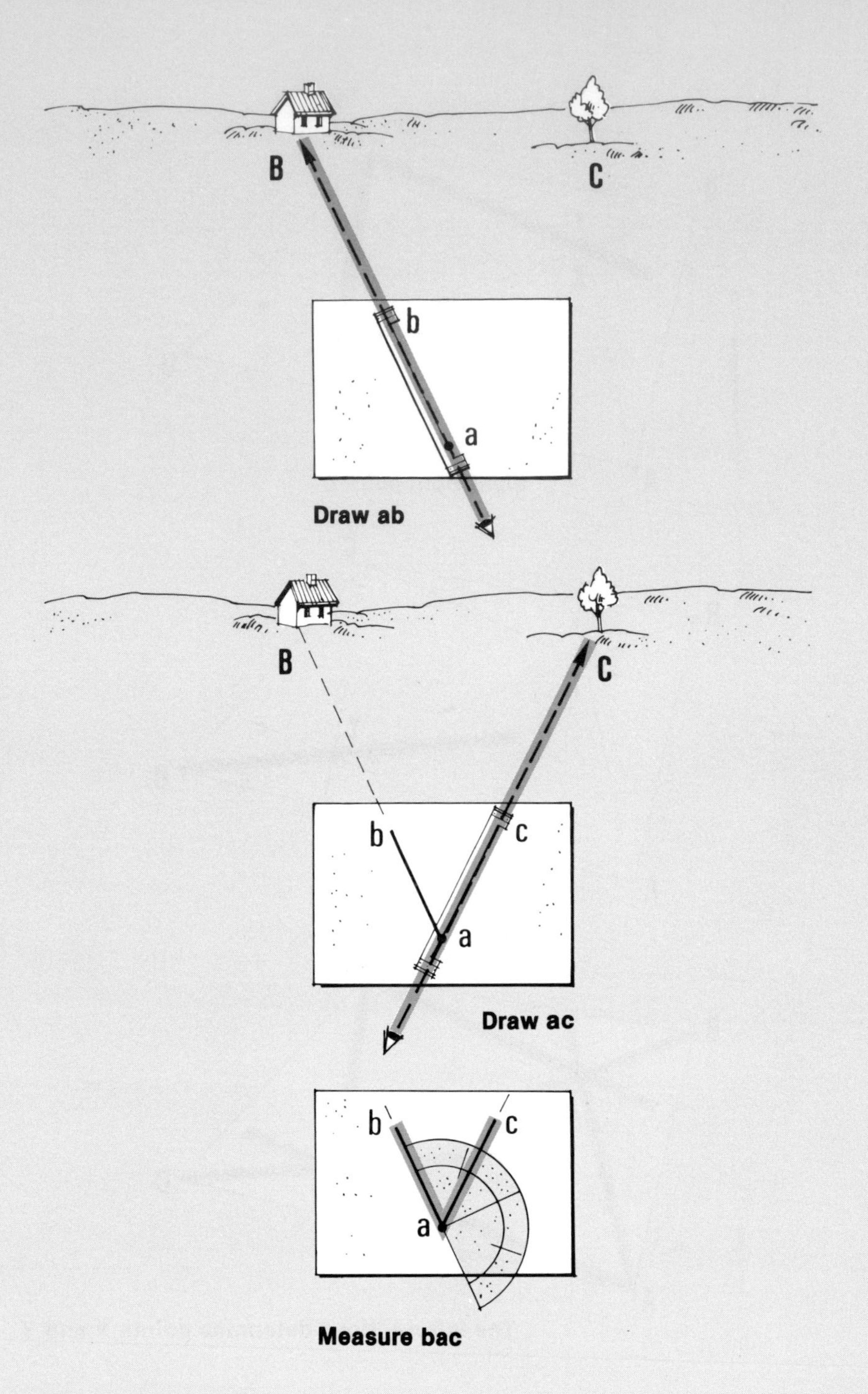

8 TOPOGRAPHICAL SURVEYS – DIRECT LEVELLING

1. In Chapters 5 and 6, you learned about various devices for measuring height differences. You also learned how to use these devices to solve **three types of problems** in measuring height differences, which you may face when you plan and develop a fish-farm (see Section 50). Now, you will learn **how to plan surveys** to solve these problems, **how to record** the measurements you make in your field-book, and **how to find** the information you need from these measurements.

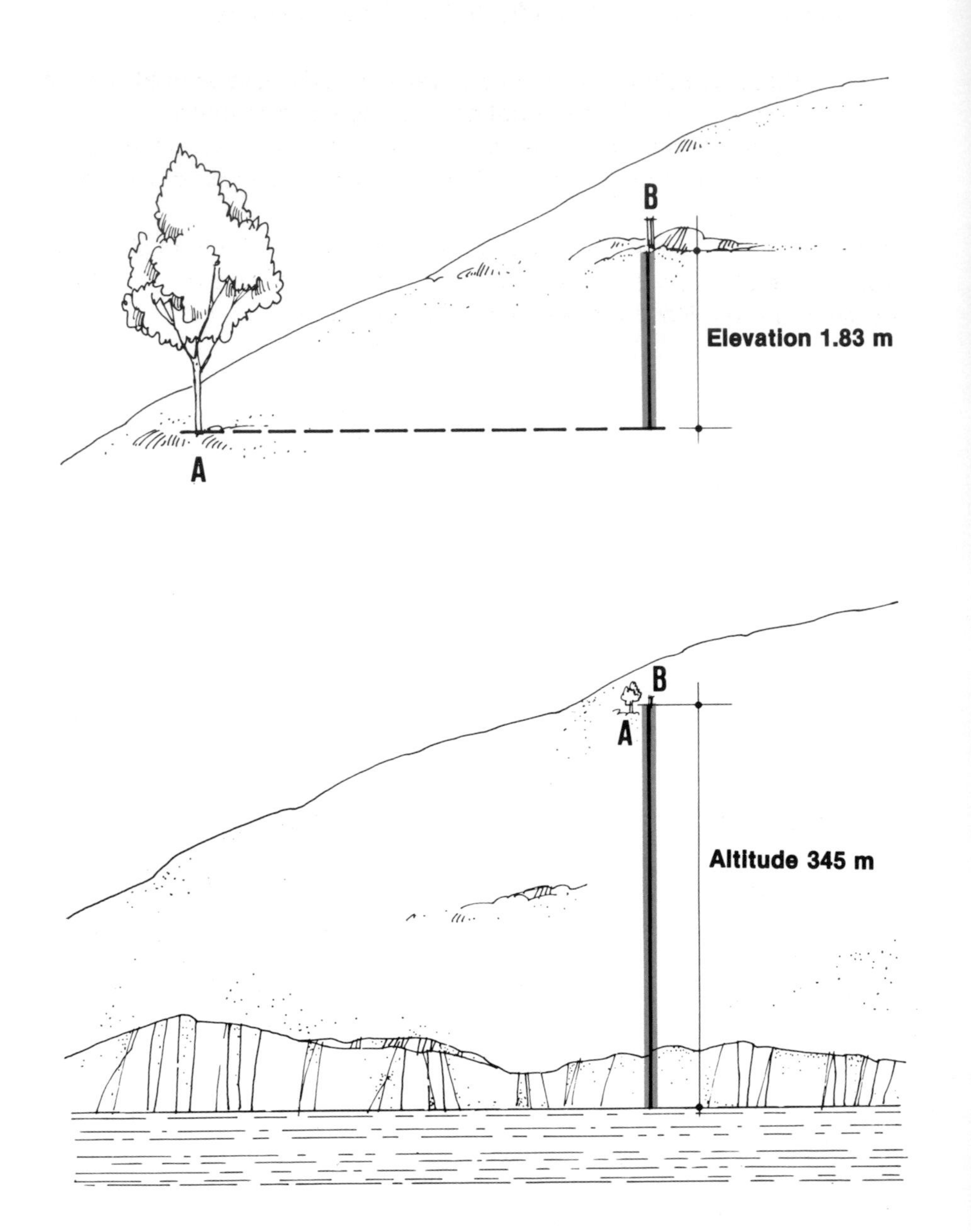

What are elevation and altitude?

2. You have learned what the height of a ground point is. Now, however, you will need to know a more accurate definition of this term.

- When the height of a point is its **vertical distance above or below** the surface of a reference plane you have selected, it is called **the elevation*** of that point.
- When the height of a point is its **vertical distance above or below mean sea level** (as the reference plane), it is called **the altitude*** of the point.

Example

Elevation of a point above a selected ground mark A	1.83 m
Altitude of the same point **above mean sea level (amsl)**	345 m

3. The vertical distance between two points is called the **difference in elevation**, which is similar to what you have learned as the difference in height (see Section 50). The process of measuring differences in elevation is called **levelling**, and is a basic operation in topographical surveys.

What are the main levelling methods?

4. You can level by using different methods, such as:

    - **direct levelling**, where you measure differences in elevation directly. This is the most commonly used method;
    - **indirect levelling**, where you calculate differences in elevation from measured slopes and horizontal distances.

You have already learned about indirect levelling in Section 50, when you learned to calculate differences in elevation from slopes or from vertical angles. Now you will learn about direct levelling.

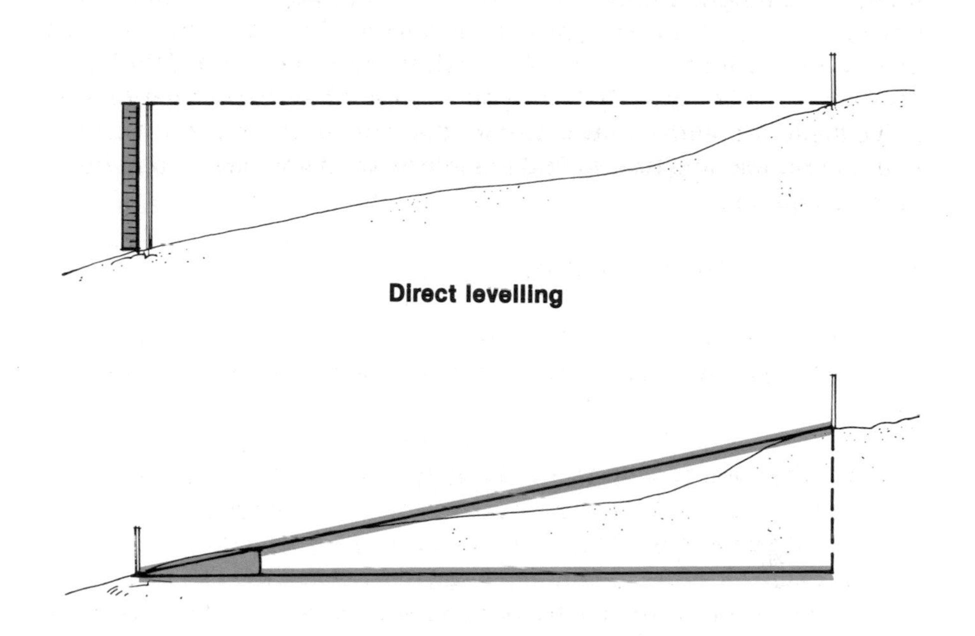

Direct levelling

Indirect levelling

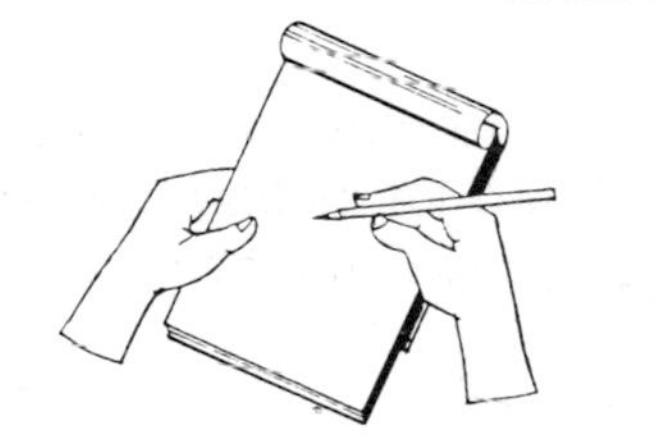

What are the kinds of direct levelling?

5. By direct levelling, you can measure both the elevation of points and the differences in elevation between points, using a **level** and a **levelling staff** (see Chapter 5). There are two kinds of direct levelling:

- differential levelling; and
- profile levelling.

6. In **differential levelling**, you find the difference in elevation of points which are some distance apart (see Section 81). In the simplest kind of direct levelling, you would survey only two points A and B from one central station LS. But you may need to find the difference in elevation between:

- either several points A, B, ... E, surveyed from a single levelling station LS; or
- several points A ... F, surveyed from a series of levelling stations LS1 ... LS6, for example.

7. In **profile levelling**, you find the elevations of points placed at short measured intervals along a known line, such as the centre-line of a water supply canal or the lengthwise axis of a valley. You find elevations for **cross-sections** with a similar kind of survey (see Section 82).

8. You can also use direct levelling to determine elevations for **contour surveying** (see Section 83), and for **setting graded lines of slope** (see Section 69), where you need to combine both differential levelling and profile levelling.

9. There are several simple ways to determine the elevations of ground points and the differences in elevation between ground points. You will use a level and a levelling staff with these methods. In the following sections, each method is fully described to help you choose between them. **Table 10** will also help you to compare the various methods and to select the one best suited to your needs in each type of situation you may encounter.

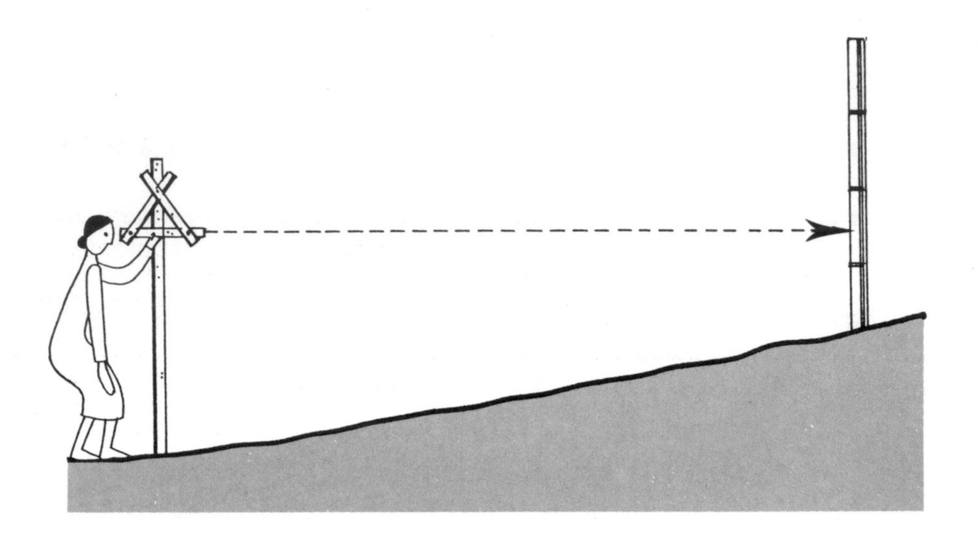

Differential levelling

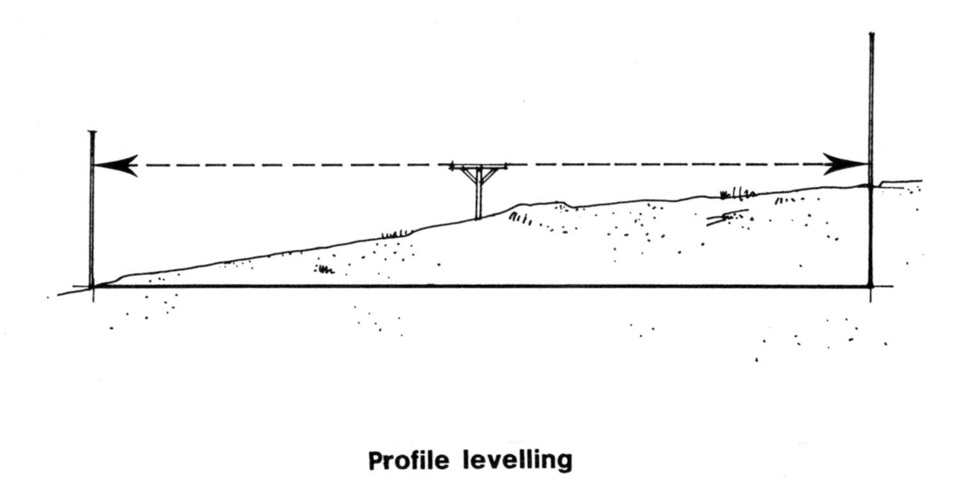

Profile levelling

TABLE 10
Direct levelling methods

Section	*Type*	*Method*	*Suitability*	*Remarks*
81	Differential levelling	Open traverse	Long, narrow stretch of land	Check for closing error
81	Differential levelling	Closed traverse	Perimeter of land area and base line for radiation	Check for closing error Combine with radiating
81	Differential levelling	Square-grid	Small area with little vegetation	Squares 10 to 20 m and 30 to 50 m
81	Differential levelling	Radiating	Large area with visibility	Combined with closed traverse
82	Longitudinal profile levelling	Open traverse	Non-sighting and sighting level	Check for closing error
82	Cross-section profile levelling	Radiating	Sighting level with visibility	
83	Contouring	Direct	Detailed mapping of small area with a sighting or a non-sighting level and target levelling staff	Slow and accurate Progress uphill
83	Contouring	Square-grid	Small area with little vegetation Especially if perimeter has been surveyed Small to medium scale mapping	Terrain, scale and accuracy depend on contour interval Progress uphill Suitable for plane-tabling
83	Contouring	Radiating	Small to medium scale mapping of large area	Fast and fairly inaccurate Progress uphill Suitable for plane-tabling
83	Contouring	Cross-sections	Preliminary survey of a long and narrow stretch of land	Fast, fairly inaccurate Progress uphill Suitable for plane-tabling

81 How to level by differential

What is differential levelling?

1. You can best understand differential levelling by first considering only two points, **A** and **B**, both of which you can see from one central levelling station, **LS**.

- Sight with a level from LS at the levelling staff on point A. The point where the line of sight meets the levelling staff is point X. Measure AX. This is called a **backsight (BS)**.
- Turn around and sight from LS at the levelling staff on point B. The point where the line of sight meets the levelling staff is point Y. Measure BY. This is called a **foresight (FS)**.
- **The difference in elevation** between point A and point B equals BC or (AX– BY) or (backsight **BS**– foresight **FS**).

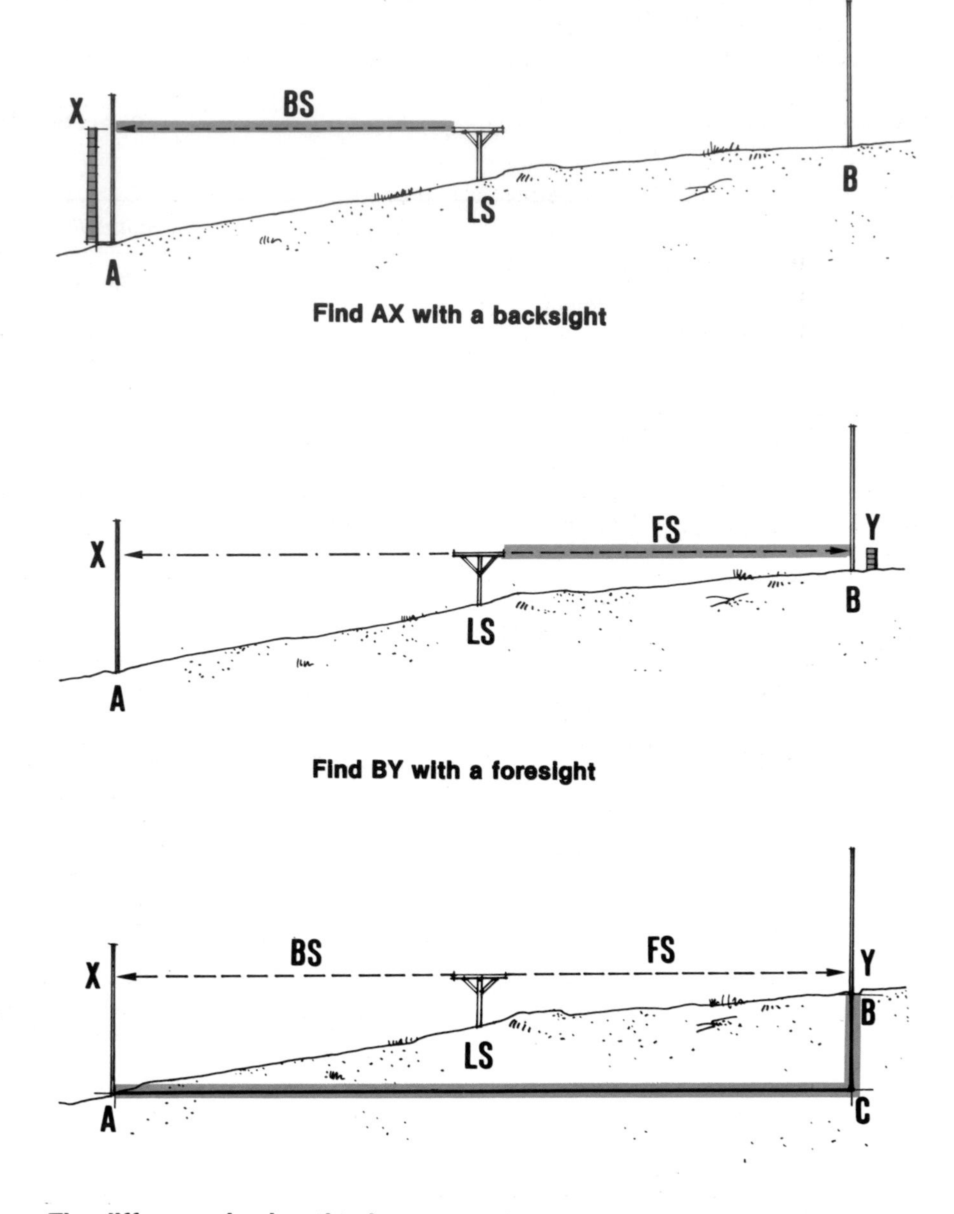

Find AX with a backsight

Find BY with a foresight

The difference in elevation between points A and B equals AX minus BY

- If you know the elevation of A, called E(A), you can calculate **the elevation of B, called E(B)**, as **BS**−**FS** + E(A).
- But BS + E(A) = **HI, the height of the instrument** or **the elevation of the line of sight** directed from the level.
- Therefore,

$$E(B) = HI - FS$$

(the elevation at point B being equal to the height of the levelling instrument, minus the foresight).

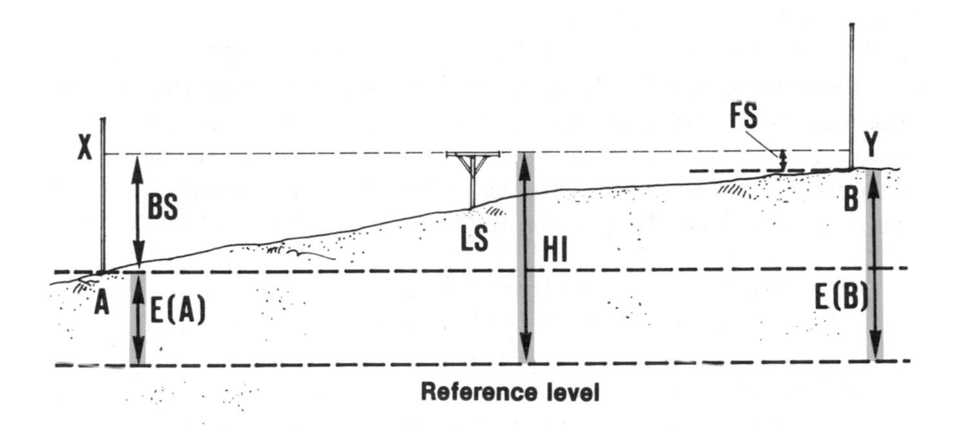

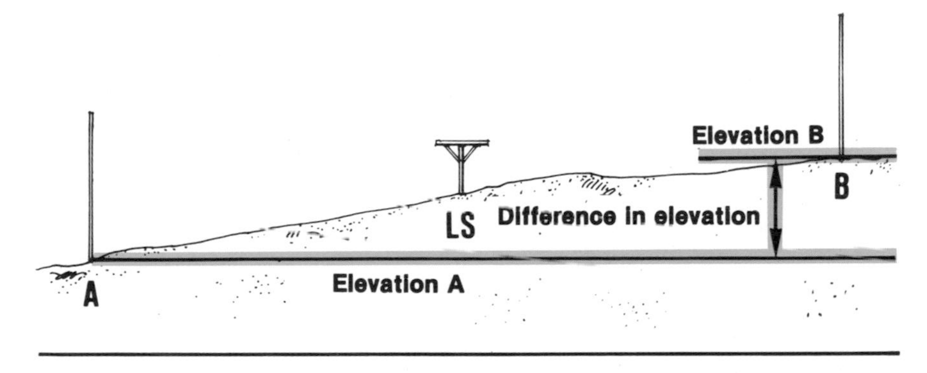

What are backsights and foresights?

It is important for you to understand exactly what **"backsight"** and **"foresight"** are in direct levelling.

2. **A backsight (BS)** is a sight taken with the level to **a point of known elevation E**, so that the height of the instrument HI can be found. A backsight in direct levelling is usually taken in a backward direction, but not always. Backsights are also called **plus sights (+ S)**, because you must always add them to a known elevation to find HI.

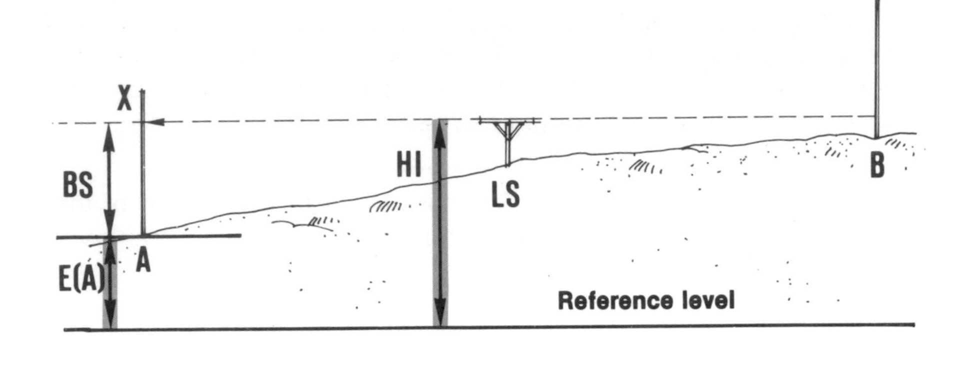

HI = BS + E

HI = BS + E

3. **A foresight FS** is also a sight taken with the level, but it can be **on any point of the sight line** where you have to determine the elevation. You will usually take it in a forward direction, but not always. Foresights are also called **minus sights (–S)**, because they are always subtracted from HI to obtain the elevation E of the point. Remember:

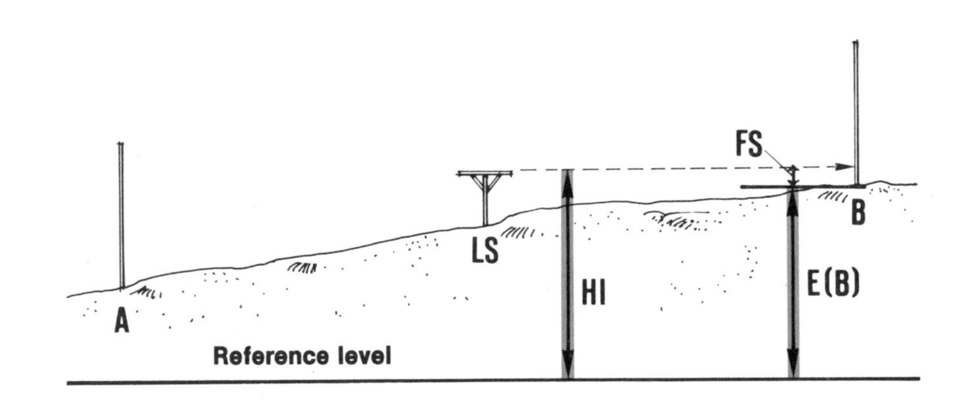

E = HI - FS

E = HI - FS

Surveying two points with one turning point

4. Often you will not be able to see at the same time the two points you are surveying, or they might be far apart. In such cases, you will need to do a **series of differential levellings**. These are similar to the type explained above, except that you will use intermediate temporary points called **turning points (TP)**.

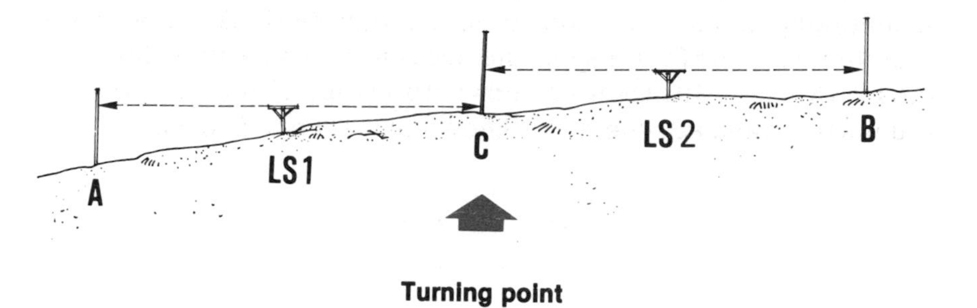

5. You know the elevation of point A, E(A) = 100 m, and you want to find the elevation of point B, E(B), which is not visible from a central levelling station. Choose a **turning point** C about halfway between A and B. Then, set up the level at LS1, about halfway between A and C.

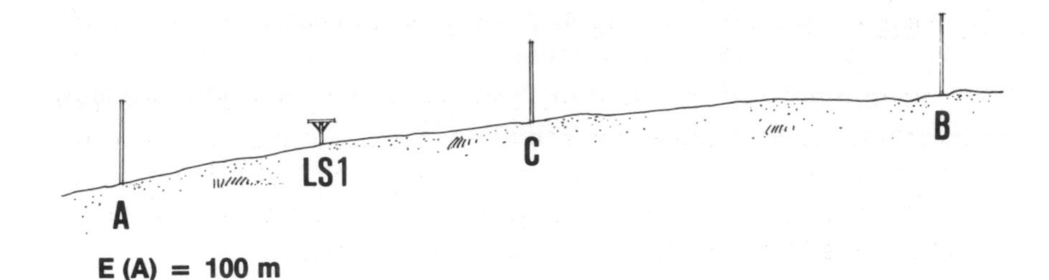

6. Measure a backsight on A (for example, BS = 1.89 m). Measure a foresight on C, FS = 0.72 m. Calculate **HI** = BS + E(A) = 1.89 m + 100 m = 101.89 m. Find the elevation of turning point C as **E(C)** = HI − FS = 101.89 m − 0.72 m = 101.17 m.

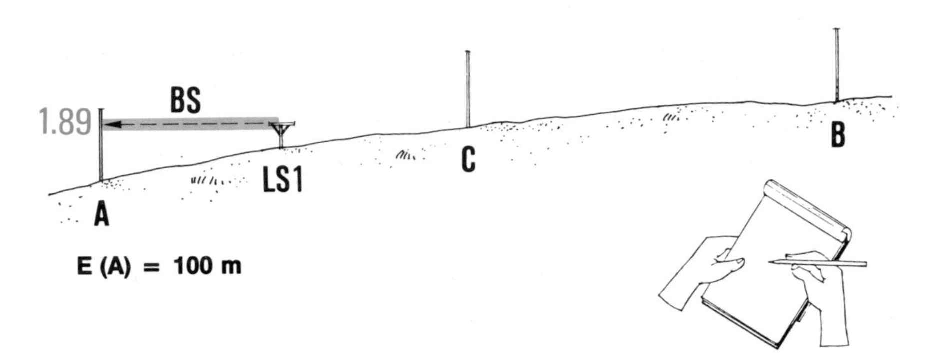

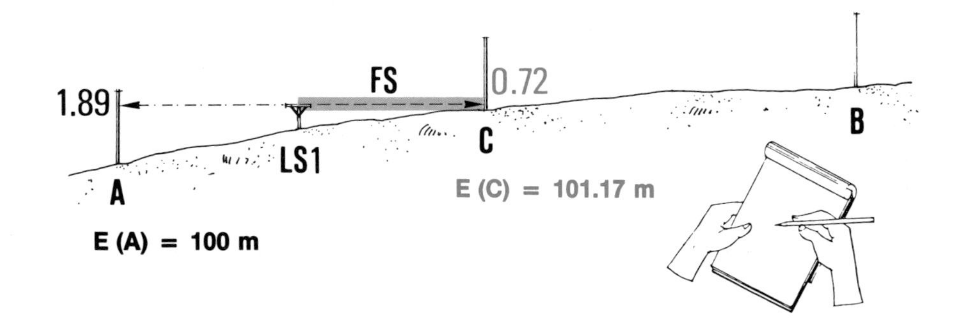

7. Move to a second levelling station, LS2, about halfway between C and B. Set up the level and measure BS = 1.96 m, and then FS = 0.87 m. Calculate **HI** = BS + E(C) = 1.96 m + 101.17 m = 103.13 m. Obtain **E(B)** = HI − FS = 103.13 m − 0.87 m = 102.26 m.

8. You can make the calculations more easily if you **record the field measurements in a table**, as shown in the example. You will not make any intermediate calculations. All BS's and all FS's must be added separately. The sum FS is subtracted from the sum BS to find the difference in elevation from point A to point B.

- **A positive difference** means that B is at a higher elevation than A.
- **A negative difference** means that B is at a lower elevation than A.

Knowing the elevation of A, you can now easily calculate the elevation of B. In this case, E(B) = 100 m + 2.26 m = 102.26 m; this is the same as the result in step 7, which required more complicated calculations. This kind of calculation is called an arithmetic check.

Example

Table form for differential levelling with one turning point.

Points	BS (+S)	FS (-S)	Elevation (m)	Remarks
A	1.89	–	100.00	Corner stone
TP(C)	1.96	0.72		property
B	–	0.87	**102.26**	
Sum	3.85	1.59		
FS(–S)	–1.59			
Difference elevation	+ 2.26			B is above A

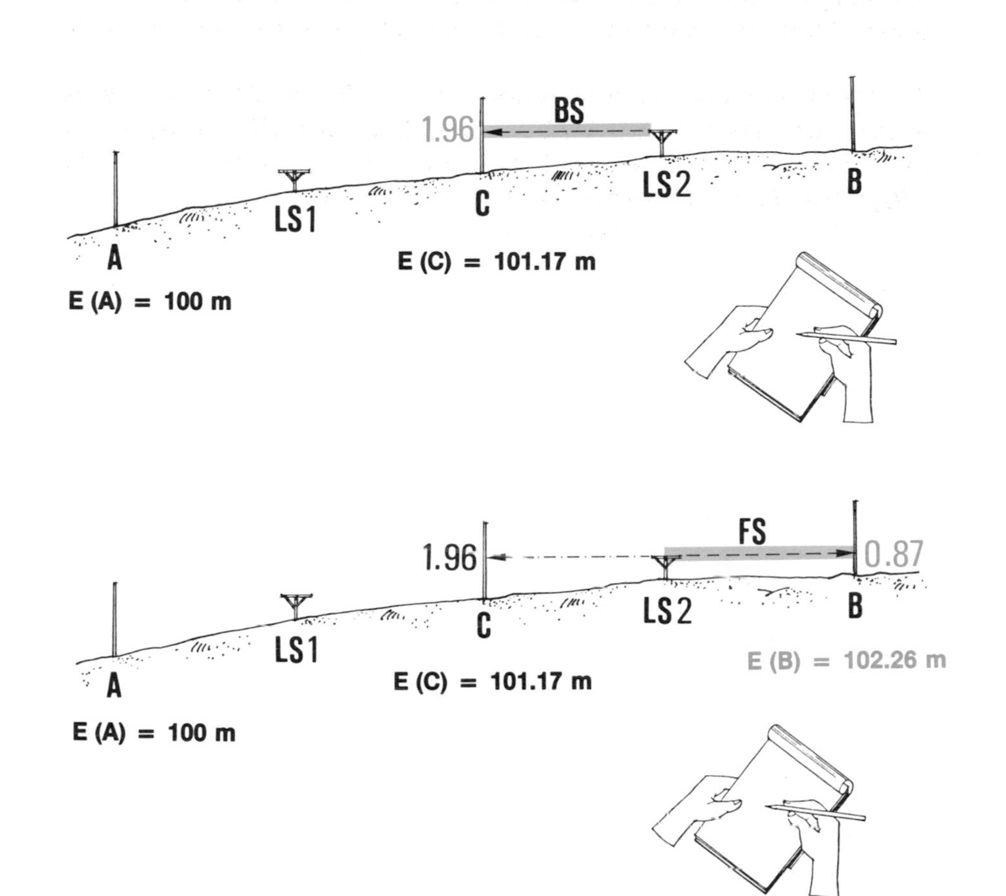

Surveying two points using several turning points

9. Often you will need to use more than one turning point between a point of known elevation and another point of unknown elevation. To do this, you can use the procedure you have just learned, but you will need to **record the field measurements in a table** to make calculating the results easier.

10. Knowing the elevation of point A, you need to find the elevation of B. To do this, you need five **turning points**, TP1 ... TP5, and six levelling stations, LS1 ... LS6.

Note: the turning points and the levelling stations do not have to be on a straight line, but try to place each **levelling station about halfway** between the two points you need to survey from it.

11. From each levelling station, measure a **backsight (BS)** and a **foresight (FS)**, except:

- **at starting point** A, where you have only a **backsight** measurement; or
- **at ending point** B, where you have only a **foresight** measurement.

Using step 8 as a guideline, enter all measurements in a table and calculate the results. You will find that point B is 2.82 m higher than point A and, therefore, that its elevation is E(B) = 100 m + 2.82 m = 102.82 m.

Example

Table form for differential levelling with several turning points.

Points	BS (+S)	FS (–S)	Elevation (m)	Remarks
A	1.50	–	100.00	Iron rod at foot of big tree
TP1	1.71	1.00		
TP2	1.85	1.15		
TP3	1.67	1.25		
TP4	1.45	1.13		
TP5	1.35	1.12		
B	–	1.06	**102.82**	
Sums	9.53	6.71		
FS(–)	–6.71			
D(E)	+2.82			B is above A

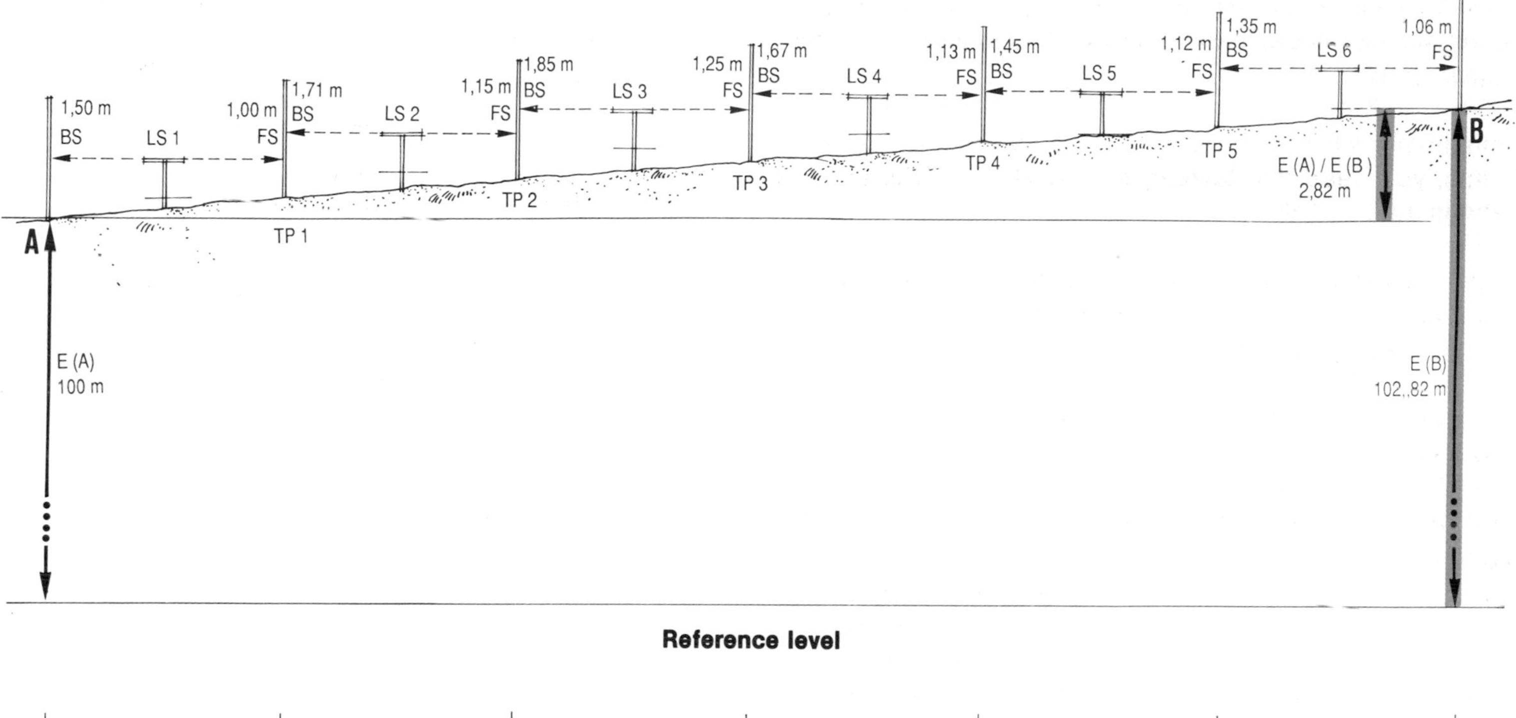
1,50 m
BS
LS 1
1,00 m
FS
1,71 m
BS
LS 2
1,15 m
FS
1,85 m
BS
LS 3
1,25 m
FS
1,67 m
BS
LS 4
1,13 m
FS
1,45 m
BS
LS 5
1,12 m
FS
1,35 m
BS
LS 6
1,06 m
FS
A
B
TP 1
TP 2
TP 3
TP 4
TP 5
E (A) / E (B)
2,82 m
E (A)
100 m
E (B)
102,.82 m
Reference level

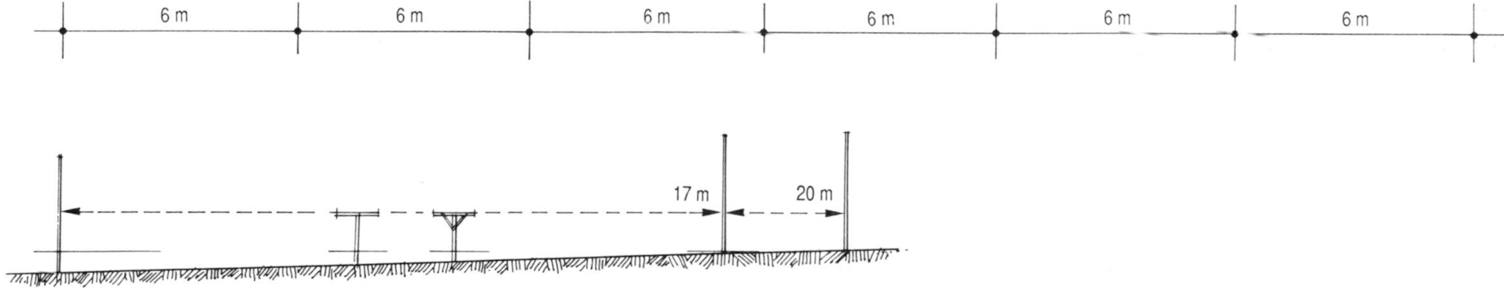
6 m
6 m
6 m
6 m
6 m
6 m
17 m
20 m

12. Even if you are careful, you may still make mistakes when you make your arithmetic calculations from the table. To reduce this kind of error, add **two additional columns** to your table that will make checking your calculations easy. In these columns, enter the difference (BS – FS), either positive (+) or negative (-), between the measurements you took at each levelling station. For example, from LS1 you measure BS (A) = 1.50 m and FS (TP1) = 1.00 m. The difference 1.50 m – 1.00 m = 0.50 m is positive, and you enter it in the (+) column on the TP1 line.

The arithmetic sum of these differences should be equal to the calculated difference in elevation D(E) = + 2.82 m. These columns will also help you to calculate **the elevation of each turning point**, and to **check** on the elevation of point B more carefully.

Example

Differential levelling with several turning points.

Points	BS	FS	(BS – FS)		Elevations	Remarks
	(+)	(–)	+	–	(m)	
A	1.50	–	–	–	100.00	Assumed elevation
TP1	1.71	1.00	0.50	–	100.50	Gate to farm
TP2	1.85	1.15	0.56	–	101.06	Paths' junction
TP3	1.67	1.25	0.60	–	101.66	Corner of maize field
TP4	1.45	1.13	0.54	–	102.20	Centre of path
TP5	1.35	1.12	0.33	–	102.53	Foot of large tree
B	–	1.06	0.29	–	102.82	Rock along path
Sums	9.53	6.71	2.82	–		
FS(–)	–6.71					
D(E)	+ 2.82					These two values should be the same

13. By now, you have learned enough to make a topographical survey of two distant points by measuring the horizontal distance between them and the difference in their elevation.

When you survey a future fish-farm site, you will use a very similar method. You can then prepare a topographic map of the site (see Chapter 9), which will become a useful guide for designing the fish-farm.

14. This is a survey method using **straight open traverses**, that is, several intermediate stations along one straight line. You know the elevation of starting point A, E(A) = 63.55 m. You want to know the distance of point B from point A, and its elevation. Because of the type of terrain on which you are surveying, you cannot see point B from point A, and you need two **turning points, TP1** and **TP2**, for levelling. Measure horizontal distances as you move forward with the level, from point A toward point B; try to progress along a straight line. If you cannot, you will need to use the **broken open traverse** survey method, which involves measuring the azimuths of the traverse sections as you move forward and change direction (see step 17).

15. **Set out a table** like the one in step 12, and add two columns to it for horizontal distances. Enter all your distance and height measurements in the main part of the table. Then, in the first additional column, record each **partial distance** you measure from one point to the next one. In the second column, note the **cumulated distance**, which is the distance calculated from the starting point A to the point where you are measuring. The last number in the second column will be total distance AB.

16. **Conclusions**. Point B is 1.55 m higher than A and its elevation is 65.10 m. It is 156.5 m distant from point A. The arithmetic check from the (BS — FS) differences agrees with the calculated difference in elevation.

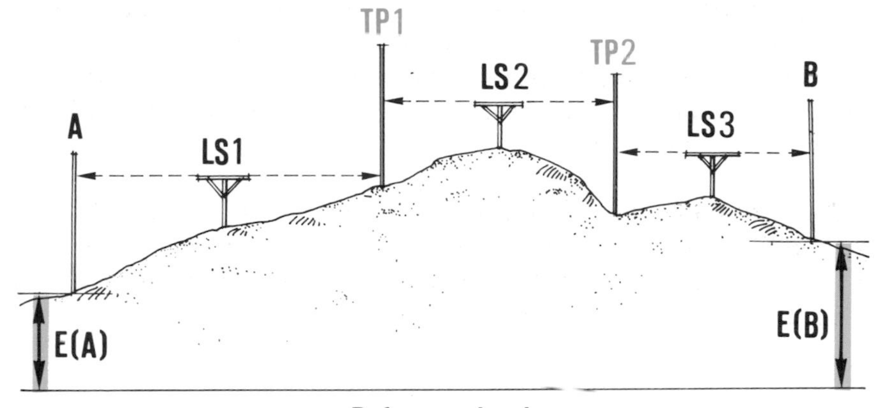

Reference level

Example

Topographical survey of a straight open traverse by differential levelling

Points	Distances		BS	FS	(BS – FS)		Elevations	Remarks
	Partial	Cumulative	(+)	(–)	+	–	(m)	
A	0	0	3.20	–	–	–	63.55	Known elevation
TP1	62.2	62.2	3.15	1.20	2.00	–	65.55	Corner cassava field
TP2	59.5	121.7	1.00	2.75	0.40	–	65.95	Centre of path
B	34.8	**156.5**	–	1.85	–	0.85	**65.10**	Foot of tall tree
Sums	–	–	7.35	5.80	2.40	0.85	65.10	
Differences	–	–	–5.80		–0.85		–63.55	
CHECKS			+ 1.55		+ 1.55		+ 1.55	

Making topographical surveys by broken open traverses

17. Remember, that if you survey by **broken open traverses** (or zigzags), you will also have to measure the azimuth of each traverse section as you proceed, in addition to distances and elevations.

18. You need to survey open traverse ABCDE from known point A. You require four turning points, TP1, TP2, TP3 and TP4. You want to know:

- the elevations of points B, C, D and E;
- the horizontal distances between these points;
- the position of each point in relation to the others, which will help you in mapping them.

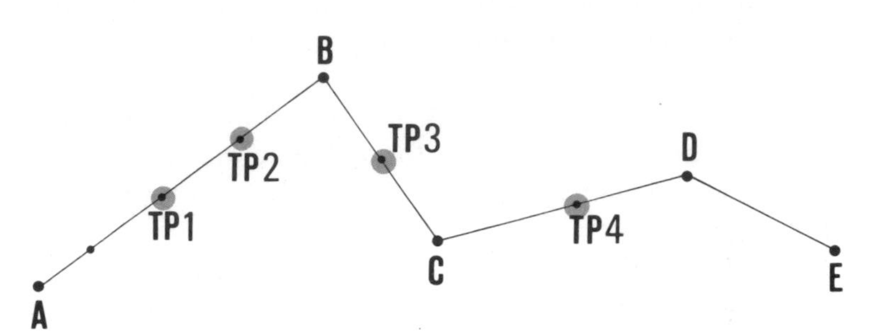

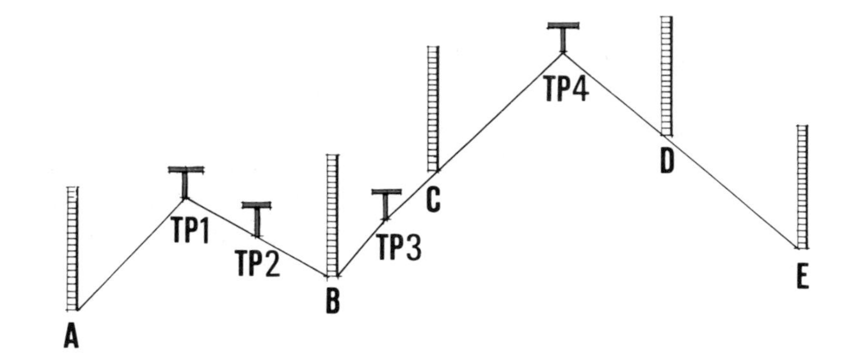

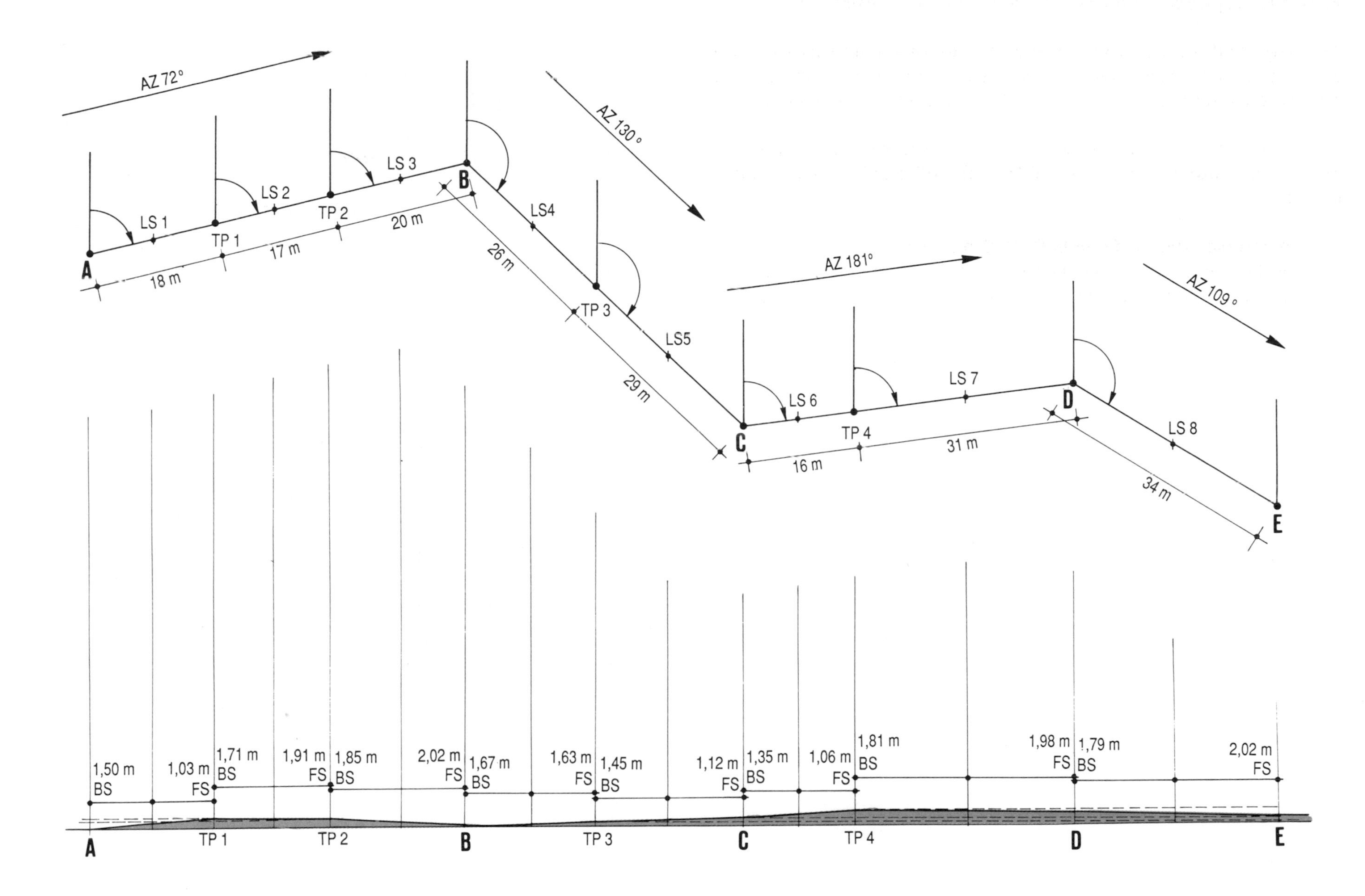
AZ 72°
AZ 130°
AZ 181°
AZ 109°
A
B
C
D
E
LS 1
LS 2
LS 3
LS4
LS5
LS 6
LS 7
LS 8
TP 1
TP 2
TP 3
TP 4
18 m
17 m
20 m
26 m
29 m
16 m
31 m
34 m
1,50 m BS
1,03 m FS
1,71 m BS
1,91 m FS
1,85 m BS
2,02 m FS
1,67 m BS
1,63 m FS
1,45 m BS
1,12 m FS
1,35 m BS
1,06 m FS
1,81 m BS
1,98 m FS
1,79 m BS
2,02 m FS

Proceed with the differential levelling as described earlier, measuring **foresights** and **backsights** from each levelling station. Measure azimuths and horizontal distances as you progress from the known point A toward the end point E. All the azimuths of the turning points of a single line should be the same. This will help you check your work.

19. Make **a table** similar to the one shown in step 15, and add **three extra columns** to it for recording and checking **the azimuth values** (see Section 71, step 17). Enter all your measurements in this table. At the bottom of the table, make all the checks on the elevation calculations, as you have learned to do them in the preceding steps.

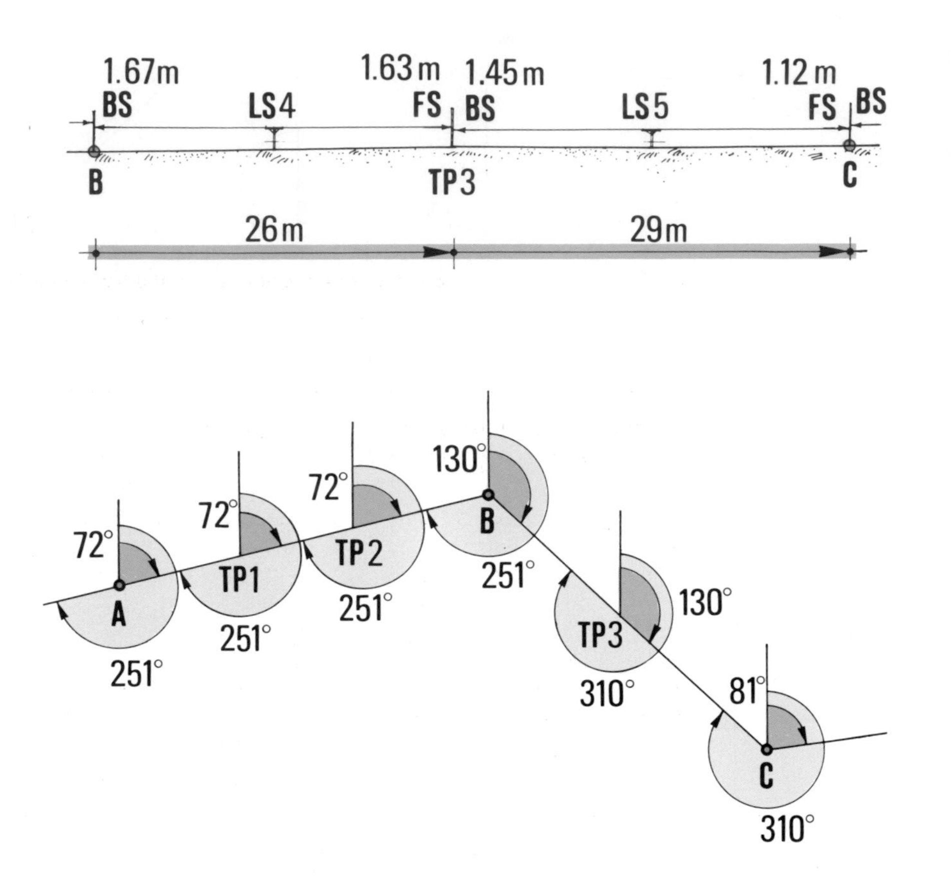

Chart 2

Example of a topographical survey of a broken open traverse by differential levelling

Points	Distances (m) Partial	Distances (m) Cumulative	Az(degrees) FS	Az(degrees) BS	Diff. Az FS/BS (degrees)	BS (m)	FS (m)	(BS − FS) (+)	(BS − FS) (−)	Elevations (m)	Remarks
	PLAN SURVEY					DIFFERENTIAL LEVELLING					
A	0	0	72	–		1.50	–	–	–	**100.00**	Assumed elevation
TP1	18	18	–	–	(72/251)	1.71	1.03	0.47	–	100.47	
TP2	17	35	–	–	179	1.85	1.91	–	0.20	100.27	
B	20	55	130	251		1.67	2.02	–	0.17	**100.10**	
TP3	26	81	–	–	(130/310)	1.45	1.63	0.04	–	100.14	
C	29	110	81	310	180	1.35	1.12	0.33	–	**100.47**	
TP4	16	126	–	–	(81/262)	1.81	1.06	0.29	–	100.76	
D	31	157	109	262	181	1.79	1.98	–	0.17	**100.59**	
E	34	191	–	289	180	–	2.02	–	0.23	**100.36**	
			Sums			13.13	12.77	1.13	0.77	100.36	Last point elevation
			Differences			−12.77		−0.77		−100.00	First point elevation
			CHECKS			+ 0.36		+ 0.36		+ 0.36	

Checking on levelling errors

20. Checking on the arithmetic calculations does not tell you how accurate your survey has been. To fully check on your accuracy, **level in the opposite direction**, from the final point to the starting point, using the same procedure as before. You will probably find that the elevation of point A you obtain from this second levelling differs from the known elevation. This difference is **the closing error**.

Example

From point A of a known elevation, survey by traversing through five turning points, TP1 . . . TP5, and find the elevation of point B. To check on the levelling error, survey by traversing BA through four other turning points, TP6 . . . TP9; then calculate the elevation of A. If the known elevation of starting point A is 153 m, and the calculated elevation of A at the end of the survey is 153.2 m, the closing error is 153.2 m − 153 m = 0.2 m.

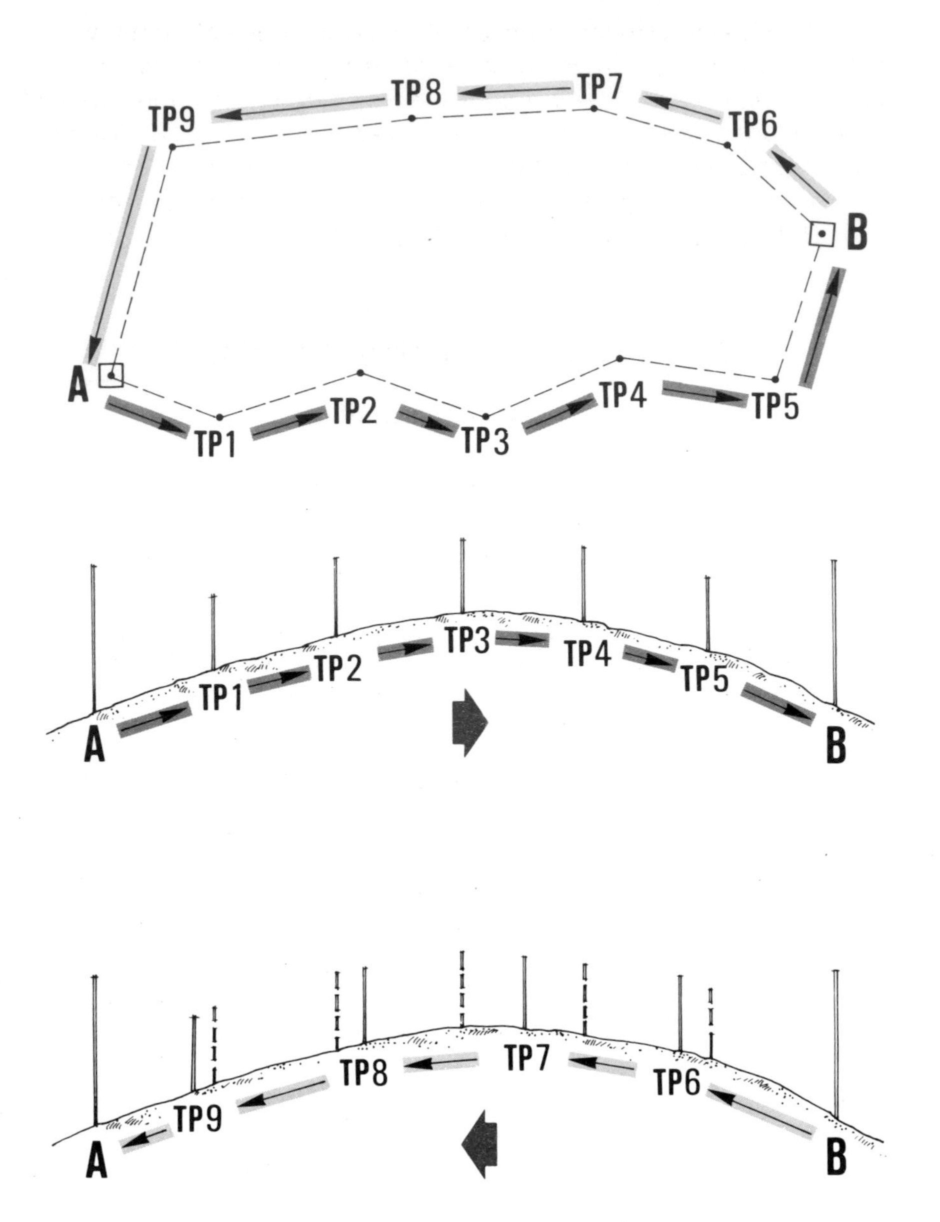

21. The closing error must be less than the permissible error, which is the limit of error you can have in a survey for it to be considered accurate. The size of the permissible error depends on **the type of survey** (reconnaissance, preliminary, detailed, etc.) and on **the total distance** travelled during the survey. To help you find out how accurate your survey has been, calculate the **maximum permissible error (MPE)** expressed **in centimetres**, as follows:

Reconnaissance and preliminary surveys:

$$MPE(cm) = 10\sqrt{D}$$

Most engineering work:

$$MPE(cm) = 2.5\sqrt{D}$$

---

where **D** is the distance surveyed, expressed **in kilometres**.

Example

You have just finished a reconnaissance survey. Your closing error was 0.2 m or 20 cm, at the closure of a traverse 2.5 km + 1.8 km = 4.3 km long. In this case, the maximum permissible error (in centimetres) equals $10\sqrt{4.3} = 10 \times 2.07 = 20.7$ cm. Since your closing error is smaller than the MPE, your levelling measurements have been accurate enough for the purposes of a reconnaissance survey.

22. In the previous section, you made a topographical survey along an open traverse joining points A and B. You can survey a **closed traverse**, such as the perimeter of a fish-farm site, in a similar way. You should use each **perimeter summit** A, B, C, D, E and F of the polygon as a survey point, and plot turning points between these summits as you need to. Make a plan survey as explained in Section 71, and use differential levelling to find the elevation of each perimeter point.

23. If you do not know the exact elevation of starting point A, you can assume its elevation, for example **E(A) = 100 m. Start the survey at point A**, and proceed clockwise along the perimeter of the area. Take levelling staff readings at TP1, TP2, B, TP3, etc., until you reach starting point A again and close the traverse. At the same time, make any necessary horizontal distance and azimuth measurements. Record your measurements either in **two separate tables**, one for plan surveying and one for levelling, or in **one table** which includes distance measurements. From the (BS — FS) columns, you can easily find the elevation of each point on the basis of the known (or assumed) elevation at point A. Make all the **checks** on the calculations as shown in steps 15 and 16. Find the closing error at point A (see step 20). This error should not be greater than the **maximum permissible error** (see step 21).

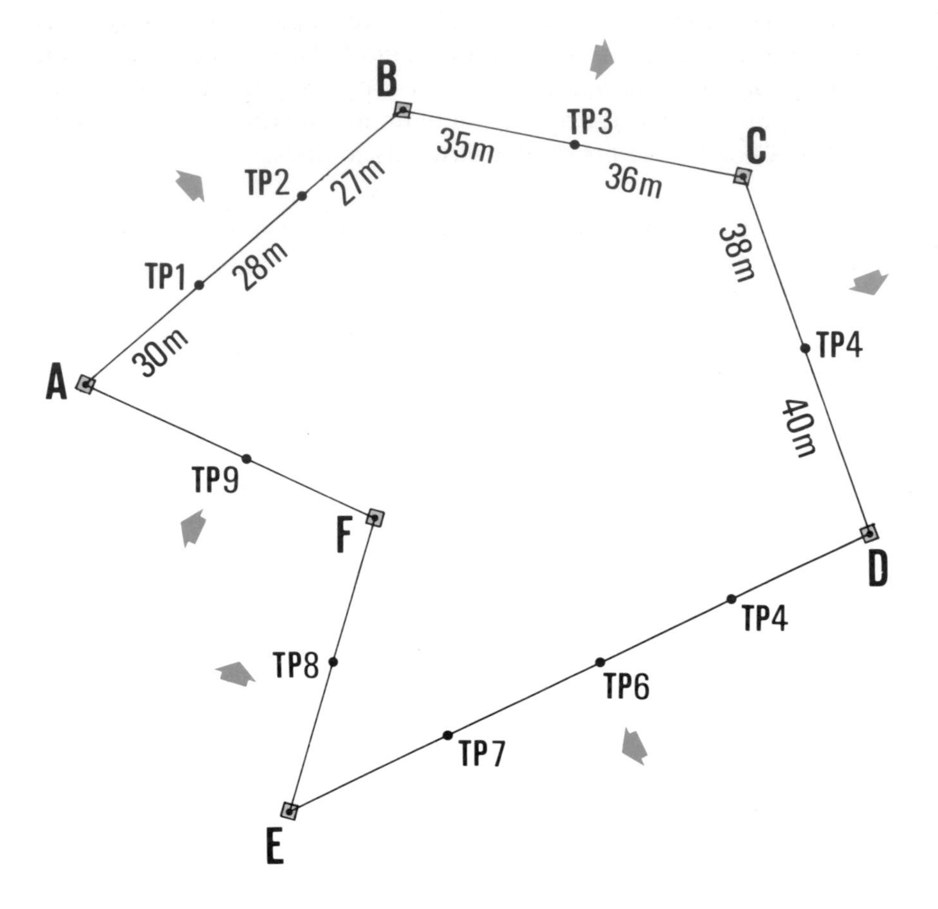

Example

Topographical survey of a closed traverse by differential levelling

Points	Distances		BS	FS	(BS – FS)		Elevations	Remarks
	Partial	Cumulative	(+)	(–)	+	–	(m)	
A	0	0	1.50	–	–	–	**100.00**	Assumed elevation
TP1	30	30	1.71	1.03	0.47	–	100.47	
TP2	28	58	1.85	1.91	–	0.20	100.27	
B	27	85	1.67	2.02	–	0.17	**100.10**	
TP3	35	120	1.45	1.63	0.04	–	100.14	
C	36	156	1.35	1.12	0.33	–	**100.47**	
TP4	38	194	1.81	1.06	0.29	–	100.76	
D	40	234	...	1.98	–	0.17	**100.59**	

Making topographical surveys by the square grid

24. The square-grid method is particularly useful for surveying small land areas with little vegetation. In large areas with high vegetation or forests, the method is not as easy or practical. To use the method, you will lay out squares in the area you are surveying, and determine the elevation of each square corner.

25. **The size of the squares** you lay out depends on the accuracy you need. For greater accuracy, the sides of the squares should be 10 to 20 m long. For reconnaissance surveys, where you do not need to be as accurate, the sides of the squares can be 30 to 50 m long.

26. In the field choose **base line AA** and clearly mark it with ranging poles. This base line should preferably be **located** at the centre of the site, and it should be parallel to the longest side of the site. When you work with a compass, you may find that it helps to orient this base line following the north-south direction.

27. Working uphill, chain along this base line from the perimeter of the area, and set stakes **at intervals** equal to the size you have chosen for the squares, such as 20 m. Clearly number these stakes 1, 2, 3, ... n.

28. From each of these stakes, lay out a line, **perpendicular to the base line**, that runs all the way across the site.

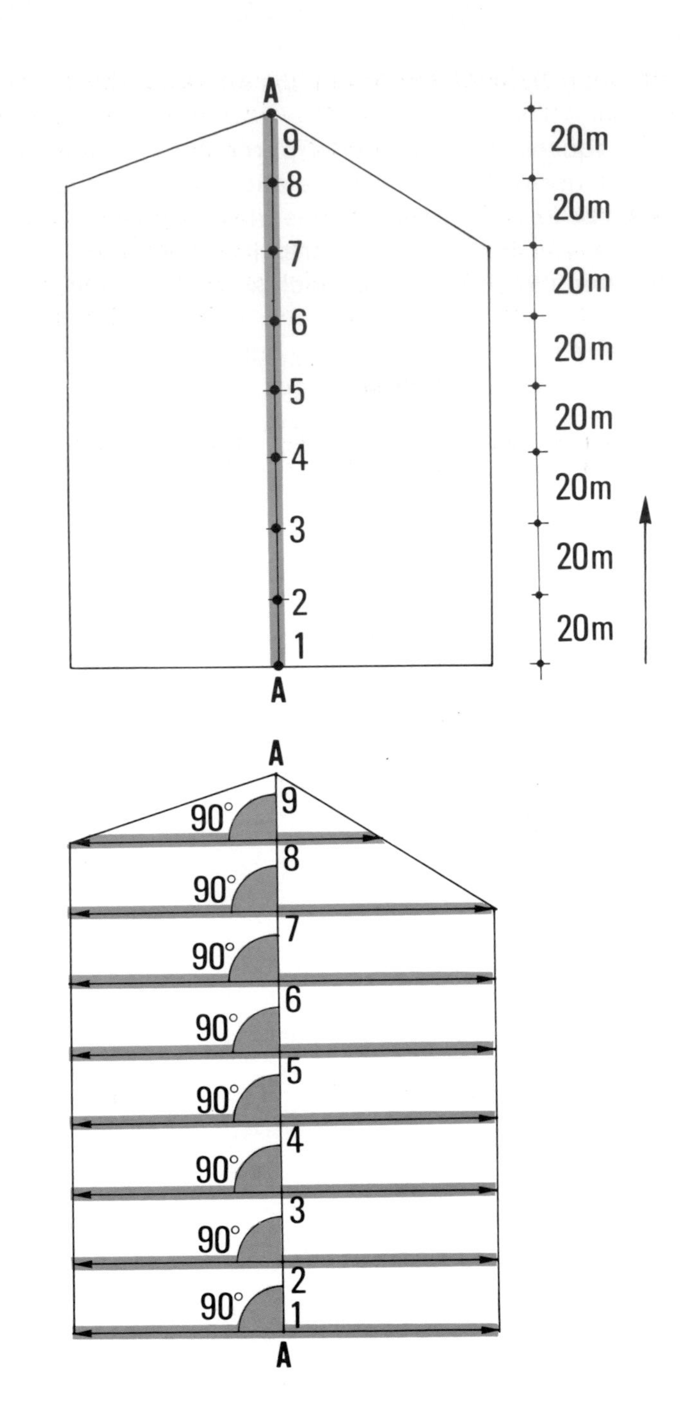

29. Proceed by chaining along the entire length of **each of these perpendiculars, on either side of the base line**. Set a stake every 20 m (the selected square size). Identify each of these stakes by:

- **a letter** (A, B, C, etc.) which refers to the line, running parallel to the base line, to which the point belongs;
- **a number** (1, 2, 3, ... n) which refers to the perpendicular, laid out from the base line, to which the point belongs.

Example

20 m from point A1, perpendicular 2 crosses line AA at point **A2**.
20 m to the left of point A2 lies point **B2**, on line BB.

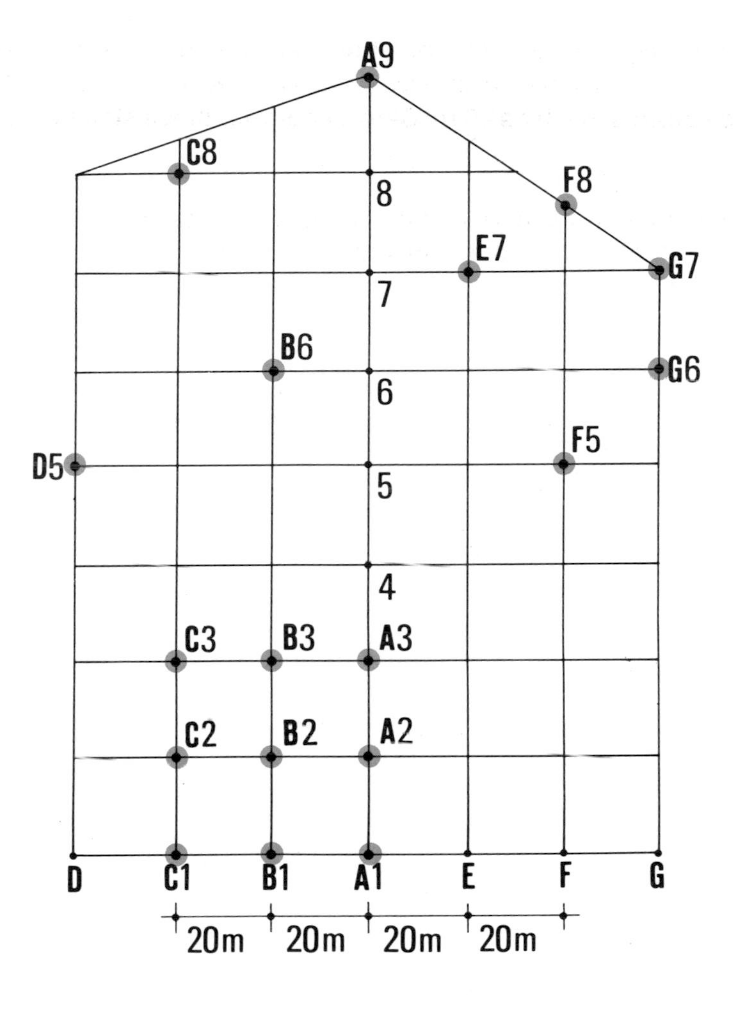

30. Now that you have laid out **the square grid** on the ground, you need to find the **elevation of each corner of the squares**, which you have marked with stakes. First establish **a bench-mark (BM) on base line AA** near the boundary of the area and preferably in the part with the lowest elevation (see steps 42-44). This bench-mark can be either at a **known elevation** (such as one point on a previously surveyed traverse), or at an **assumed elevation** (such as 100 m) (see step 45).

31. You will level **the square grid points in two stages**.

- (a) Starting from the bench-mark, measure the differences in elevation for **all the base points** A1, A2, A3, ... An. This is called **longitudinal profile levelling** (see Section 82).
- (b) Then, starting at these base-line points with known elevations, measure the differences in elevation for all points of each of the perpendiculars, on each side of the base line (for example, B2, C2 and D2 followed by E2, F2 and G2). This is called **cross-section levelling** (see Section 82).

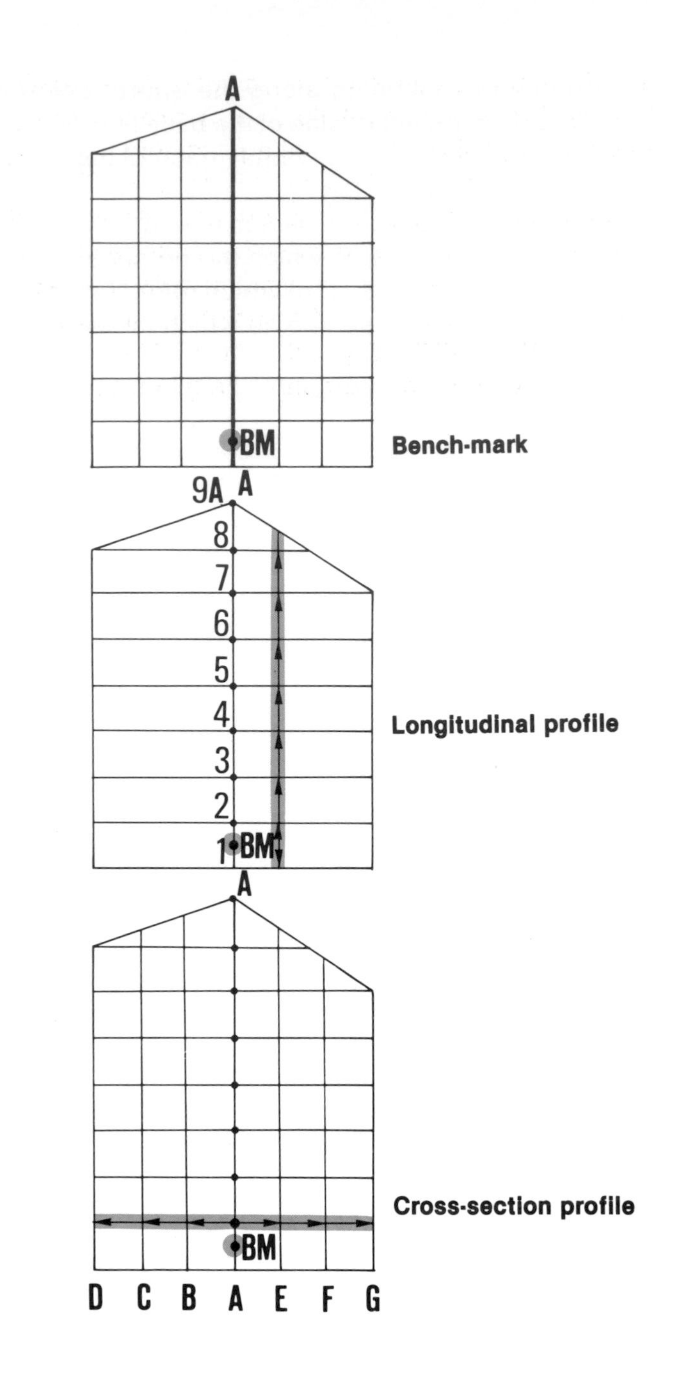

32. If you use a sighting level you can make **a radiating survey** (see step 34). Set up your level at LS1 and take a backsight reading on the bench-mark (BM). Then, take foresight readings on as many base-line points as possible. From this, find the height of the instrument (HI) and point elevations, with HI = E(BM) + BS and E (point) = HI — FS. When necessary, change the levelling station and find a new HI on the last known point, which is used as a turning point. Then measure a series of foresights. Since the distances of the square grid are all fixed, you do not need to measure them any more. Note down your measurements in a table, as shown in the example.

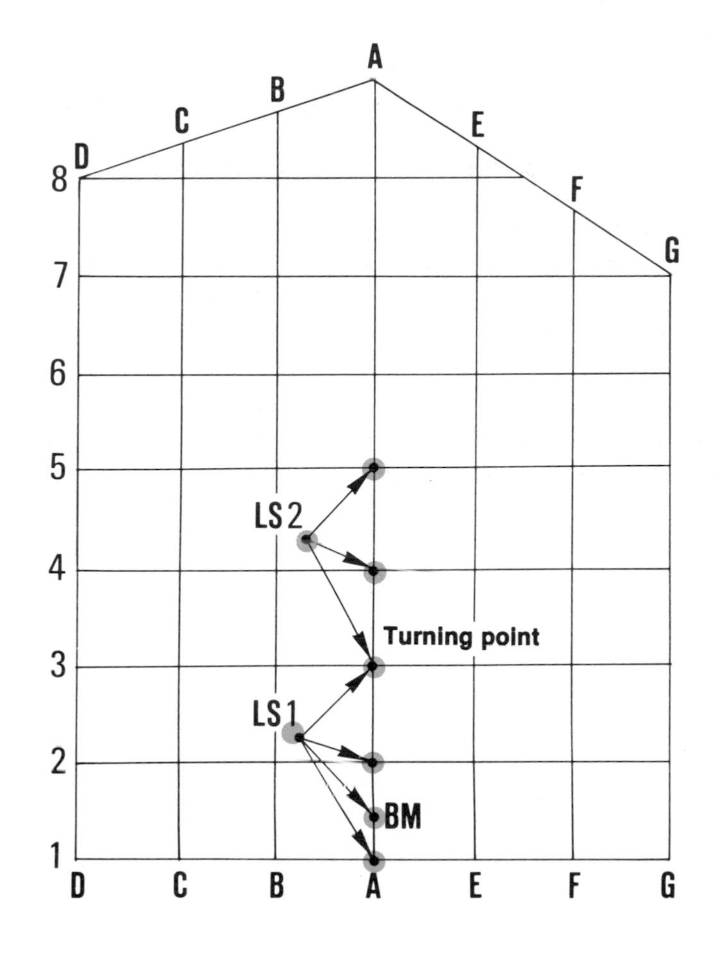

Example

Topographical survey by square grid with a sighting level

Levelling station	Point	BS	HI	FS	Elevation	Remarks
1	BM	1.53	101.53	–	100.00	Assumed elevation
	A1	–	101.53	1.25	100.28	
	A2	–	101.53	1.20	100.33	
	A3	–	101.53	1.15	100.38	
2	A3	1.48	101.86	–	100.38	Turning point
	A4	–	101.86	1.41	100.45	
	...	...	...	...	...	
	A9	...	...	...	...	
5	A1	1.20	101.48	–	100.28	Known from A1 above
	B1	–	101.48	0.23	100.25	
	C1	–	101.48	0.25	100.23	
	D1	–	101.48	0.28	100.20	
6	D1	1.30	101.50	–	100.20	Turning point
	E1	–	101.50	0.35	100.15	
	...	...	...	...	...	
	G1	–	101.50	0.47	100.03	
9	A2	1.35	101.68	–	100.33	Known from A2 above
	B2	...	...	–	...	
...		...	...	...	...	
12	F2	...	...	–	...	
	G2	...	...	–	...	
13	A3	...	...	–	100.38	Known from A3 above
	...	...	...	...	...	

Example

Topographical survey by square-grid with a non-sighting level

Points	Heights (m)		Difference (Rear-Forward)		Elevation (m)	Remarks
	Rear	Forward	(+)	(—)		
BM	1.06	–	–	–	1.41	Nail in tree root
A1	1.07	1.13	–	0.07	1.34	
A2	1.10	1.16	–	0.09	1.25	
...	...	...	–	...	...	
A10	1.04	1.11	–	0.02	1.20	
A1	1.50	–	–	–	1.34	Known from above
B1	...	1.52	–	0.02	1.32	
C1	...	...	...	...	...	
...	...	...	...	...	...	
G1	...	...	...	...	...	
A2					1.25	Known from above
B2						
...						
G2						
A3	...	...	...	...	...	...
...	...	...	...	...	...	...

33. If you use **a non-sighting level**, first follow base line AA. Start with the bench-mark as a reference point, and survey all its points A1, A2, ... A9. Then, repeat this surveying procedure along each of the perpendiculars, starting with the known base-line points as the reference points.

Enter all your measurements in a table, and find the elevation of each point of the square grid (see steps 38-41 for a further explanation).

You can check calculations and survey measurements at the bottom part of the table (see this Section, step 41).

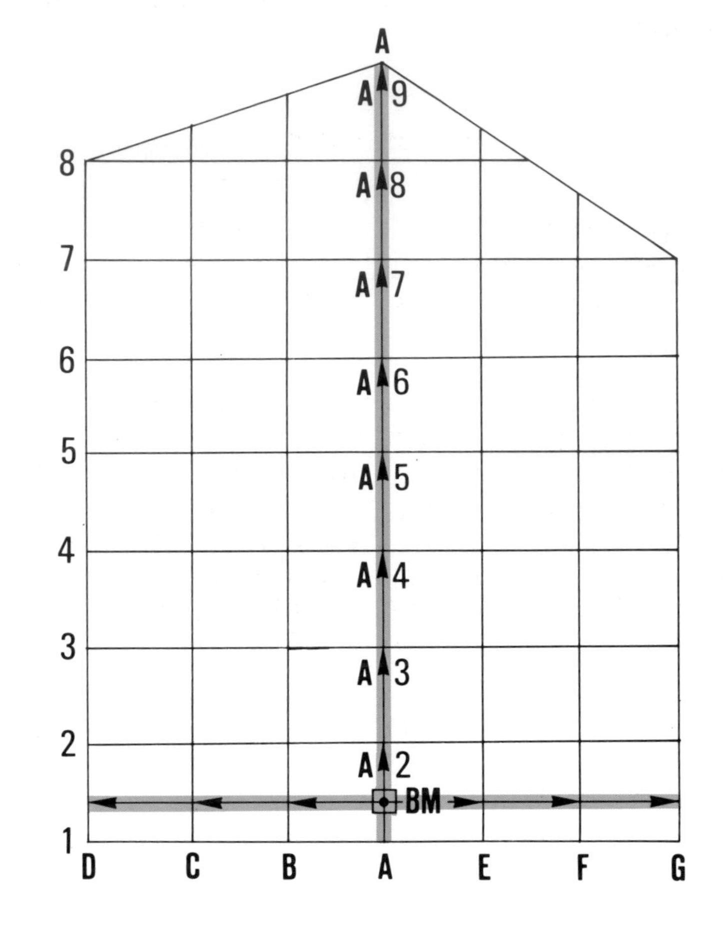

34. When you make a radiating survey (see Section 72), you first need to determine **the height of the instrument HI** at levelling station O. Sight at a point X of known elevation E(X), and find a **backsight (BS)**. Then,

$$\mathbf{HI = BS + E(X)}$$

35. Then you need to find the elevation of each of the points A, B, C and D. Sight at each of them in turn. You will find a **foresight (FS)** for each. Calculate their elevations as E (point) = HI — FS.

36. Record all your measurements in a table. This table may also include plan-surveying information, such as azimuths and horizontal distances. You might also use two different tables as explained in step 23. The first line of the table will refer to **the known point X**. This point can be one of the perimeter points which you have already determined, or it can be a benchmark (see step 42). You find **the position of point O** from the azimuth of line OX and the horizontal distance OX.

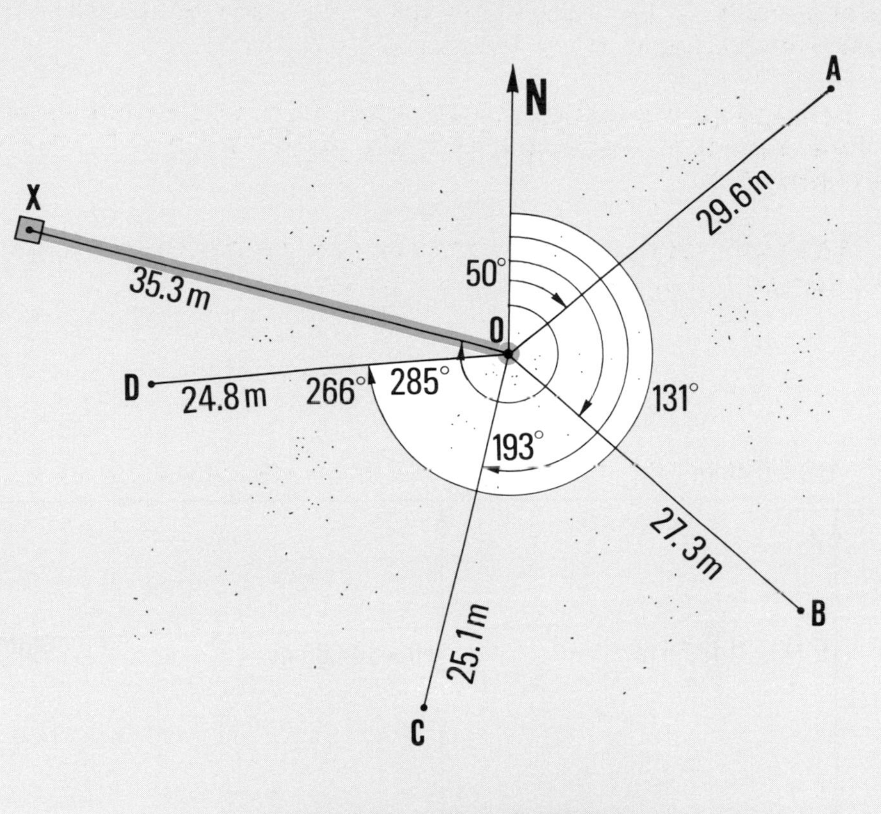

Use X as a point of reference

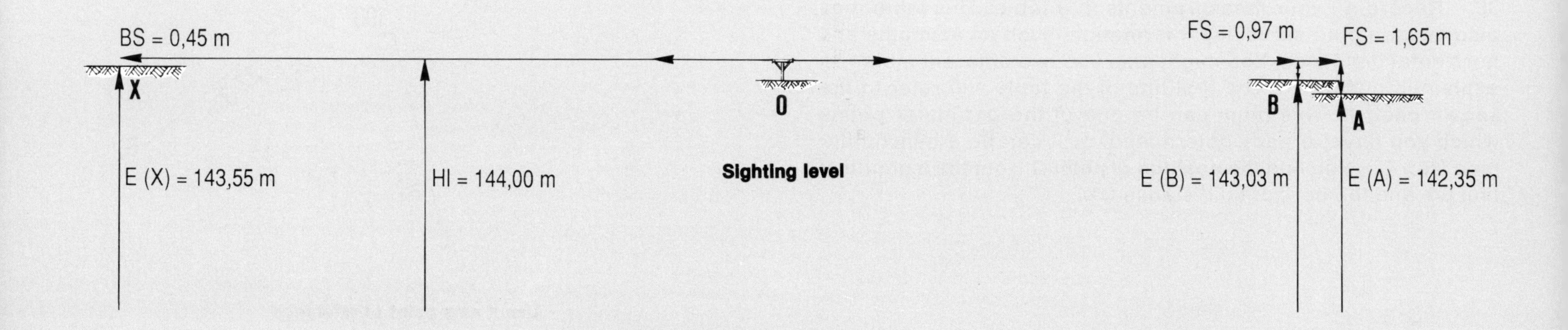
BS = 0,45 m
FS = 0,97 m
FS = 1,65 m
X
O
B
A
E (X) = 143,55 m
HI = 144,00 m
Sighting level
E (B) = 143,03 m
E (A) = 142,35 m

Example

Topographical radiating survey

Levelling station	Point	BS (m)	HI (m)	FS (m)	Elevation (m)	Azimuth (degree)	Distance (m)	Remarks
O	X	0.45	144.00	–	143.55	285	35.3	Known elevation
O	A	–	144.00	1.65	142.35	50	29.6	
O	B	–	144.00	0.97	143.03	131	27.3	
O	C	–	144.00	0.60	143.40	193	25.1	
O	D	–	144.00	1.12	142.88	266	24.8	

Combining traversing and composite radiating

37. This method combines radiating with a closed traverse. You can use it to gather the information you need to make a **topographical map of a land area** such as a fish-farm site (see Chapter 9). Using what you have learned so far about surveying, do the following:

(a) With a closed traverse, plan survey **the boundaries of the area** ABCDEA. Find the lengths and directions of all of its sides (see Section 71).

(b) In the interior of the site, choose **a series of levelling stations** 1, 2, 3, . . . 6, from which you can survey the surrounding area by radiating.

(c) **Fix the position of levelling station 1** by measuring it in relation to known boundary points such as A and B. You can use the plane-tabling and triangulation methods (see Sections 74 and 75).

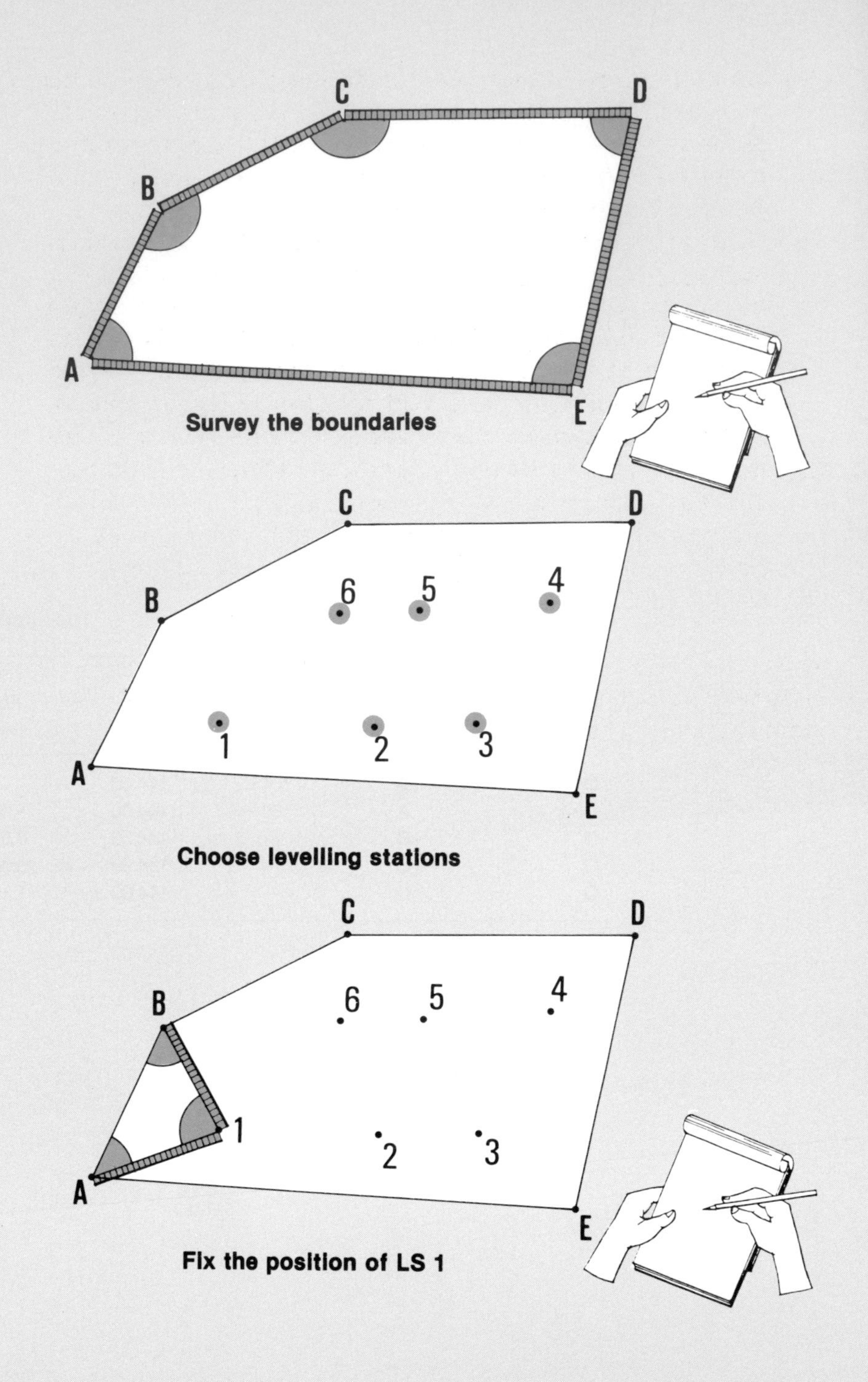

Survey the boundaries

Choose levelling stations

Fix the position of LS 1

(d) Join all the selected levelling stations by straight lines **to form a closed traverse**. Survey it, using turning points as necessary, to fix **the position of each station** and to determine **its elevation**. Check for the closing error (see Section 71 and this section, step 22).

(e) Now you are ready to start the detailed topographical survey, proceeding from each known levelling station in turn. From station 1, set up a series of **radiating straight lines at a fixed-angle interval** (such as 20°). This means that each radiating line will be 20° from the next. Use your magnetic compass and ranging poles or stakes. Mark on the ground the **north-south line**. You will call this the **zero-degree line**. Standing on this line at station 1, measure and mark a line with a 20° azimuth. Then, moving around in a clockwise direction on the same point, measure and mark in turn lines with azimuth 40°, 60°, . . . 340°.

Note: **the fixed-angle interval** you use depends on how accurate a survey you need. Smaller angles will help you make a more accurate map of the site.

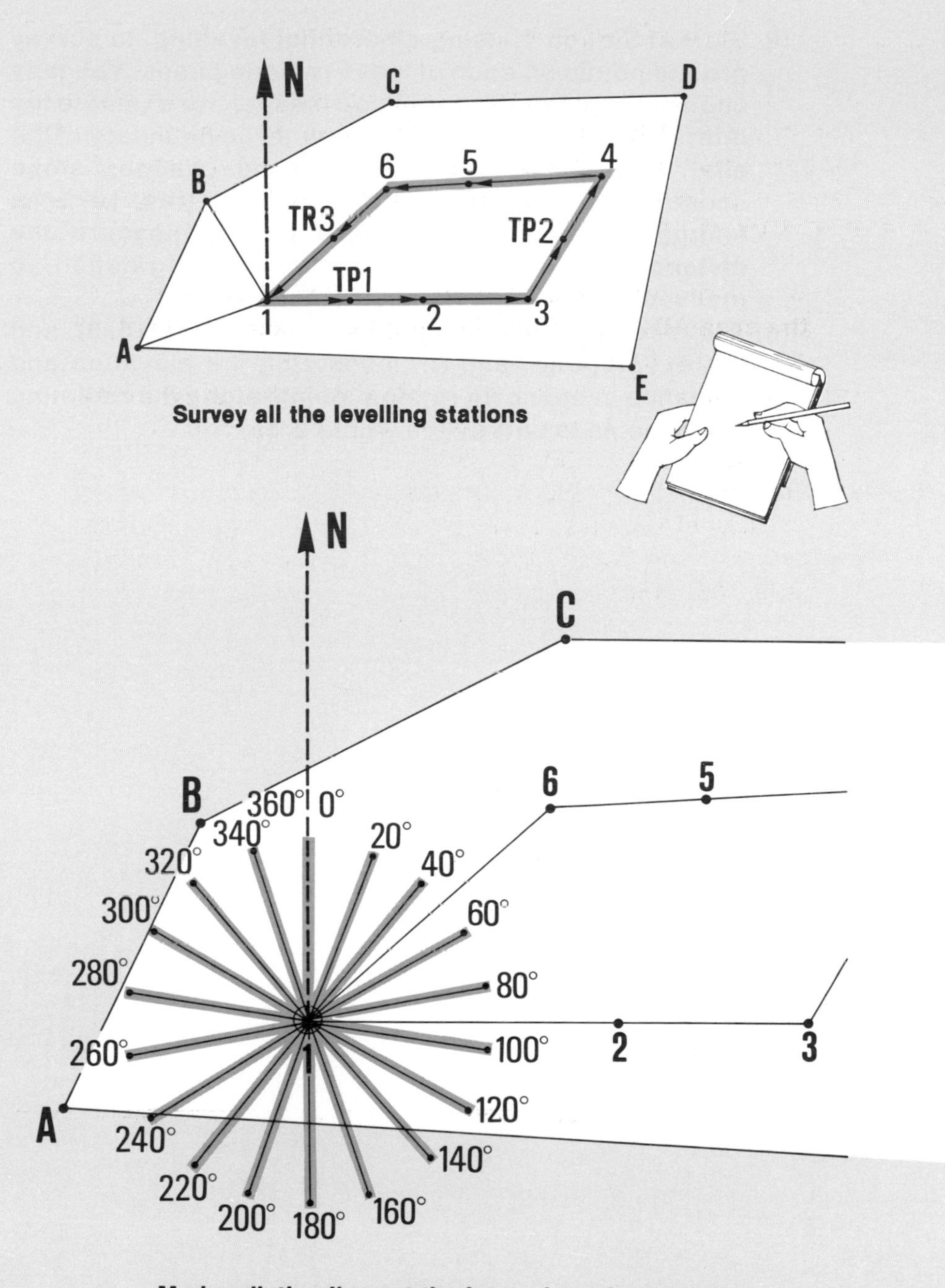

Survey all the levelling stations

Mark radiating lines at the interval you have chosen

(f) Start at Station 1, using **differential levelling**, to survey ground points on each of these radiating lines. You may choose any points you want to measure, for example the intersection of the radiating line with the boundary of the site, or a point where the ground changes slope suddenly, or the location of a rock or tree. Besides finding **the elevation** of these points, measure **the distance** between each point and the levelling station, so that you will be able to map them later on.

(g) Move to each levelling station in turn (2, 3, 4, 5, 6), and repeat steps (e) and (f), measuring the elevation and distance of **unknown random points along the radiating lines**, so as to survey the whole area.

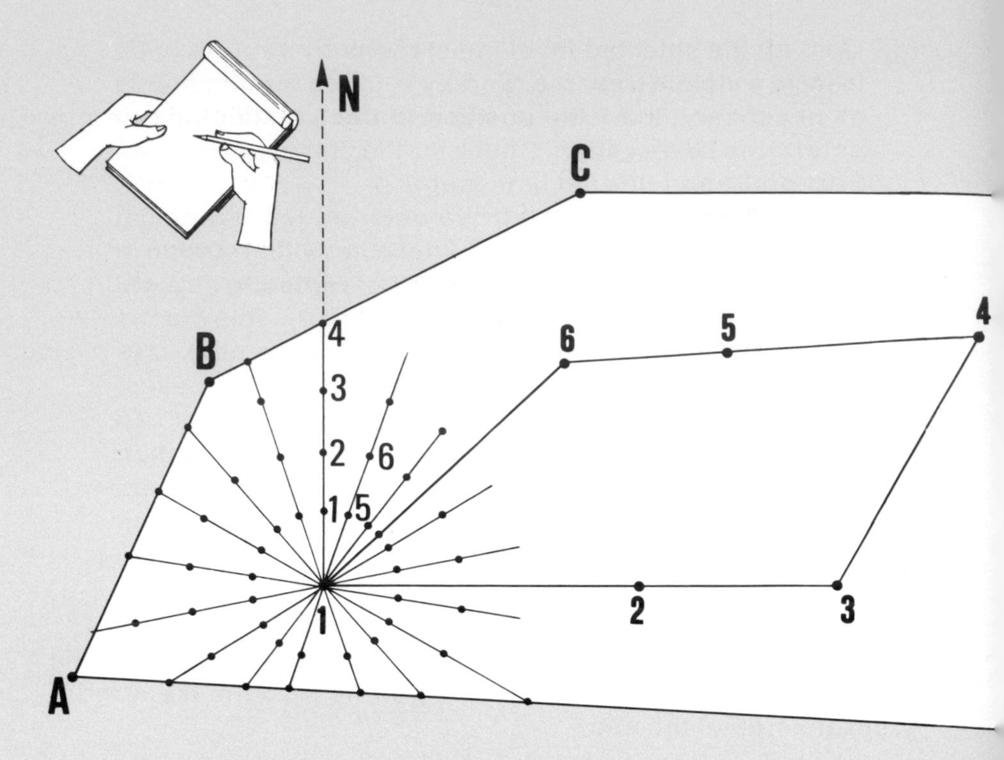

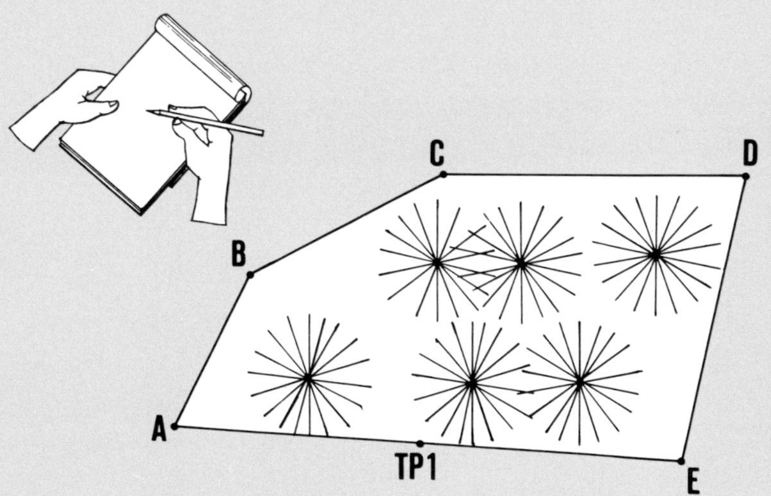

(h) Record all the measurements in a table, and calculate the elevations of all the surveyed points (see this section, step 36). You will need **two additional columns** in this table:

- one column marked **"Line"**, where you will record the **azimuth** of each line;
- one column marked **"Cumulative distances"**. In this space, you can **separately** calculate the distance from the levelling station to the points surveyed for each line.

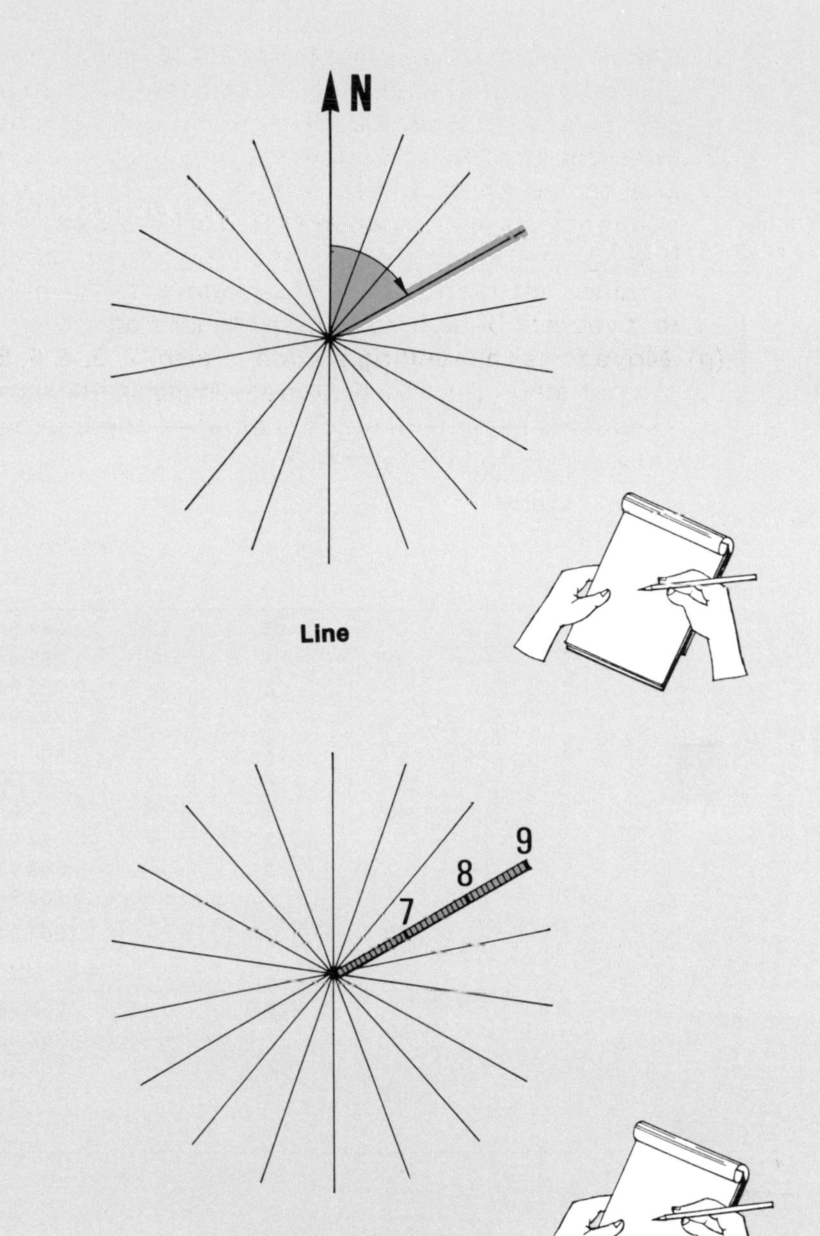

Line

Cumulative distance

Example

Topographical survey of partial area by composite radiating

Levelling station	Line	Points	BS	HI	FS	Elevation (m)	Distances (m)		Remarks
							Partial	Cumulative	
1		B	1.32	283.94	–	282.62	44.6	–	Known elevation
1	0°	1	–	283.94	0.86	283.08	15.2	15.2	
1		2	–	283.94	0.93	283.01	16.8	32.0	
1		3	–	283.94	0.97	282.97	15.9	47.9	
1	20°	1	–	283.94	0.74	283.20	19.7	19.7	
1		2	–	283.94	0.89	283.05	23.2	42.9	
1	40°	1	–	283.94	...	...	...	...	
1		2	–	283.94	...	...	...	...	
1		3	–	283.94	...	...	...	...	
1		4	–	283.94	...	...	...	...	
1	60°	1	–	283.94	...	...	...	...	
...	...	...	...	...	...	...	...	...	
2		TP1	1.44	282.28	–	280.84	–	–	Known elevation
2	0°	1	–	282.28	1.12	281.16	18.9	18.9	
2		2	–	282.28	1.34	280.94	21.6	40.5	
2	20°	1	–	282.28	...	...	...	...	
...	...	...	...	...	...	...	...	...	

Making topographical surveys with non–sighting levels

38. You can also make topographical surveys along straight lines by **using non–sighting levels**, such as **the line level** (see Section 52) or **the flexible–tube water level** (see Section 53). You have already learned how to measure height differences by using the square–grid method with such levels (see this section, step 33).

Remember, when you lay out your grid, that **the distance between points** cannot be more than the length of your level.

39. Work in a team of two or three with this method. Both the **rear person** and the **front person** will take measurements in the field, but **only one person** should be responsible for noting down these measurements in the field book.

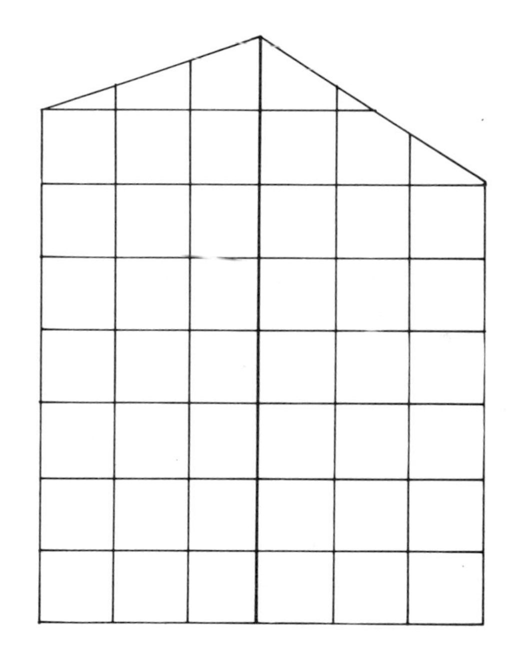

40. Record the measurements in a table for each levelled section. You will be measuring **horizontal distances** from one point to the next, and **differences in elevation** between one point and the next. At both the starting point and the last point, there is only one height measurement. The rear person will measure it on the starting point, and the front person will measure it on the last point.

41. Find the **cumulated distances** from the starting point and **the elevations** of each point, as shown in the example. There are **three possible checks**, which you make at the bottom part of the table.

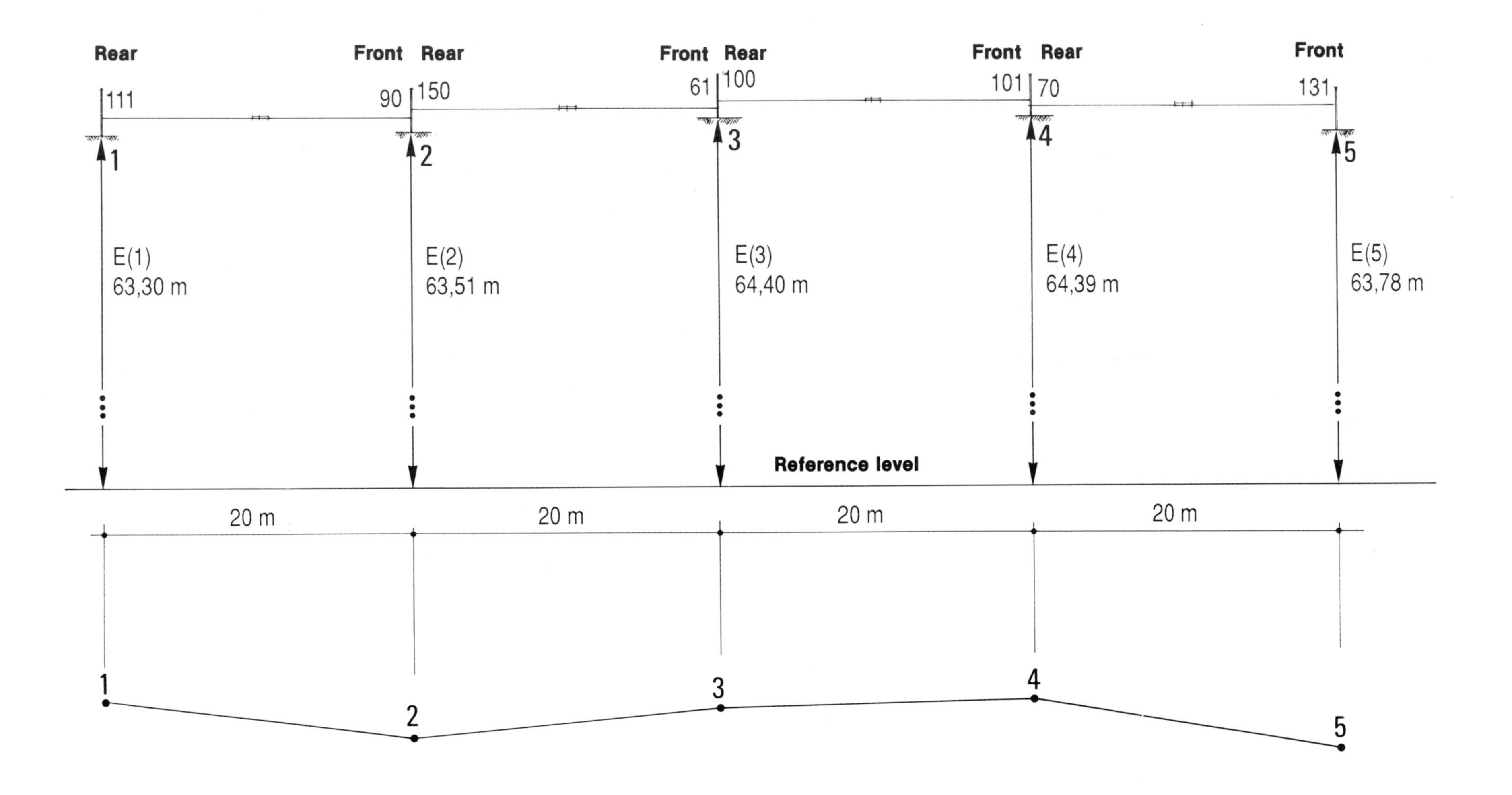

Example

Topographical survey with a line level (20 m)

Points	Distances (m)		Heights (m)		Differences (Rear – Forward)		Elevation	Remarks
	Partial	Cumulative	Rear	Forward	(+)	(–)		
1	0	0	1.11	–	–	–	63.30	Known elevation
2	19.8	19.8	1.50	0.90	0.21	–	63.51	
3	19.9	39.7	1.00	0.61	0.89	–	64.40	
4	19.9	59.6	0.70	1.01	–	0.01	64.39	
5	19.8	79.4	–	1.31	–	0.61	63.78	
Sums			4.31	3.83	1.10	0.62		
Differences			–3.83		–0.62		–63.30	
Checks			0.48		0.48		0.48	

Making bench-marks for topographical surveys

42. As you have just learned, you will always start differential levelling surveys by measuring a height on **a ground point of known or assumed elevation**. This point becomes **a bench-mark (BM)**. The elevation of this bench-mark will form the basis for finding the elevation of the other points you need to survey in the area.

43. A bench-mark should be **permanent**. You should always establish at least one bench-mark near the construction site of a fish-farm to act as a fixed reference point or object. You may also use a bench-mark as a turning point during topographical surveys.

44. A bench-mark should be a very **well-defined point**. You should be able to find and recognize it easily. It should be easy to reach, so that you can hold a levelling staff on it. You can establish a bench-mark:

- on wooden or bamboo stakes set near the construction site;
- by driving a nail into a tree or tree stump, near the ground line, where it will remain even when the tree is cut down;
- by fixing a piece of iron rod in a concrete block near ground level;
- on permanent objects or structures which are unlikely to settle, move or be disturbed, such as a bridge, a large rock or the wall of a building.

Note: it is best to paint the bench-mark, or set several signs near it, to show its location.

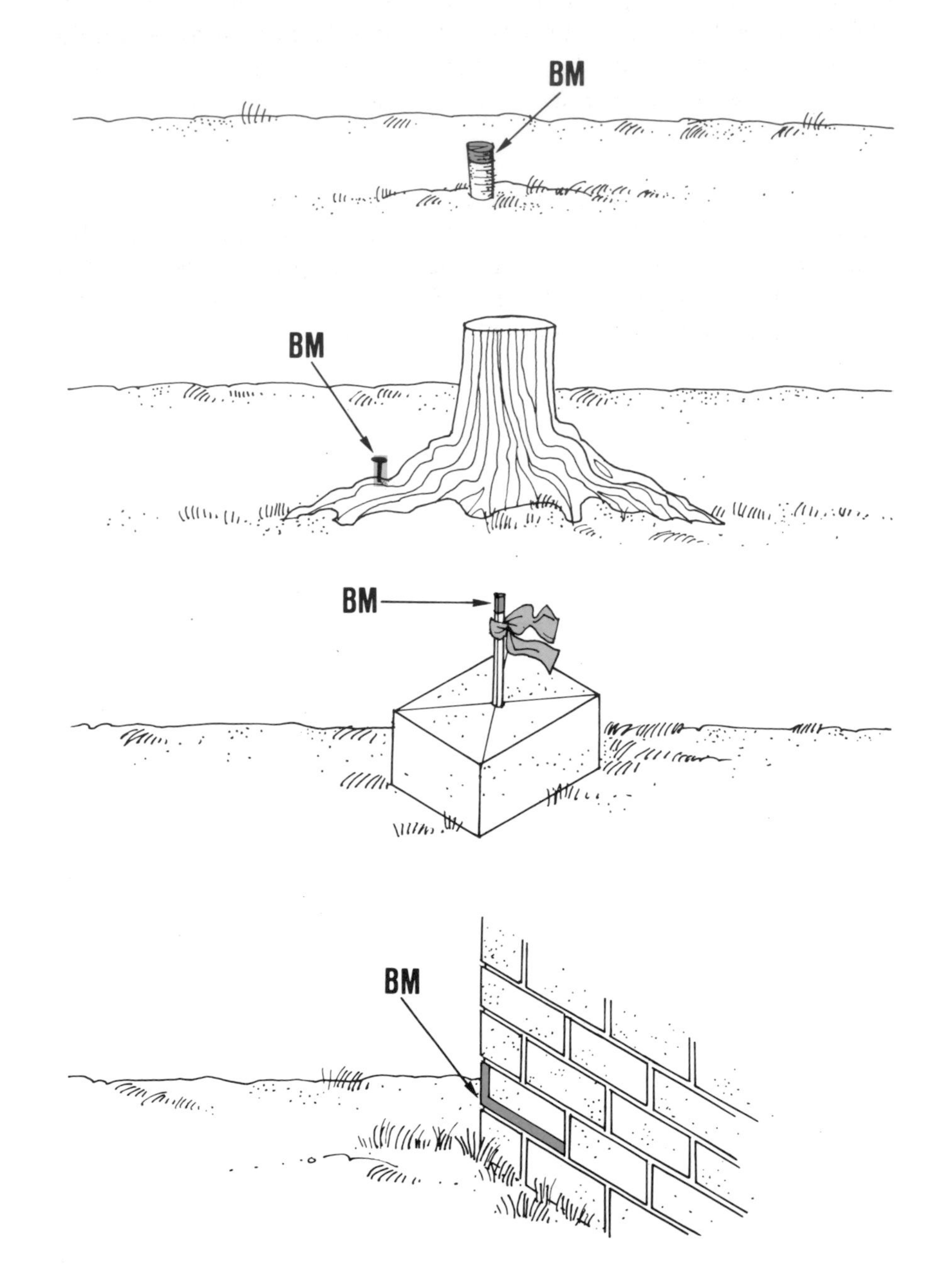

45. Generally, **the elevation of a bench-mark** E(BM) **is not known but is assumed**. When you have established the first bench-mark for a building project, you give it an elevation that is a **convenient whole number**, such as 100 m. The number you choose should be large enough to prevent any point in the surveyed area from having a negative elevation.

Note: you have seen in previous examples that some surveys are related to previously surveyed points. This means that the measurements in the survey are based on these points. These points then become **turning-point bench-marks**. You find their elevations by levelling, and these then become known elevations.

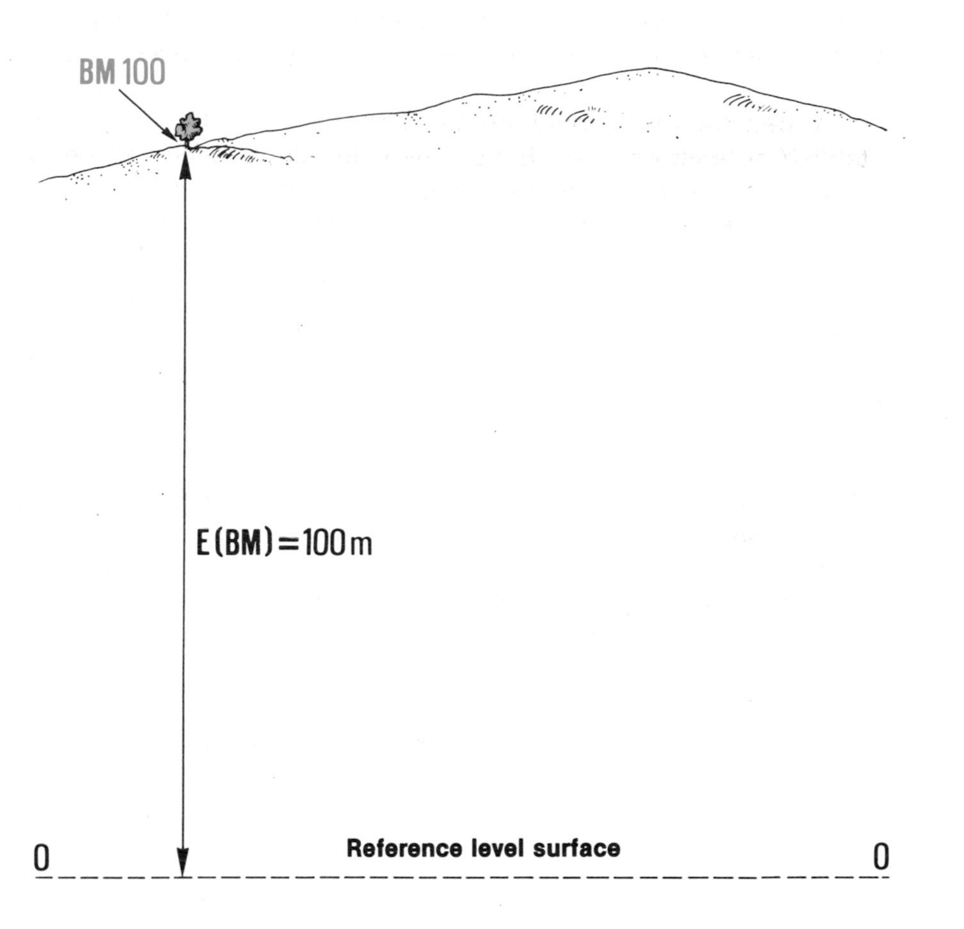

82 How to level by profile

What is the purpose of profile levelling?

1. The purpose of profile levelling is to determine the changes in the elevation of the ground surface **along a definite line**. (You have already learned about profile levelling used with the square-grid method in Section 81, step 31.) This definite line AB might be the centre-line of a water-supply canal, a drainage ditch, a reservoir dam, or a pond dike. This line might also be the path of a river bed through a valley, where you are looking for a dam site, or it might be one of several lines, perpendicular to a river bed, which you lay out across a valley when you are surveying for a suitable fish-farm site.

2. You will usually transfer the measurements you obtain during profile levelling onto paper, to make a kind of diagram or picture called **a graph**. This will show changes in elevation, and how they are related to horizontal distances. This kind of graph is called a **ground profile**. You will learn how to make one in Sections 95 and 96.

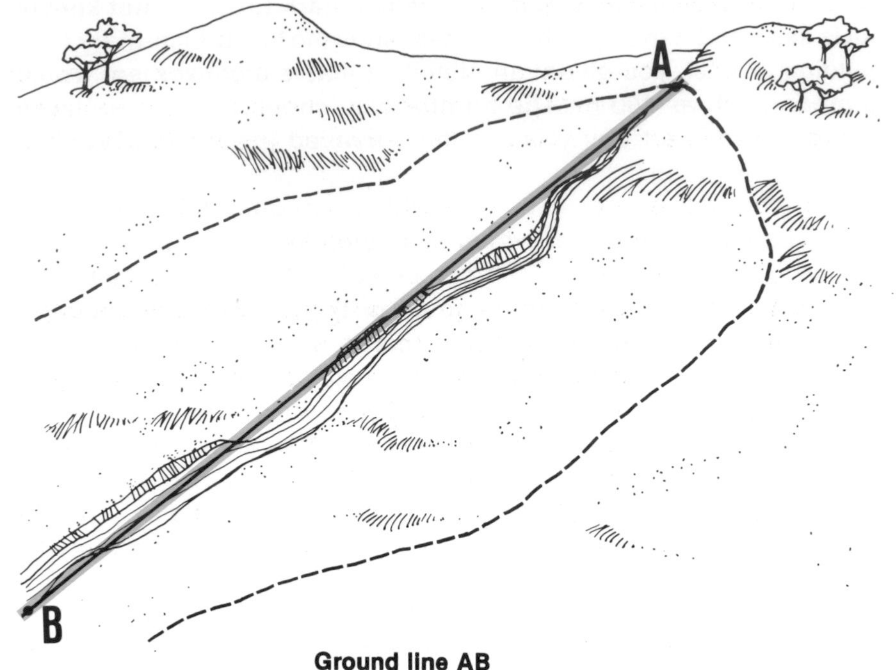

Ground line AB

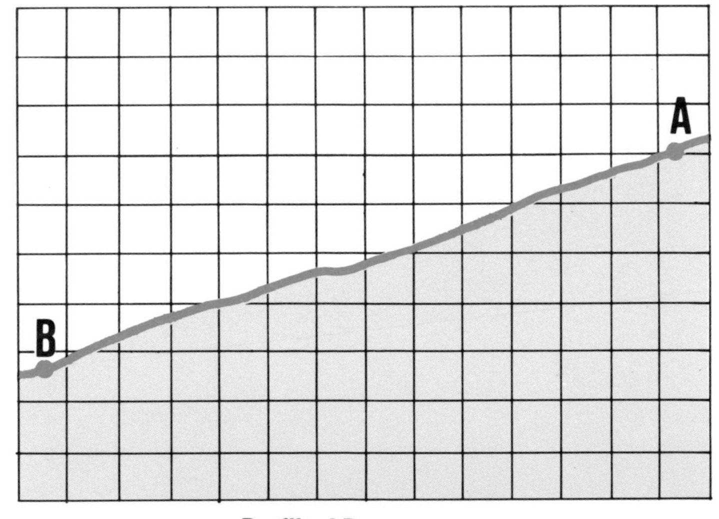

Profile AB

What does profile levelling consist of?

3. When you profile level, you are determining a series of elevations of points which are located **at short measured intervals along a fixed line**. These elevations determine the profile of the line.

4. There are two kinds of profiles which are commonly used in fish culture: longitudinal and cross-section profiles.

- You survey **longitudinal profiles** by profile levelling along a line which is the main axis of the survey. This can be the centre-line of a water canal or the base line of a square grid.
- You usually survey **cross-section profiles** along a line which is perpendicular to a surveyed longitudinal profile, using its points of known elevation as bench-marks. Cross-sections of valleys are useful in helping you locate a good fish-farm site. On a smaller scale, you can also survey cross-sections for water-supply canals, for dam construction, and for pond construction. You have already learned how to use cross-section profiles when surveying by the square-grid method (see Section 81, step 31).

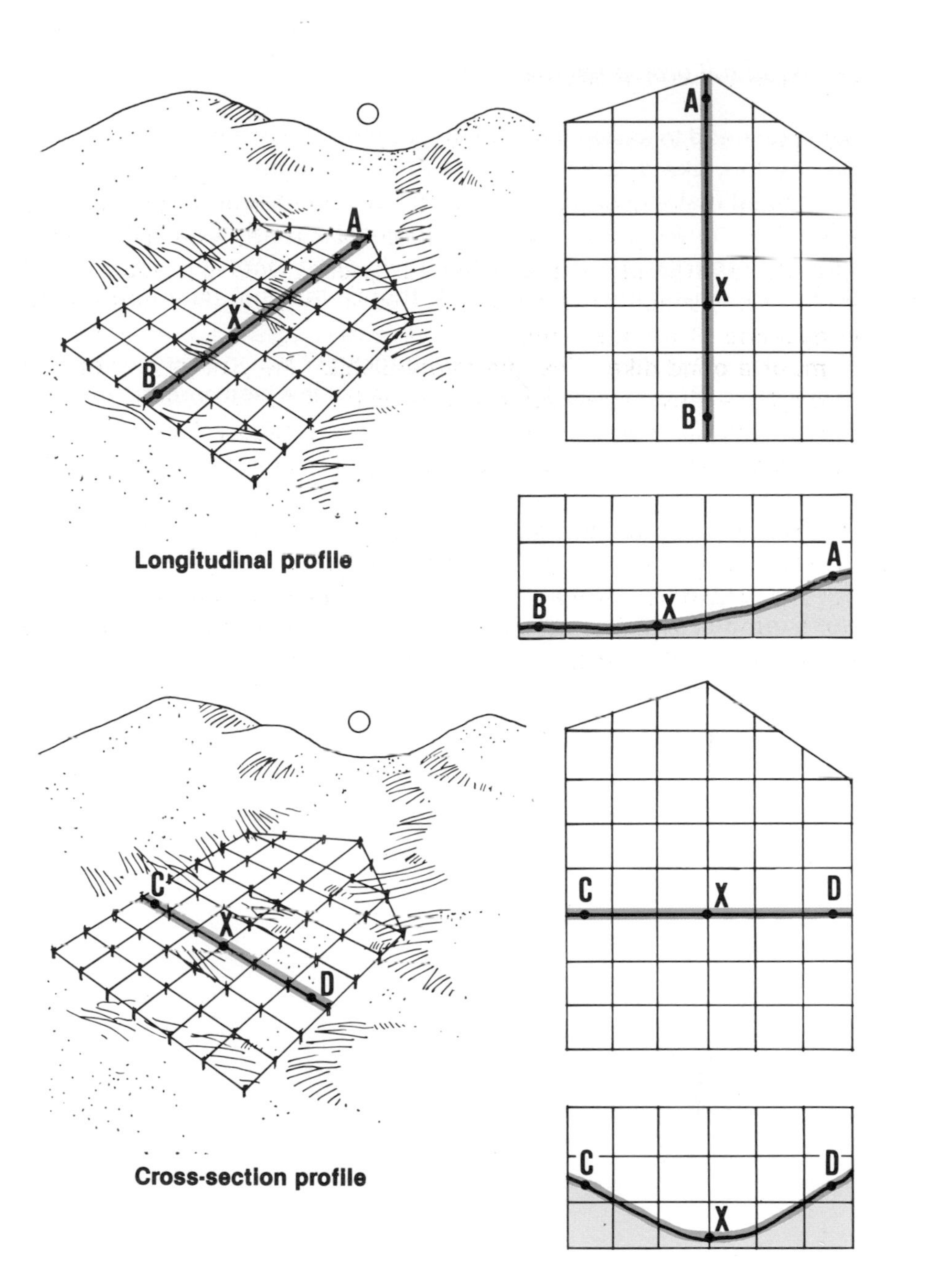

Longitudinal profile levelling by radiating

5. You need to survey line AB, the centre-line of a water canal. You decide to make a radiating survey using **a sighting level**. Measure horizontal distances and mark every 25 m of the line with a stake, from its initial to its final point. Add points between the stakes where there are **marked changes of slope**. On each stake, clearly indicate its distance from the initial point A, that is, the cumulated distance.

6. Set up your level at LS1. Take a backsight BS on a **bench-mark** of elevation E(BM) to determine **the height of the instrument HI**, as:

$$\mathbf{HI = BS + E(BM)}$$

7. From levelling station LS1, read foresights FS on as many points (for example, six) of line AB as possible, starting from the initial point A.

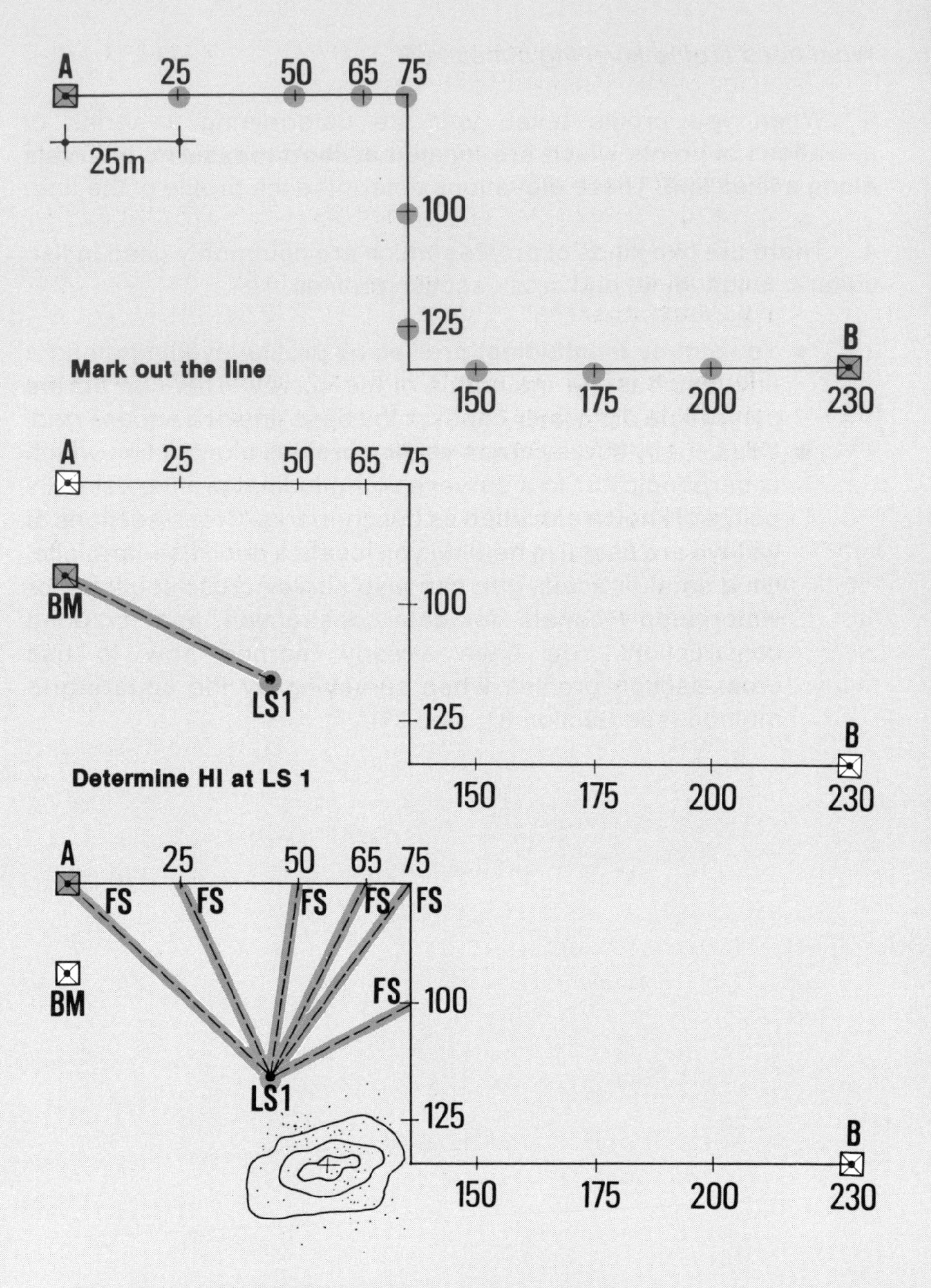

Mark out the line

Determine HI at LS 1

Take foresights at the points you have marked

8. When you need to move the level to a new station so that you can take readings on the points ahead:

- first, choose a **turning point** TP and take a foresight FS to find its elevation from LS1;
- move to the next levelling station LS2, from which you can see the turning point TP;
- take a backsight BS on this turning point to find the **new height of the instrument** HI.

9. Read foresights FS on as many points as possible until you reach the end point of AB. If necessary, use another turning point and a new levelling station as described in step 8.

10. Note down all your measurements in a field book, using a table similar to the ones you have used with other methods. Find the elevations of the points (except for the turning point) by subtracting each FS from its corresponding HI. In the example of the table shown here, cumulated horizontal distances (in metres) appear as point numbers 00, 25, 50, 65, etc. in the first column.

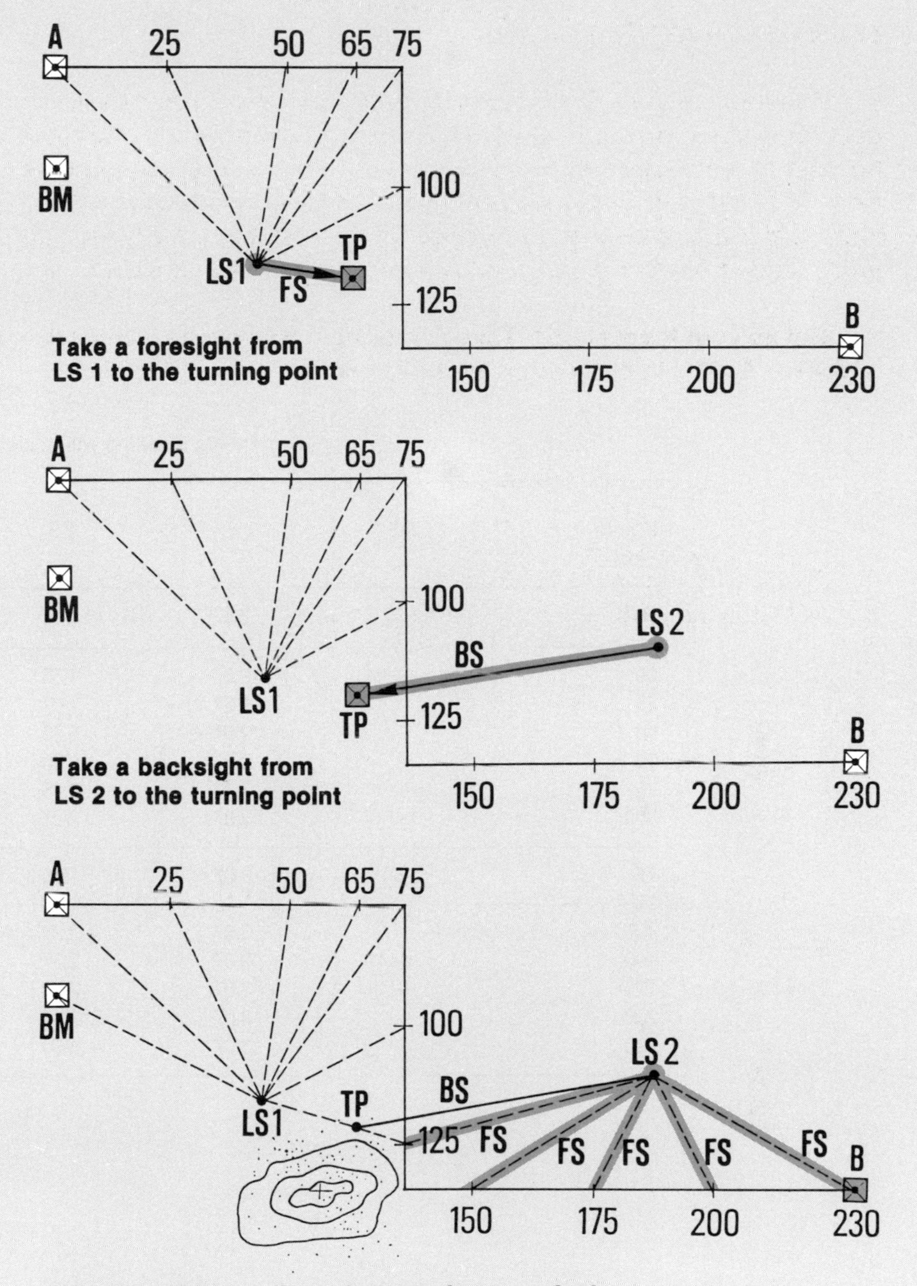

Take a foresight from LS 1 to the turning point

Take a backsight from LS 2 to the turning point

Take foresights at the points you have marked

Example

Profile levelling with a sighting level in a radiating survey

Points (m)	BS	HI	FS	Elevation (m)	Remarks
BM	1.37	2.87	–	1.50	Nail at foot of tree stump
00	–	2.87	1.53	1.34	Beginning of canal
25	–	2.87	1.67	1.20	
50	–	2.87	1.73	1.14	
65	–	2.87	1.90	0.97	Marked change of slope
75	–	2.87	2.05	0.82	
100	–	2.87	2.22	0.65	
TP	1.80	3.07	1.60	1.27	On stone
125	–	3.07	2.27	0.80	
150	–	3.07	2.37	0.70	
175	–	3.07	2.57	0.50	
200	–	3.07	2.77	0.30	
230	–	3.07	3.00	0.07	End of canal

Longitudinal profile levelling by traversing

11. You need to survey the same line AB, the centre-line of a water canal, for profile levelling. You will use a non-sighting level, such as the **flexible tube water level** (see Section 53). Since you are using this kind of level, you will survey by traversing. **Mark the line** AB with stakes driven into the ground at regular intervals. The length of these intervals depends on the working length of your level (in this case, 10 m). Where there are marked changes in slope, add intermediate stakes. On each stake, mark its distance from the initial point A.

12. Level a **tie-in line** between bench-mark BM and the initial point A (see Section 52, steps 6-12). This will give you the elevation of point A, through intermediate point 1.

13. Proceed with **the levelling** of the marked points along the line, using this method. At each point, you will make two scale readings, one rear and one forward, except at the final point where you will take only one height measurement.

14. One person should be responsible for **recording the measurements** in a field book, using a table similar to the one in Section 81, step 41. But, in this case, you will not need to enter the distances in the table, since they identify the surveyed points. Checks are made at the bottom of the table as usual. Remember that in this type of survey there is no need for turning points.

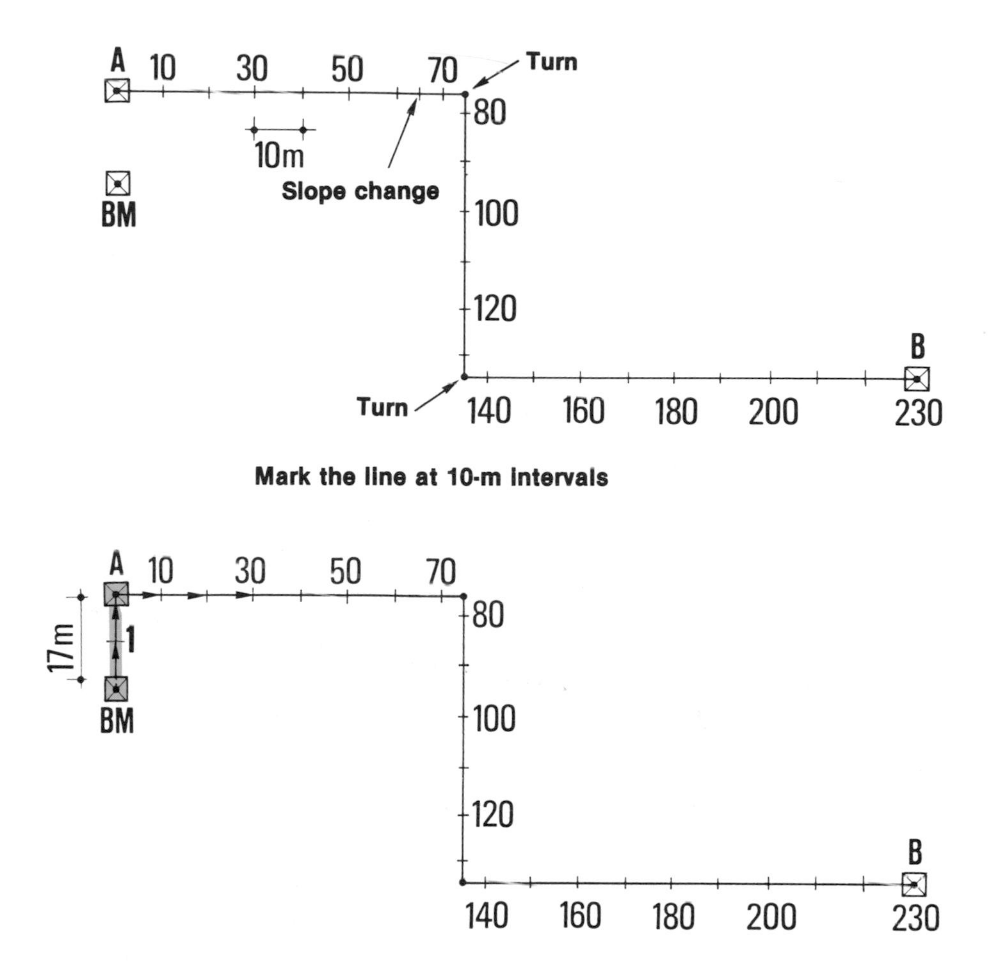

Mark the line at 10-m intervals

Level a tie-in from the bench-mark, then level the points on the line

Example

Profile levelling by traversing with a flexible tube water level (10 m)

Points and cumulative distances	Heights (m)		Differences (Rear – Forward) (m)		Elevation (m)	Remarks
	Rear	Forward	(+)	(–)		
BM	1.08	–	–	–	1.50	Nail in fence post
1	1.06	1.17	–	0.09	1.41	
A	1.07	1.13	–	0.07	1.34	Beginning of canal
10	1.10	1.16	–	0.09	1.25	
20	1.11	1.12	–	0.02	1.23	
30	1.09	1.12	–	0.01	1.22	
40	1.04	1.11	–	0.02	1.20	
...	...	...	...	...	...	(points 50,60,. . . 210)
220	1.08	0.96	–	0.08	0.12	
230	–	1.13	–	0.05	0.07	End of canal
Sums	27.09	28.52	0.00	1.43		
Differences	–28.52		–1.43		0.07	
					–1.50	
Checks	–1.43		–1.43		–1.43	

Cross-section profile levelling

15. After you have found the elevations of points along a longitudinal profile, you can proceed with the survey of **perpendicular cross-sections**. These cross-sections can pass through as many of the points as necessary. Cross-sections are commonly used for contouring long, narrow stretches of land (see Section 83).

16. You will need to have more information on some of the longitudinal profile points. Choose these points and mark them. Then, **set out and mark perpendicular lines** at these points (see Section 36), and extend these perpendiculars **on both sides** of the traverse as far as you need to. In this type of levelling, such perpendiculars are called the **cross-section lines.**

Note: at points where the traverse changes direction (for example, at point 175 in the drawing), you should set out two perpendicular lines E and F; each line will be perpendicular to one of the traverse sections.

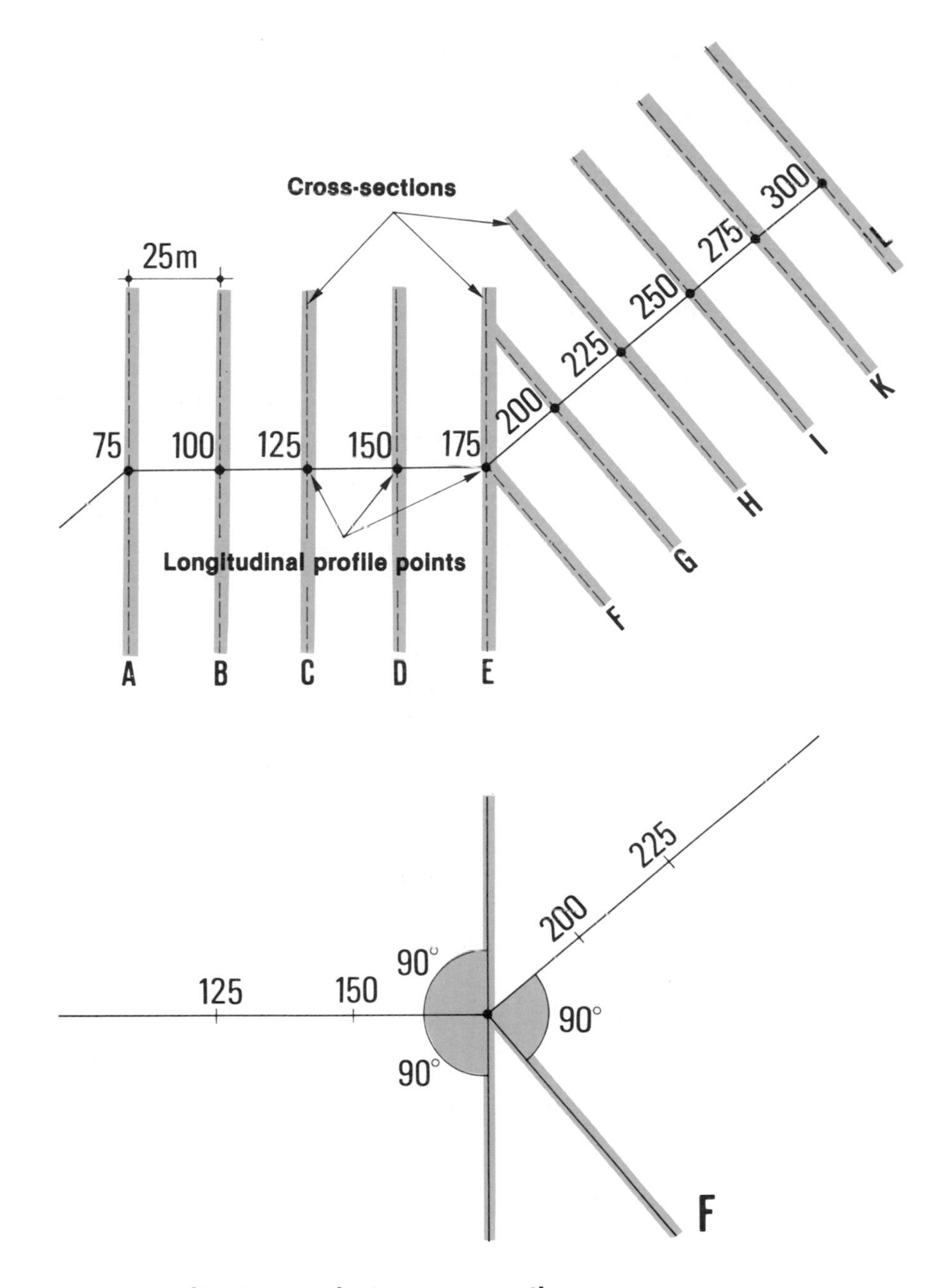

At a turn, make two cross-sections

17. Choose and clearly mark the points you want to survey on each cross-section line. In this case, these points do not have to be regularly spaced. Rather, they should be at places **where the terrain changes** since they should mark changes in slope.

18. As you know the elevations of the traverse points from a previous survey, you may treat these points as bench-marks. Proceed with the profile levelling of selected points along the cross-section lines as explained earlier. You may survey them:

- by radiating, with a **sighting level** (see this section, steps 5-10); or
- by traversing, with **a non-sighting level** (see this section, steps 10-14).

Note: you can also survey by traversing using a **simple sighting level** such as a bamboo sighting level (see Section 56) or a hand level (see Section 57).

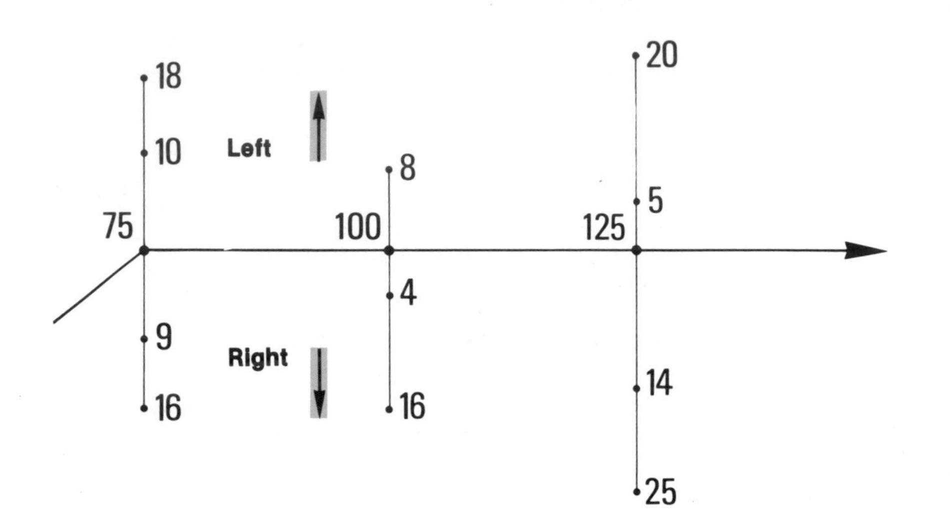

19. Your field notes will be similar to those shown in either step 10 or 14, depending on the levelling method you use. You will identify the points differently, however. You identify each cross-section line by **the number of the traverse point** of known elevation. To do this, identify the surveyed points along each cross-section line according to whether they are **to the left or the right of the traverse**. Also use their distance (in metres) from the traverse points as identification. The following example is of field notes and calculations for a radiating survey, where each cross-section was surveyed from a single levelling station.

Example

Cross-section profile levelling

Traverse Point	Point		BS (m)	HI (m)	FS (m)	Elevation (m)	Remarks
	Left	Right					
50	...	...	...	...	...	...	
75	–	–	0.54	40.94	–	40.40	Edge of existing path
	10	–		40.94	1.09	39.85	
	18	–		40.94	1.15	39.79	
	–	9		40.94	0.85	40.09	
	–	16		40.94	0.68	40.26	
100	–	–	1.15	38.96	–	37.81	Edge of maize field
	8	–		38.96	1.23	37.73	
	–	4		38.96	1.11	37.85	
	–	16		38.96	0.78	38.18	
125	–	–	0.97	36.64	–	35.67	Edge of small forest
	5	–		36.64	1.12	35.52	
	20	–		36.64	1.55	35.09	
	–	14		36.64	1.03	35.61	
	–	25		36.64	0.89	35.75	
150	...	...	...	...	...	...	

83 How to contour

What is a contour?

1. **A contour** is an imaginary continuous line or curve which joins ground points of an equal elevation. The elevation of the ground points must be measured from the same reference plane.

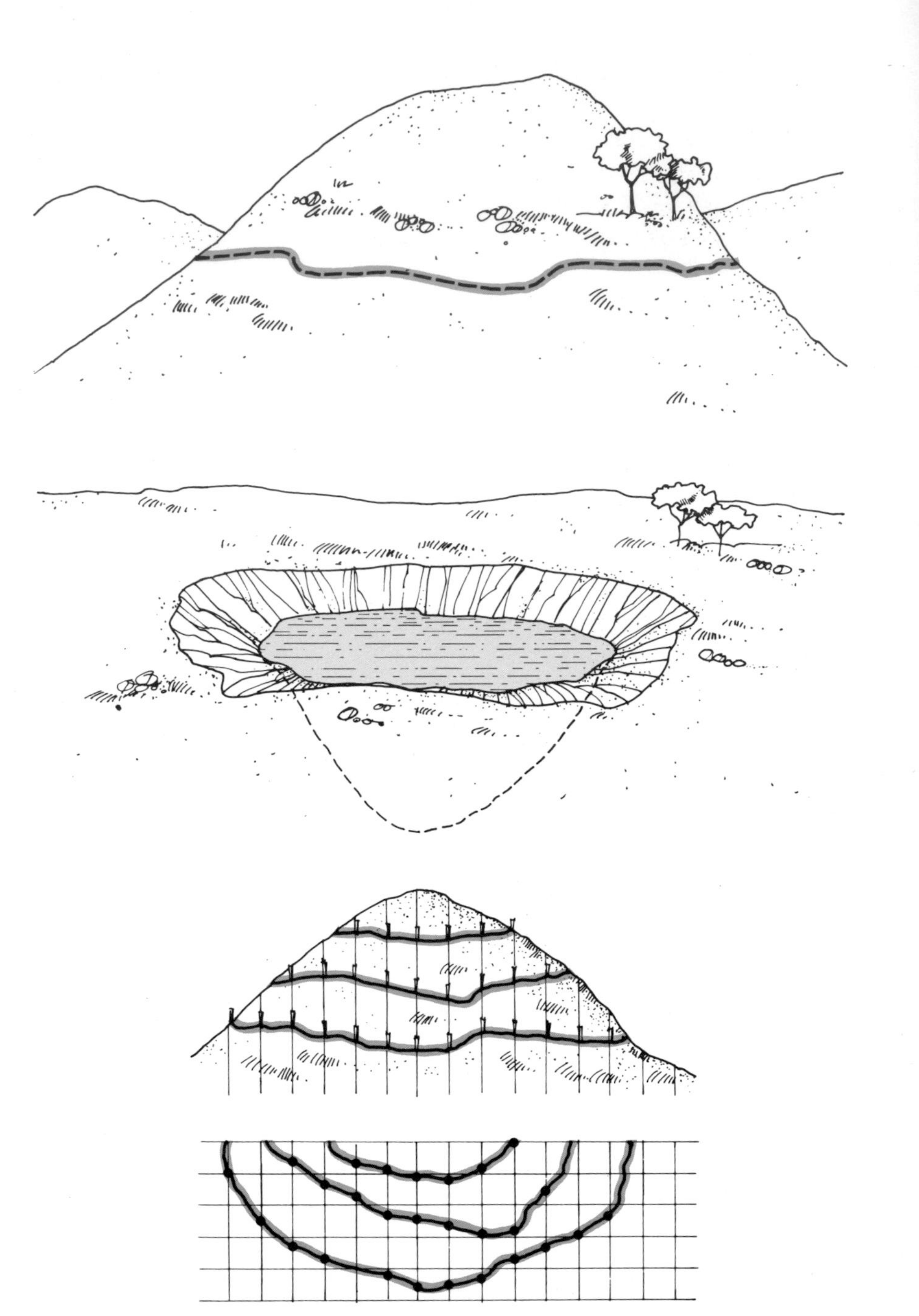

Example

When you pour water into a hole in the ground, you will see that the surface of the water forms a continuous line made up of the water's points of contact with the sides of the hole. This line shows one contour for this particular water depth in the hole. A lake or a reservoir also has a surface contour which depends on its water level.

What is contouring?

2. **Contouring** means surveying to identify the contours on the ground, lay them out with markers, and plot them on a plan or map. You will learn more about planning and mapping contours in Section 94.

3. Contouring is used in fish culture to solve two kinds of problem:

- if you have fixed the location of a point, you may have **to identify the contour** passing through that point;

Example

You have chosen the end-point of your water-supply canal on a fish-farm site. Now you have to identify the canal's centre-line, which usually follows a contour back to the water source (which may be a point along a river, or the outlet pipe of a pump).

- If you need to prepare a plan or map showing the ground relief of an area, you must find out the **location of contours on the ground** and be able to transfer them onto paper.

Example

You have chosen a fish-farm site. Before you can plan, design and build the farm, you will need to make a topographical map showing the location of a series of contours from which you will be able to define the ground relief of the site.

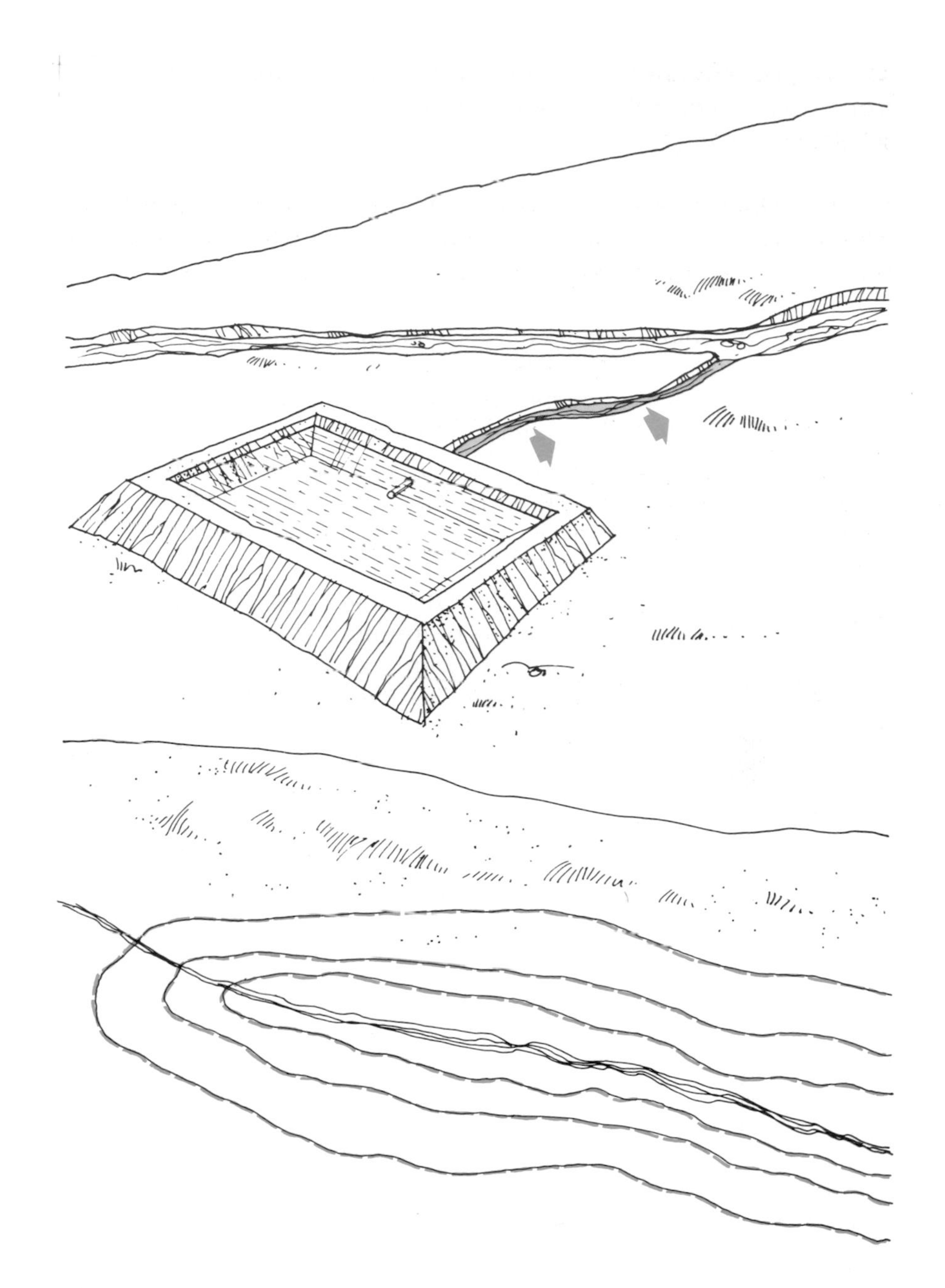

4. You have already learned how to find a contour on the ground from a fixed point, in the sections on contouring devices (see Sections 62-68).

5. In the following steps, you will learn how **to survey contours over a land area** so that you can prepare a topographical map (see Section 94).

What are the main methods for contouring?

6. It would be an impossible task to identify all the contours in one area. Therefore, you will have to decide how many contours you need to identify in each area. You will have to fix **the difference in elevation between contours** which are next to each other. This is called the **contour interval**.

7. Choosing which contour interval to use depends mainly on the accuracy you need, on the scale of the map you will prepare (see Section 91) and on the kind of terrain you are surveying. Contour intervals usually **vary from 0.25 m to 1 m**. This range of intervals allows good accuracy, and makes it possible to produce large-scale topographical maps for flat or slightly sloping ground (which is usually the type of ground used for fish-culture sites). Since smaller contour intervals make contouring much more difficult, you will usually make reconnaissance and preliminary surveys with a contour interval greater than the one you use for later, more detailed surveys.

Example

Relationship between the size of contour intervals and factors

Factor	Contour intervals	
	Smaller	Larger
Required accuracy	High	Low
Mapping scale (Section 9)	Large-scale	Small-scale
Type of terrain	Flat	Sloping

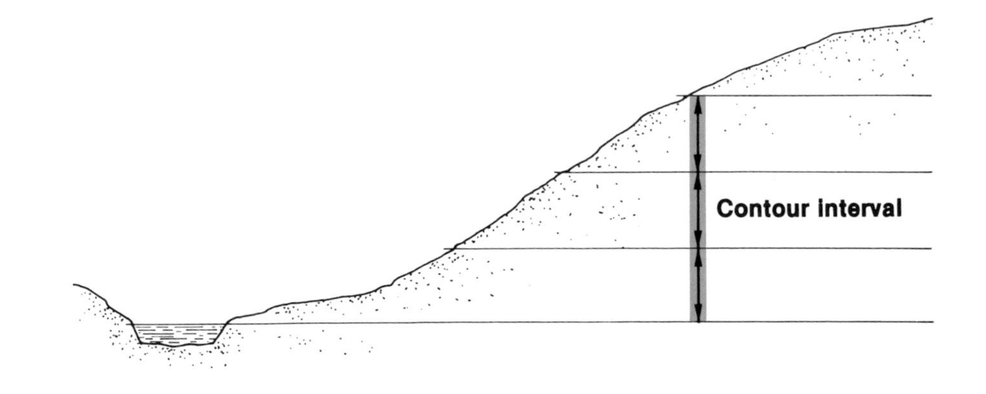

8. There are two main methods of surveying contours:

- **a direct method**, in which you trace and mark the line of each contour on the ground, and then plan survey these lines so that they can be mapped;
- **an indirect method**, in which you make a topographical survey of the area to find a series of points of known elevation. Then you enter them on a map and determine the contours from this map.

Selecting the contouring method

9. When selecting the method you will use for contouring, remember that:

- **the direct method** is much slower, but is more accurate. Use it only to contour a relatively small area which you need to map in detail, on a large scale;
- **the indirect method** is faster, but it is not so accurate. Use it to contour large areas that you will map on a medium or small scale. It should preferably be combined with plane-tabling (see Section 75).

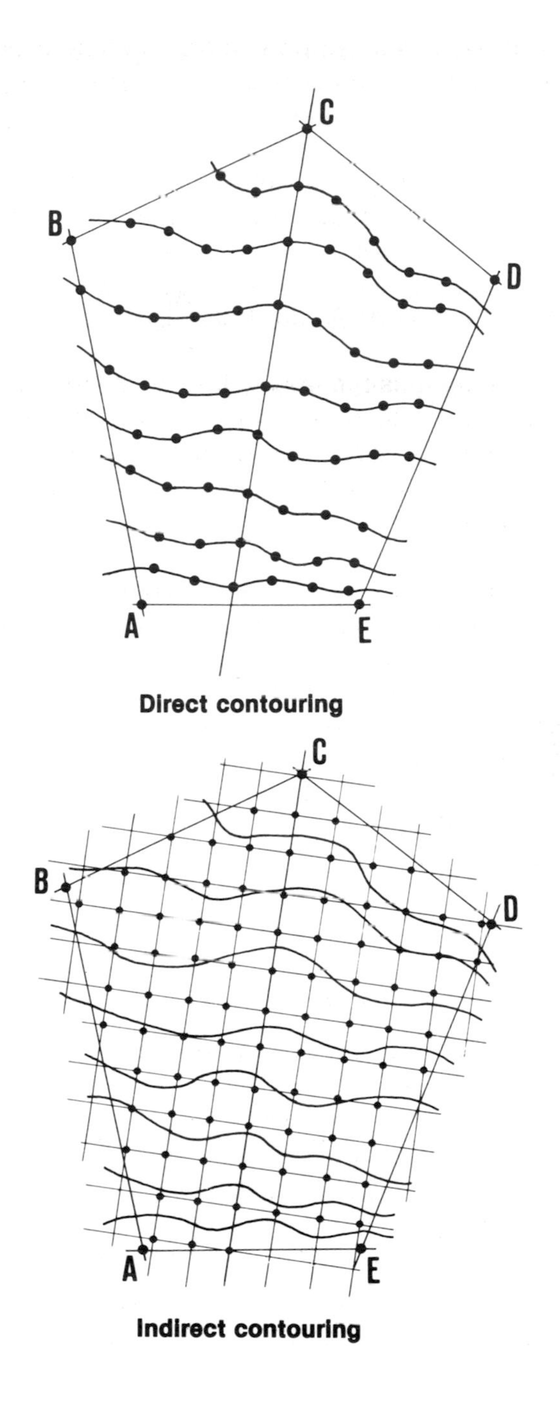

Laying out contours on the ground with a sighting level

You will now learn the direct method of contouring which will enable you to lay out a number of points on the ground which have exactly the same elevation.

10. Start your contouring survey of site ABCDEA at a point of known elevation, such as **an existing bench-mark BM**. If there is no such point of known elevation in the area, you can establish one:

- either by **differential levelling** from a bench-mark outside the area to a point within the area;
- or by **assuming a convenient elevation** for your bench-mark (such as 100 m) so that you will not have points with negative elevation later.

Note: try to establish this bench-mark in the middle of the lowest ground of the area, so that you can survey uphill.

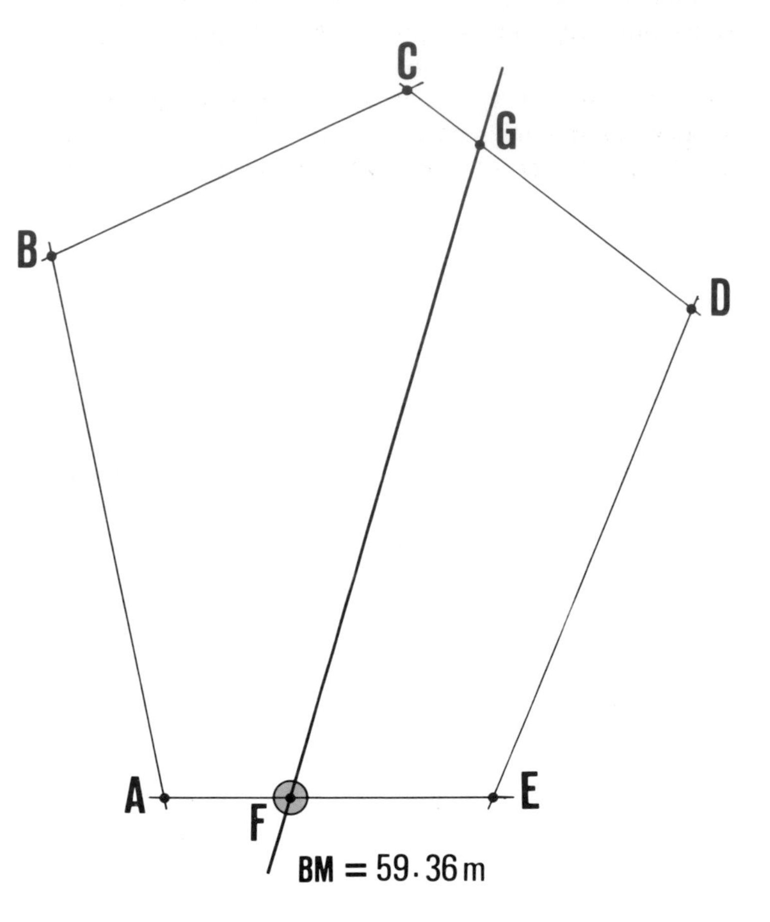

Establish a bench-mark in the lowest part of the site

11. Through this bench-mark **BM** at point F, lay out and mark **a straight line FG**. Make sure you follow the direction of **the greatest ground slope**. The line should cross the entire site.

12. At regular intervals, set out **a series of lines parallel** to FG. To choose the interval between parallels, use:

- **10 m or less**, if the contour interval is to be 0.25 to 0.50 m;
- **25 to 30 m**, if the contour interval is to be 1 to 1.5 m;
- **50 m**, if the terrain has a very gentle or regular slope.

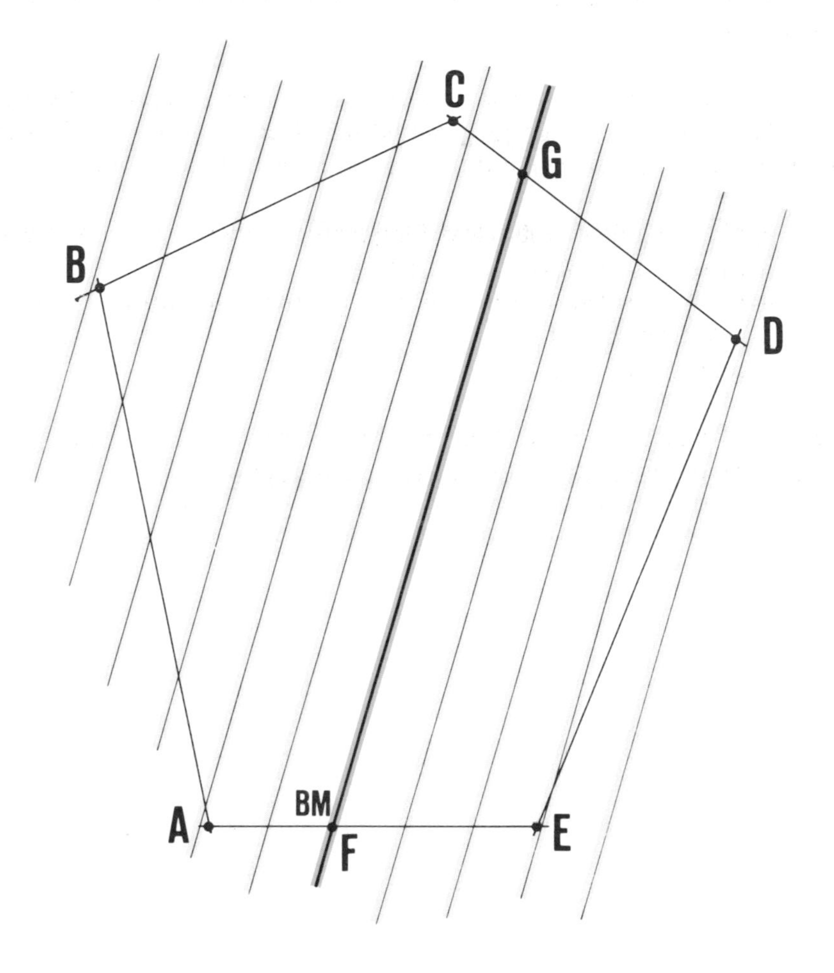

Lay out line FG from the bench-mark, and parallel lines at regular intervals

13. If you know the elevation E(BM) of the benchmark BM from a previous survey, first find the point on the line with an elevation that corresponds to **a multiple of the contour interval** you have selected. You can use **a sighting level** together with a **target levelling staff**. The method will enable you to set the target on the staff in the right position for identifying the first contour on the ground.

Example

- BM is at elevation 59.36 m.
- With a sighting level set up at LS1 and a levelling staff held on BM, read **BS = 3.23m**.
- Choose the contour interval, for example **CI = 0.25m**.
- Calculate the multiple of CI (= nCI) closest to E(BM) = 59.36m as follows:
  (a) E(BM) ÷ CI = 59.36 m ÷ 0.25 m = 237.44 ... or the round number **n = 238**;
  (b) n × CI = 238 × 0.25 m = **59.50 m**.
- The difference between E(BM) and n(CI) equals 59.50 m – 59.36 m = **0.14 m**.
- Set the target on a **target levelling staff at the height of** BS minus this difference or 3.23m – 0.14 m = **3.09 m**.
- Find the position of the first contour at the elevation 59.50 m.

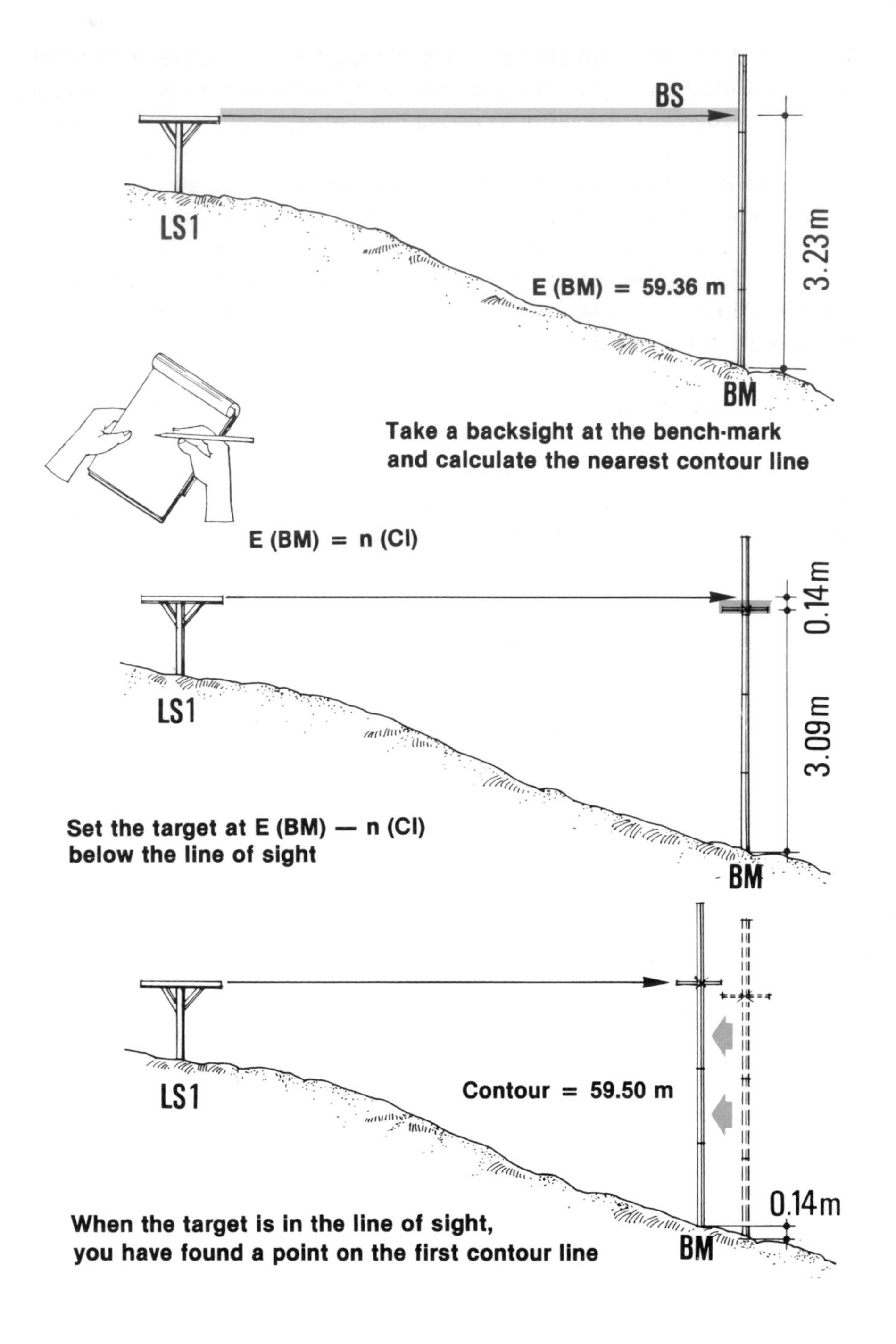

14. You will need an assistant for this method. At LS1, the point from which you can survey as many surrounding points as possible, set up the level. Holding **the adjusted target levelling staff**, your assistant walks slowly uphill from the bench-mark **along the central line FG**. Sight with the level at the target, and signal to your assistant to stop when the sighting line lines up with the target line. The ground point X where the levelling staff stands should be at elevation 59.50 m. This is the first point of the 59.50 m contour. Direct your assistant to mark this point with a stake. Remember also to indicate clearly **the elevation of the point** on the stake.

15. Your assistant then moves with the levelling staff to another parallel line, where you determine and mark a second point Y at elevation 59.50 m in the same way. This procedure is repeated on all the parallel lines, until you have marked **contour 59.50 m** completely on the ground across the site.

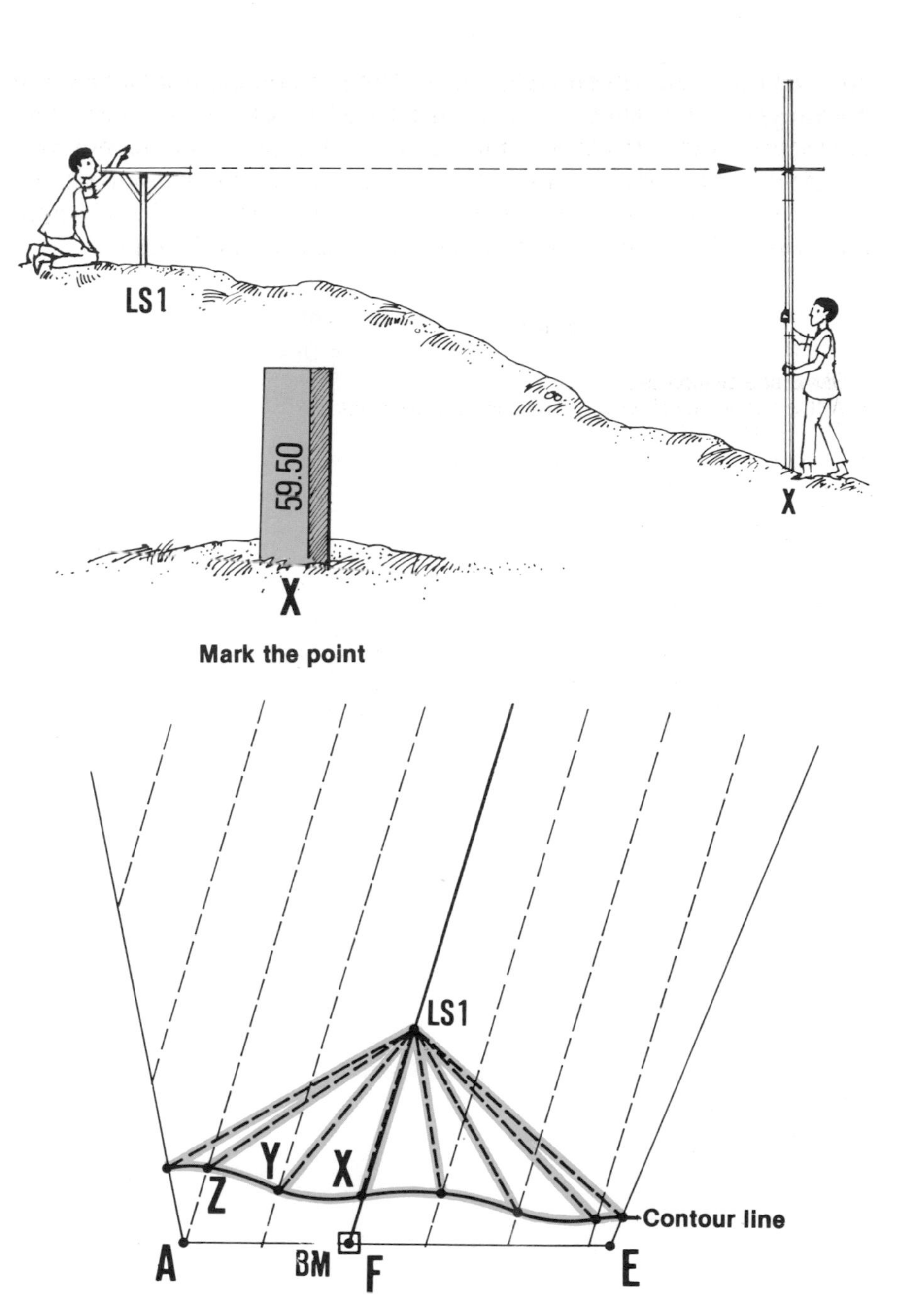

Mark the point

Survey other points on the same contour

16. To determine the next contour, you must change the position of **the target** on the staff. As you are **moving uphill**, using a selected contour interval of 0.25 m, you will **lower the target** by 0.25 m to a height of 3.09 m — 0.25 m = 2.84 m. In this position, the target will show the ground points at elevation 59.50 m + 0.25 m = 59.75 m, if you continue surveying **from the same levelling station LS1**.

17. From LS1, find all the points **on the parallel lines** at elevation 59.75 m, and mark a second contour on the ground. **Again lower the target by 0.25 m** to the height of 2.84 m – 0.25 m = 2.59 m to determine points at the next elevation of 60 m.

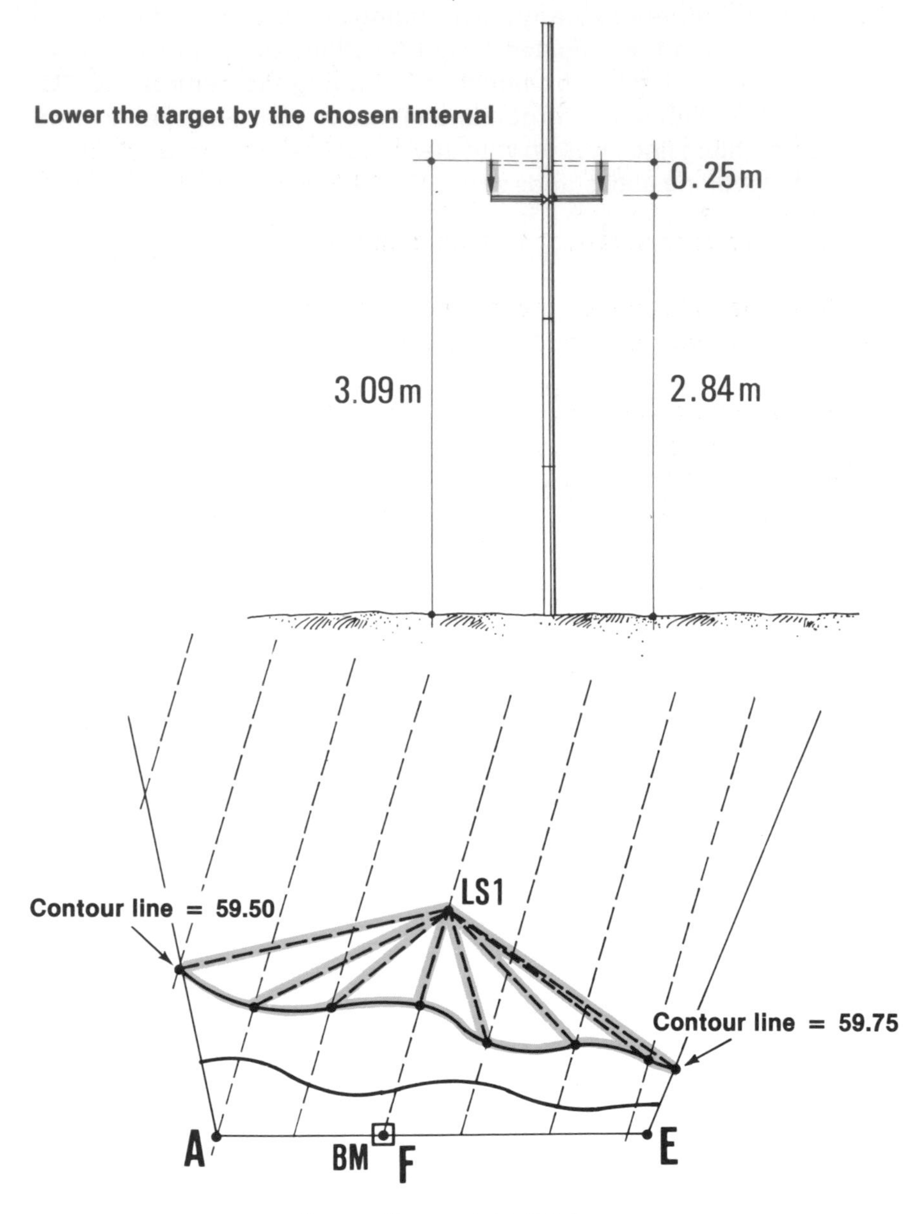

Lower the target by the chosen interval

Survey the next contour

18. If you need to change the levelling station but continue to survey the same contour:

- ask your assistant to hold the levelling staff on one of the points of that contour;
- move the level to a new, more convenient levelling station;
- tell your assistant to adjust the target height until it lines up with the line of sight of the level;
- continue to survey the same contour.

19. If you need to change the levelling station at the same time you are ready to determine another contour:

- ask your assistant to keep the levelling staff on a point of the last surveyed contour;
- move the level to its new station;
  adjust the target height to the new line of sight;
- change this target height to determine the new contour (by lowering it 0.25 m, for example, see step 16).

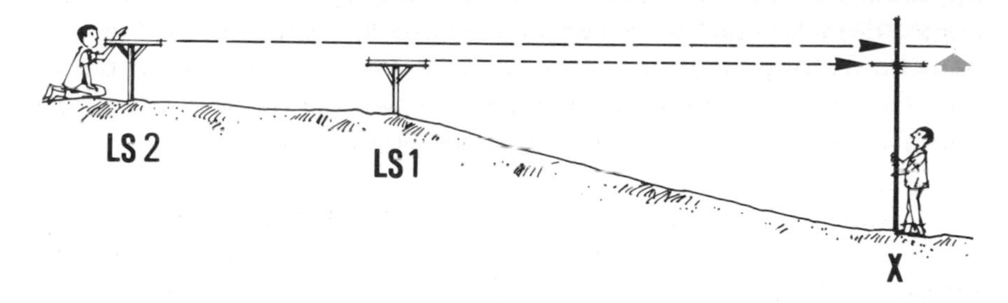

To continue on the same contour, move the level, then adjust the target

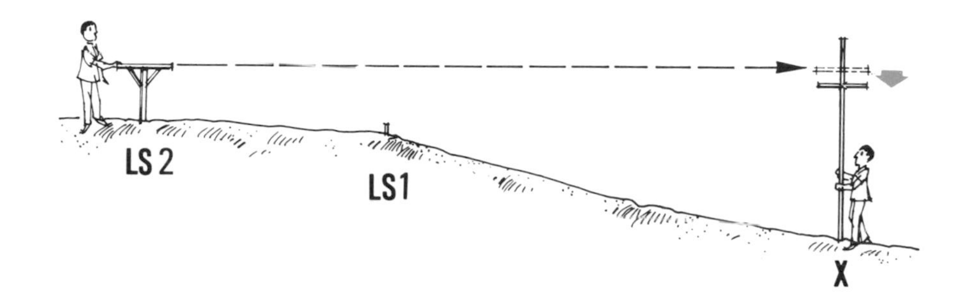

For a new contour, set the target lower than the line of sight

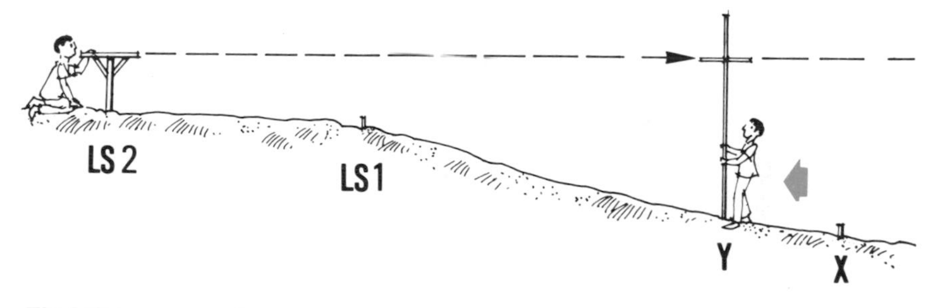

Find the new contour

20. When you have determined the various contours at their intersection with each parallel line, you will have to measure the horizontal distances between all the marked points. To do this, you can chain along the parallel lines starting from the area boundaries (see Section 26). These measurements will help you to prepare a topographical map of the area (see Section 94).

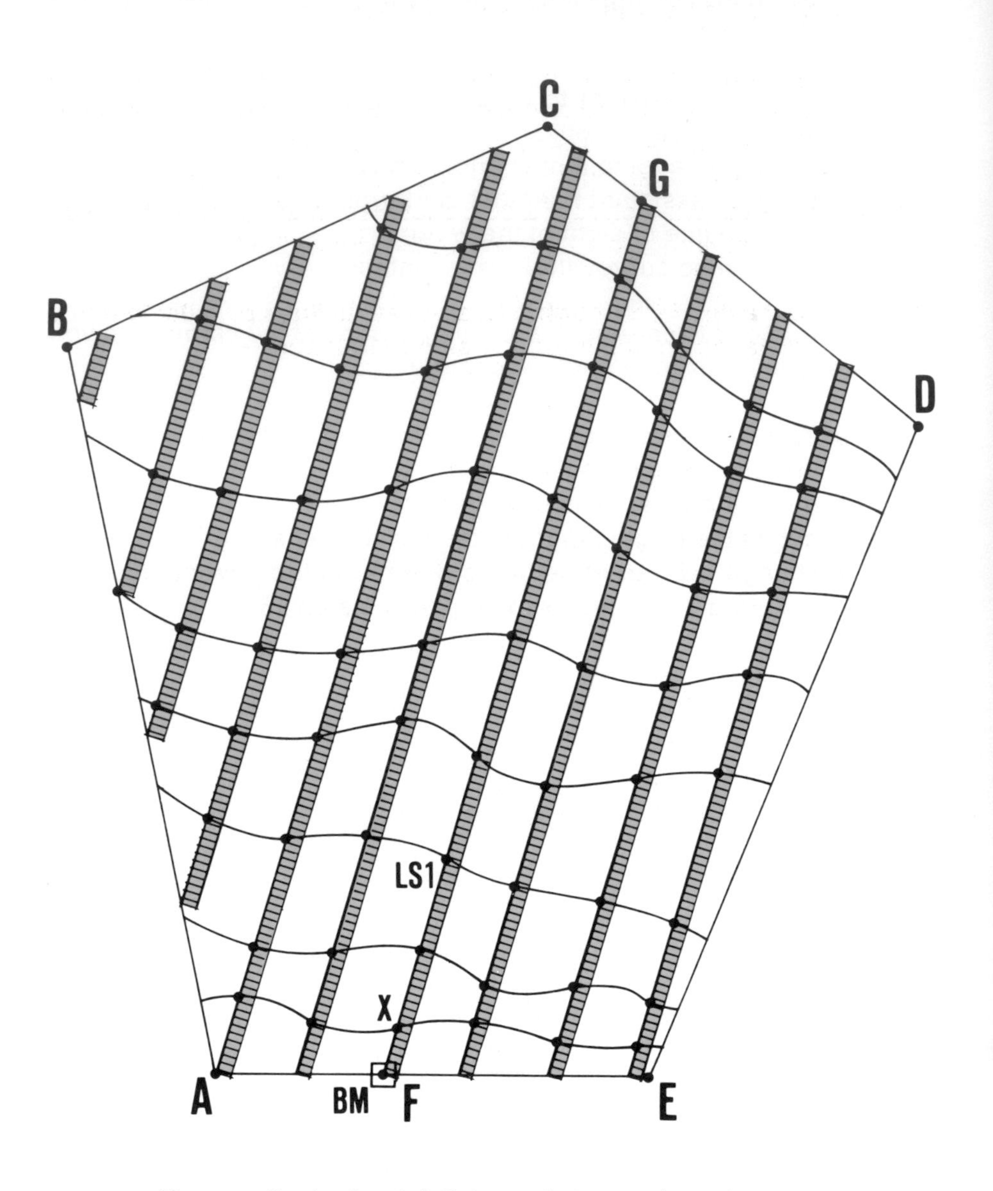

Measure the horizontal distance between the points

Laying out contours with a non-sighting level

21. When you use a non-sighting level (such as a line level or an A-frame level) to lay out contours over an area of land, you first need to establish a **bench-mark BM** near the boundary of the area. As usual, this bench-mark may be either of known elevation or of assumed elevation. It should also be located in the part of the area with the lowest elevation (see Section 81, steps 42-44).

22. Set out **a line FC through BM**, and set out **lines parallel to it** at a selected distance, as described in steps 11-12 above.

Example

Selected distance between parallels = 10 m.

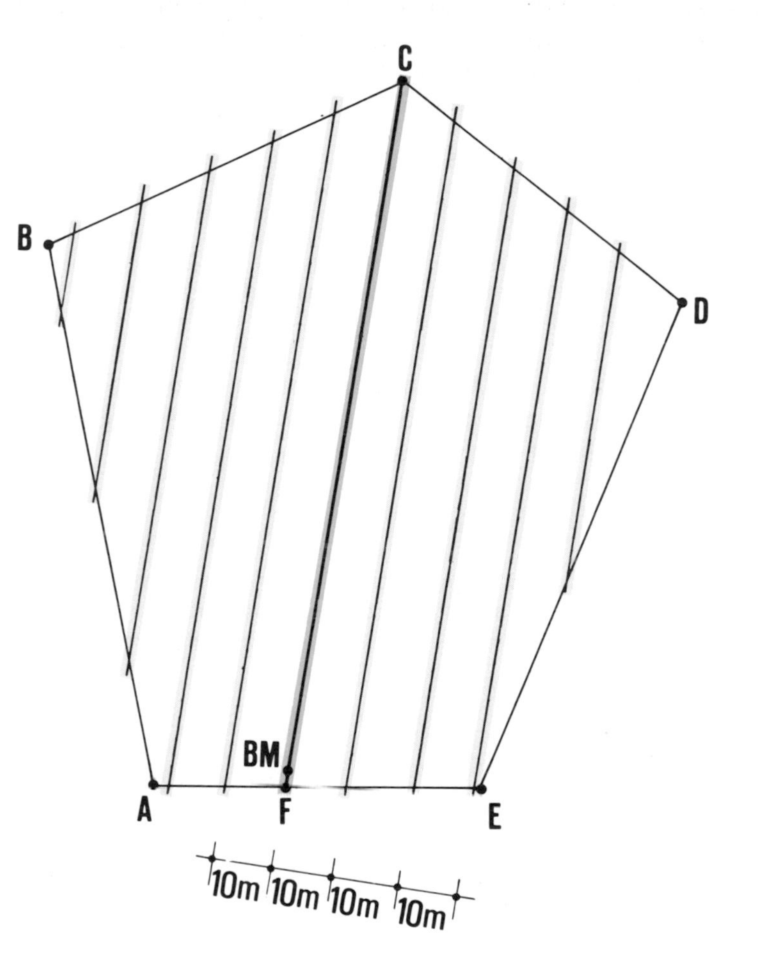

Set out a line through the bench-mark, and parallels at regular intervals

23. If you are using a bench-mark with a **known elevation**, proceed as shown above in step 13 to calculate the elevation of the first contour you will survey near the bench-mark. Also calculate the **difference** between the elevation of this first contour and the elevation of the bench-mark.

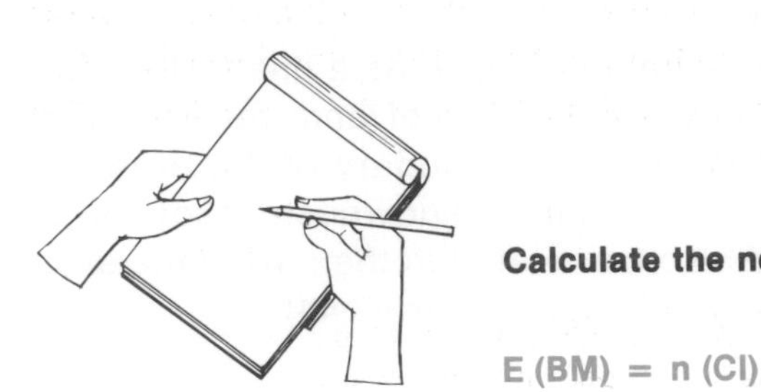

Calculate the nearest contour line

E (BM) = n (CI)

Example

- BM elevation E(BM) = 127.85 m
- Selected contour interval = 0.50 m
- Multiple of E(BM): 127.85 m ÷ 0.50 m = 255.7 and therefore you choose n = 256
- **First contour** will be at elevation 256 × 0.50 m = 128 m
- **Difference in elevation** between E(contour) and E(BM) = 128 m – 127.85 = **0.15 m**.

24. Then, **next to the bench-mark**, place some objects (such as bricks, stones, wooden planks, a tin or a box) that will provide the elevation calculated for the first contour.

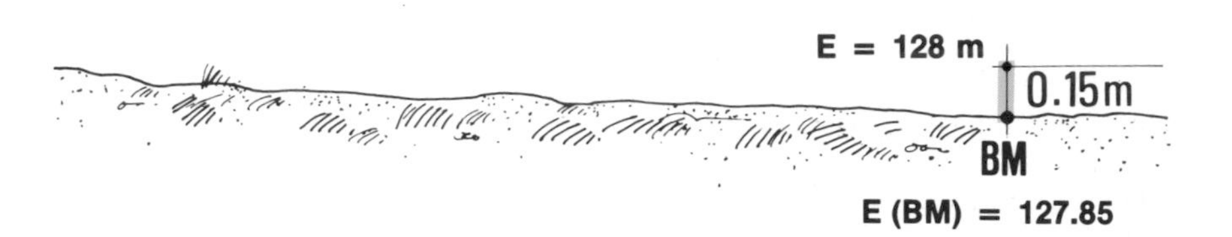

Find the difference in height

Example

Next to BM, place some bricks and adjust their top height at 0.15 m higher than E(BM), using a straight-edge and a mason's level (see Section 51). The top of these bricks will be at the 128 m elevation.

Use bricks to make up the height difference at BM

25. Find a ground point X which is near BM, is located on the line CF passing through BM, and has the same elevation as the objects piled near BM. To do this, use one of the methods described earlier (see Sections 51, 62-64 and 66). This ground point X is the first point of the contour 128 m.

Example

Using a straight-edge level, transfer the level 128 m from the top of the bricks to a ground point X on the line CF passing through BM.

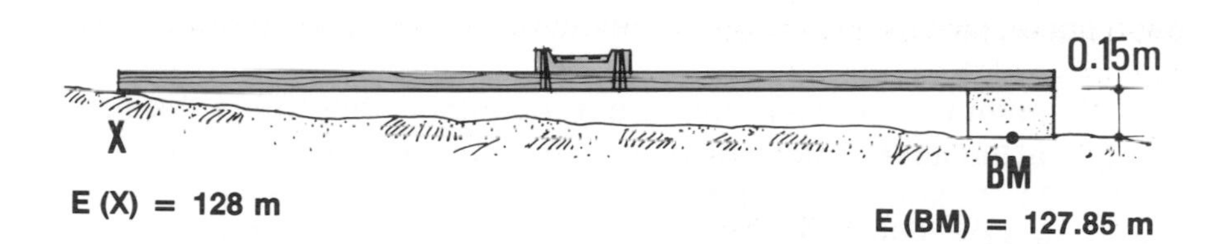

Finding the contour from a known bench-mark

26. If you are using a bench-mark with an **assumed elevation**, and are working uphill, determine the point X of the line passing through BM in the same way. The elevation of this point will equal assumed E(BM) plus the contour interval CI.

Example

- If E(BM) = 100 m and CI = 0.50 m, pile bricks 0.50 m high at BM.
- Locate nearby point X where E(X) = 100 m + 0.50 m = 100.50 m.

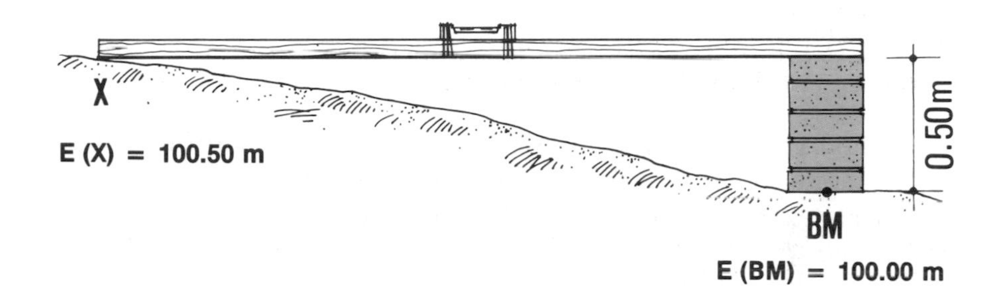

Finding the contour from an assumed bench-mark

27. Start contouring from point X using one of the methods described in Chapter 6. With **a stake**, mark each point where the contour you are following intersects with one of **the parallel lines**. On each stake, clearly indicate the elevation of the ground point.

28. Each time you finish laying out a contour, determine **the first point**, Z, **of the next contour** by using a method like the one described in step 24. At known point X, where the last contour line crosses central line CF, place **objects with a total height equal to the contour interval**. Transfer this new level horizontally along line CF to point Z on the next contour. If the contour interval is large, you may have to use intermediate points to do this in stages.

Example

- CI = 0.50 m.
- Transfer first E(contour) by + 0.25 m, from X to Y.
- Repeat again from Y to Z, to total + 0.50 m = 2 × 0.25 m.

29. When you have laid out all the contours on the ground with stakes, measure, from stake to stake, **the horizontal distances** along the parallel lines. This will help you to prepare a topographical map (see Section 94).

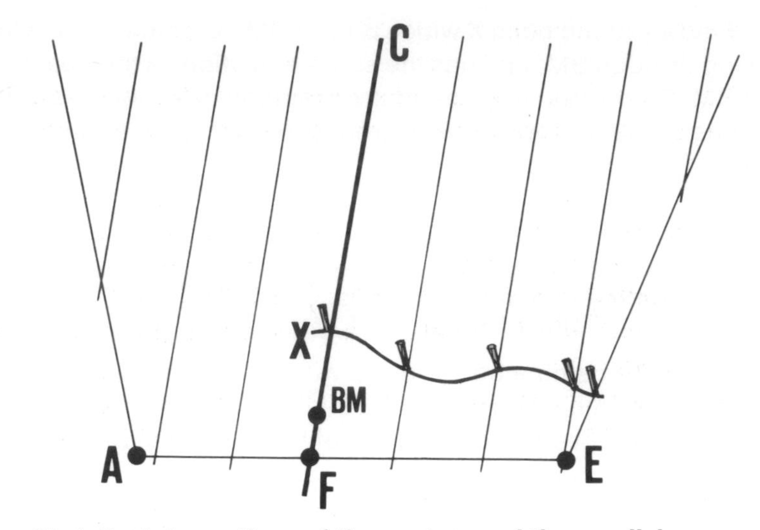

Mark the intersections of the contour and the parallels

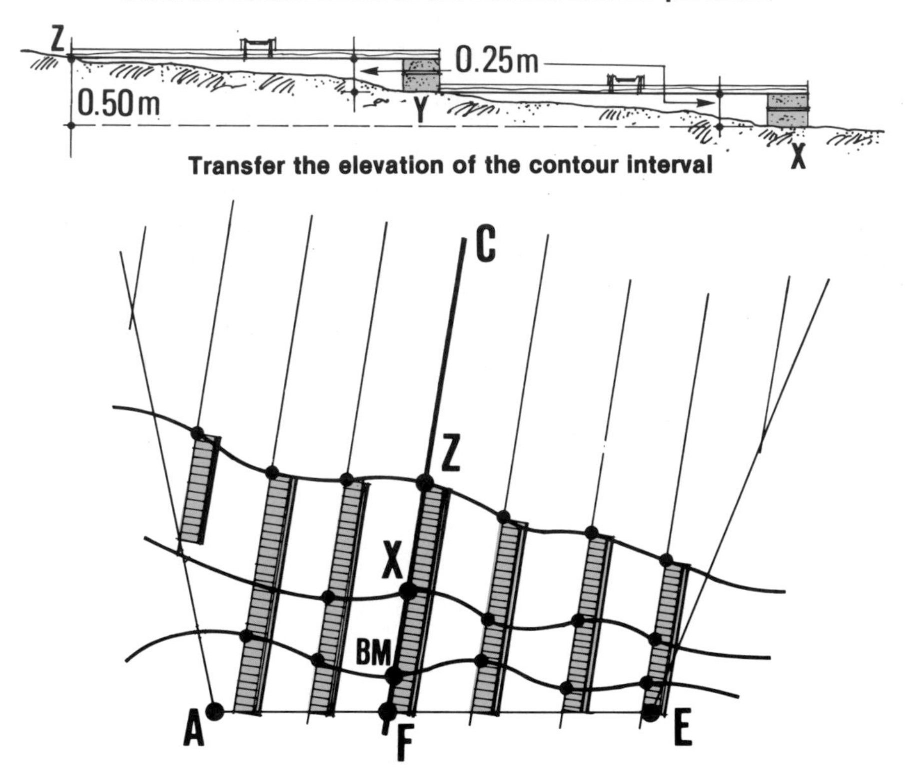

Transfer the elevation of the contour interval

Measure the horizontal distances between the points

Contouring by the indirect method

30. You can also contour by the **indirect method**. In this method, you make a topographical survey of the area, using a definite pattern, such as:

- **a square grid** to determine elevations for points located at the intersections of a grid made of square or rectangular blocks;
- **radiating** to determine elevations for random points located on lines which radiate at a selected angle interval from a known point;
- **cross-sections** to determine elevations for points located on short lines laid out at right angles to a surveyed base line.

31. You learned earlier that **the square-grid pattern** is commonly used to contour relatively small areas, particularly if their perimeters have already been surveyed (see Section 81, steps 24-33).

32. You also learned about **the radiating pattern**, which is particularly useful for large areas (see Section 81, steps 34-36).

33. Finally, you learned about **cross-sections**. These are commonly used in preliminary surveys, where you need a contoured plan of a long narrow stretch of land to select the best possible route for your purpose. You lay out lines about 30 to 100 m apart and about 50 to 100 m long on either side of a main compass traverse, and at right angles to it. Then you can find elevations of points along these cross-sections (see Section 82, steps 15-19).

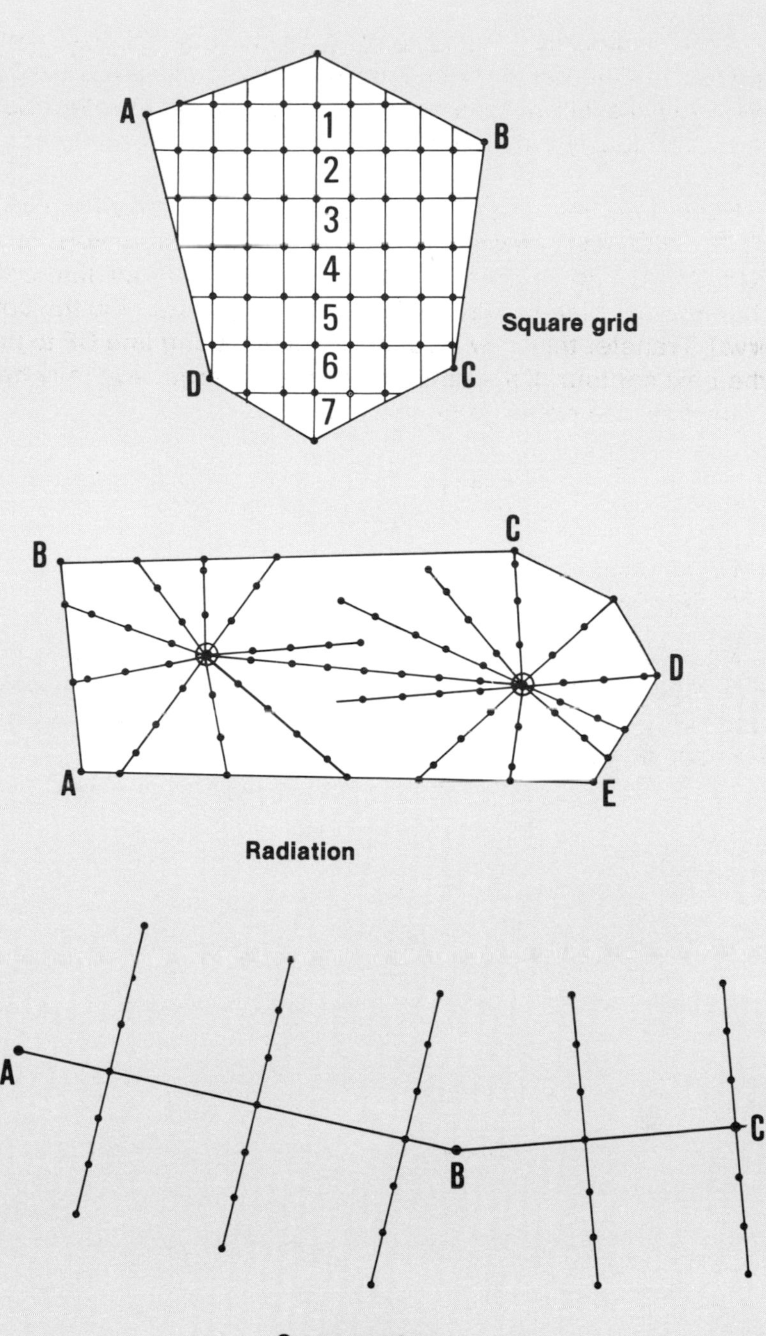

9 TOPOGRAPHICAL PLANS AND MAPS

What are topographical plans and maps?

1. Topographical plans and maps are drawings which show **the main physical features** on the ground, such as buildings, fences, roads, rivers, lakes and forests, as well as the changes in elevation between land forms such as valleys and hills (called vertical relief). You base these plans and maps on the information you collect from topographical surveys.

2. **Plans** are usually large-scale drawings; **maps** are usually small-scale drawings. Depending on the scale you use to make the drawing (see Section 91):

- it is a **plan** if the scale is **larger** than 1 cm for 100 m (1 : 10 000), for example 1 cm for 25 m;
- it is a **map** if the scale is **equal to or smaller** than 1 cm for 100 m (1 : 10 000), for example 1 cm for 200 m or 1 cm for 1 000 m.

Example

- An engineering plan could show information you need for building fish-farm features such as dikes, ponds, canals or outlet structures, at the scale of 1 cm for 25 m (1 : 2 500).
- A topographical map could show a fish-farm site (scale 1 cm for 200 m or 1 : 20 000) or a region of a country (scale 1 cm for 1 000 m or 1 : 100 000).

3. Plans and maps have **two main purposes** in fish-farm construction. They help guide you in choosing a site, planning the fish-farm, and designing the structures that are needed for the farm. Plans and maps also guide you as you lay out marks on the ground, so that you can follow the plan you have made of the fish-farm, and build the structures on it correctly.

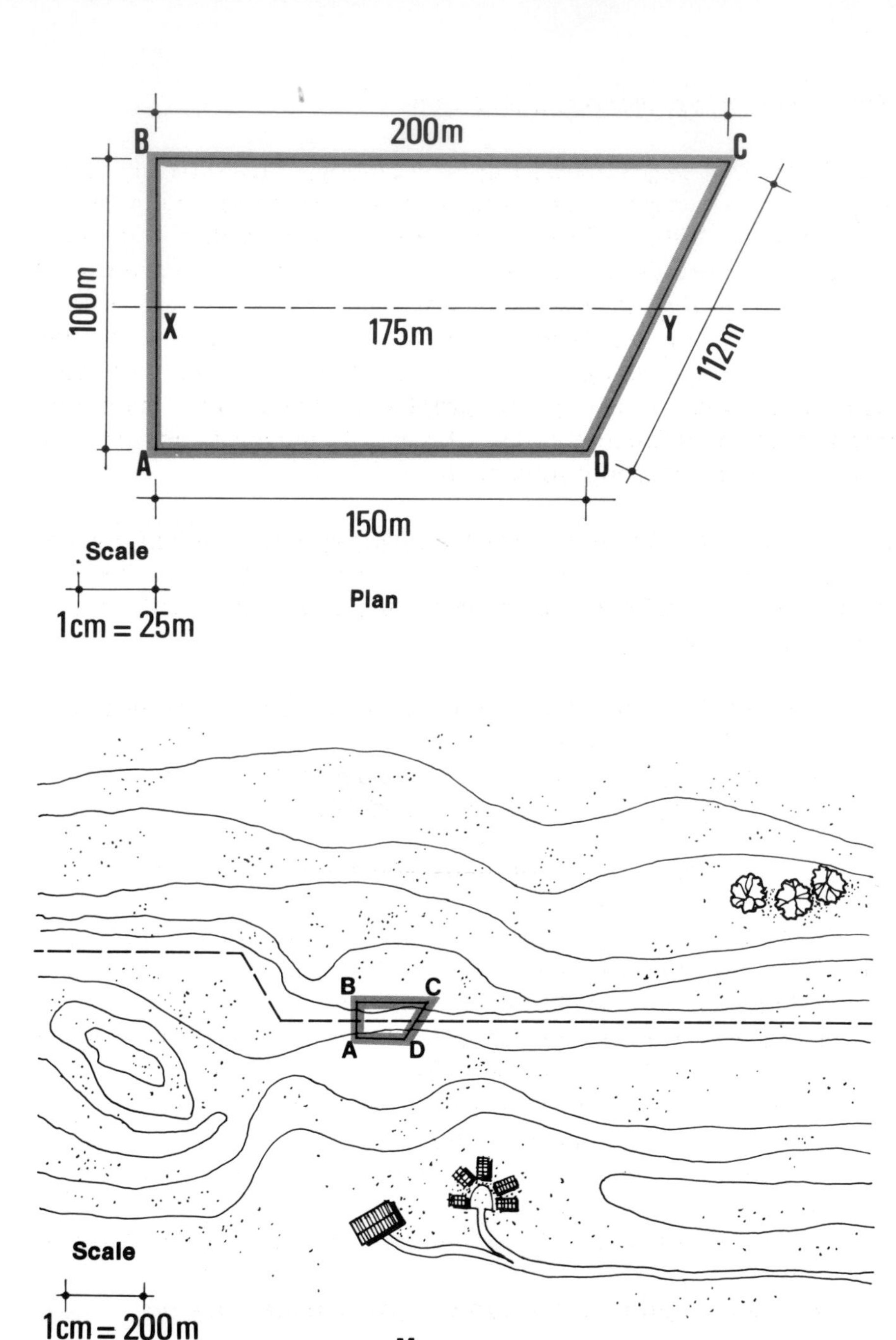

4. Before you begin a topographical survey, you should try to get any available topographical plans and maps of the area, even though they may not be exactly the kind of plan or map that you need. **General topographical maps** are available from governmental organizations which are responsible for geological surveys or land surveys, for example. National geographical institutes, soil survey departments and agricultural development agencies can also usually provide existing topographical maps. The cadastral department (that calculates land taxes) of your local government may provide local topographical plans.

5. You will often have to make the topographical plans and maps yourself, however. You will base them on a plan survey (see Chapter 7) and direct levelling (see Chapter 8). In the following sections, you will learn how to:

- make the plan or map directly in the field by plane-tabling (see Section 92); or
- make the plan or map from the field measurements recorded in your notebook (see Sections 93-96).

6. On topographical plans or maps, you should always look for:

- **the name** of the area or piece of land mapped, and/or the name of the type of project for which it is used;
- **the exact location** of the piece of land;
- **the name of the person or people** who made surveys on which the plan or map is based;
- **the date(s)** on which the surveys were made;
- the direction of **magnetic north**;
- **the scale** at which the plan or map was drawn (see Section 91);
- **the contour interval**, if the vertical relief is shown (see Section 93);
- **a key, or guide**, to the symbols used in the drawing.

This information is often located in one corner of the map. It is called the **legend**.

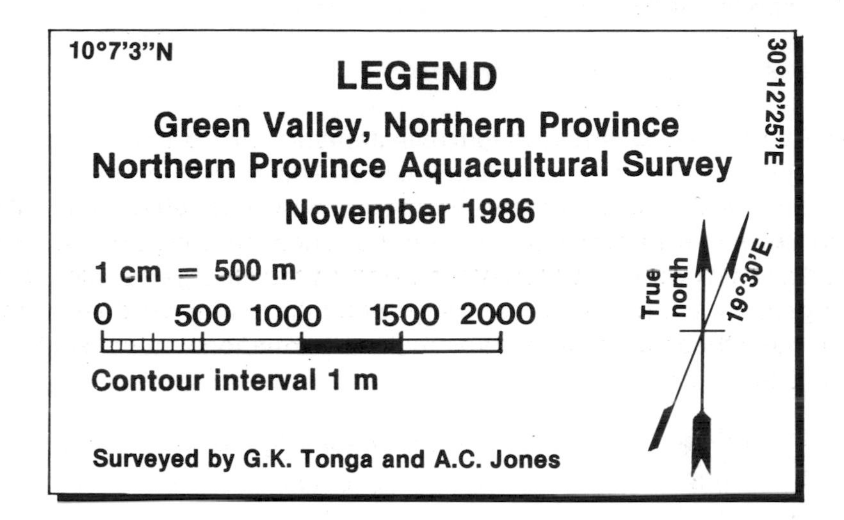

SYMBOLS FOR PLANS AND MAPS

Deciduous trees	Evergreen trees
Woods	Orchard
Marsh or Swamp	Rough pasture
Rocks	Sand and shingle
Embankment	Cutting
Railway single line or Double line	Path or
Road (fenced)	Road (unfenced)
Bridge	Wall and gate
House (brick)	Green house
Shed with open sides	Shed with closed sides
Fence or	Hedge
Concrete or brick drain	Earth drain or
River	Canal with lock
Boundaries	Lake or pond

91 How to make scales for plans and maps

What is the scale of a plan or map?

1. To represent distances you have measured in the field on a piece of paper, you need to **scale them down**. This means that you must reduce the size of the distances proportionally according to a scale. **The scale** expresses the relationship which exists between the distance shown on a drawing or map and the actual distance across the ground.

Example

- 1 cm on the plan represents 20 m on the ground, or scale 1:2 000.
- 1 cm on the map represents 100 m on the ground, or scale 1:10 000.
- 1 cm on the map represents 1 250 m on the ground, or scale 1:125 000.

Note: a ratio with a **smaller** number is a larger scale, that is, 1:500 is a larger scale than 1:1 000.

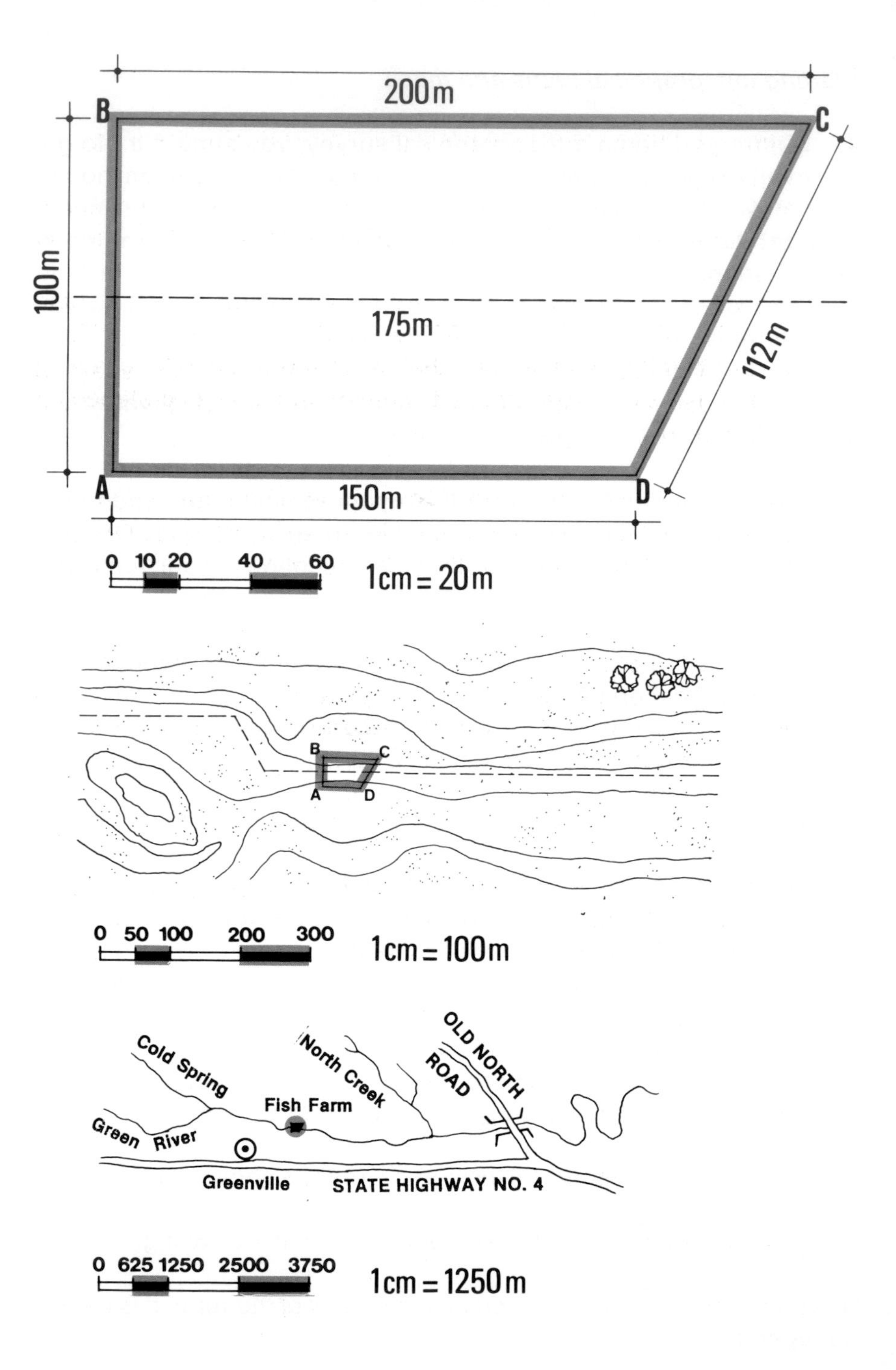

Expressing a scale

2. There are three ways of expressing the scale of a drawing:

- as **a numerical equivalent** such as "1 cm = 20 m", which you should read as "1 cm on the plan represents 20 m on the ground";
- as **a ratio** such as "1 : 2 000" which you should read as "1 cm on the plan represents 2 000 cm = 20 m on the ground";
- **graphically**, with a line that is marked off into drawing distances that correspond to convenient units of distance on the ground.

3. **Table 11** gives **the numerical equivalents** of the most common scales, expressed as fractions. Scales for both distances (in metres) and surface areas (in square metres) are shown.

Choosing a scale

4. General topographical maps usually have scales ranging from 1 : 50 000 to 1 : 250 000. These are **small-scale maps**. In most countries, 1 : 50 000 maps are now available. You can use these for general planning of aquaculture development, including the planning of your fish-farm.

5. To show greater detail, plans are drawn to **a larger scale**, showing individual structures or land areas. The scales most often used in plans are 1 : 500, 1 : 1 000, 1 : 2 000, 1:2 500 and 1:5 000. Detailed engineering drawings use scales much larger than 1 : 500, for example 1 : 100 or 1 : 10.

Note: special rulers, called "Kutsch" scales or **reduction scales**, make it easy to transfer ground distances onto drawings.

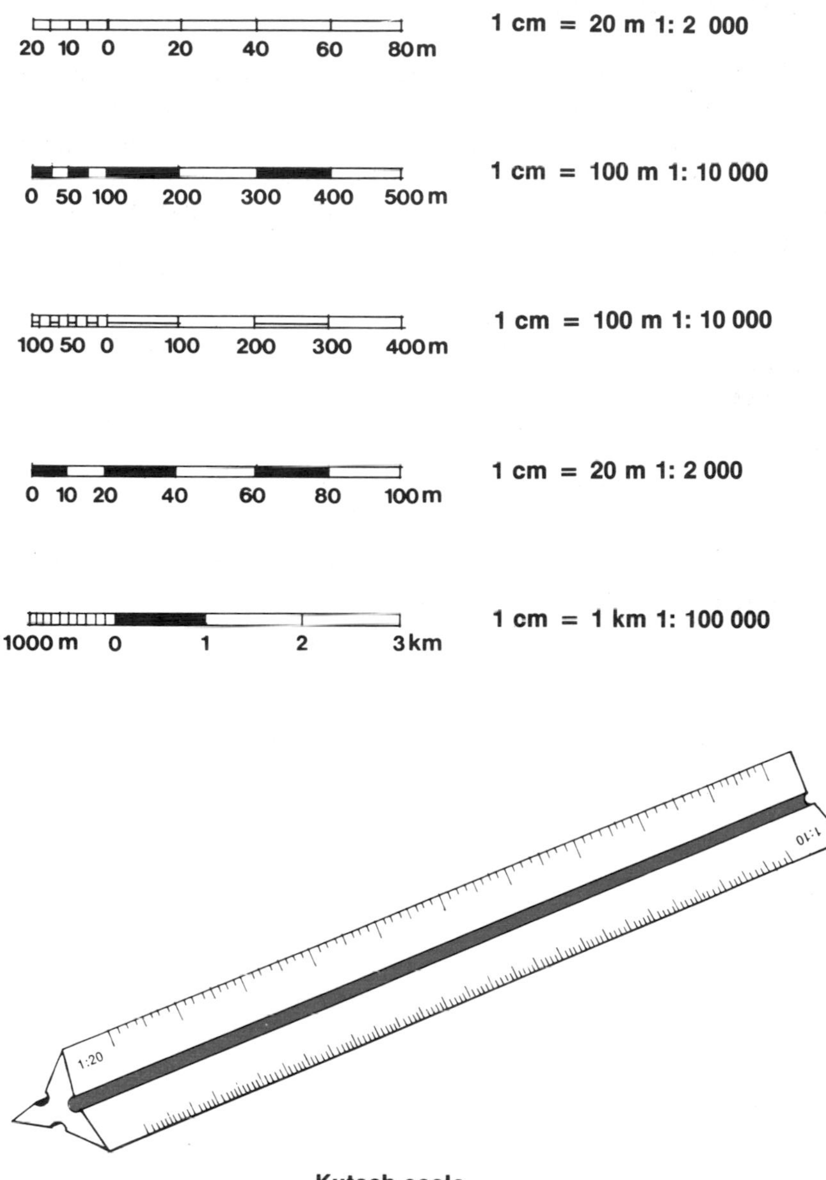

Kutsch scale

TABLE 11

Distances and surface areas expressed by scales

Scale		*Distance* *1 cm equals (m)*	*Surface area* *1 cm ^{2} equals (m ^{2})*	
1:	300	3	9	
	500	5	25	
	600	6	36	
	1 000	10	100	
	1 200	12	144	
	1 500	15	225	
	2 000	20	400	
	2 500	25	750	
	5 000	50	2 500	(0.25 ha)
1:	10 000	100	10 000	(1 ha)
	25 000	250	62 500	(6.25 ha)
	50 000	500	250 000	(25 ha)
	100 000	1 000	1 000 000	(100 ha)
	125 000	1 250	1 562 500	(156.25 ha)
	200 000	2 000	4 000 000	(400 ha)
	250 000	2 500	6 250 000	(625 ha)

92 How to make a map by plane-tabling

1. In Section 75, you read that you can use a **plane-table** to make a reconnaissance survey and to plot details. In this section, you will learn how to do this. It is best to use an **alidade** for this method (see Section 75, steps 21-28), but you can use a simple ruler, and a series of tailor's pins to show the observed directions, instead.

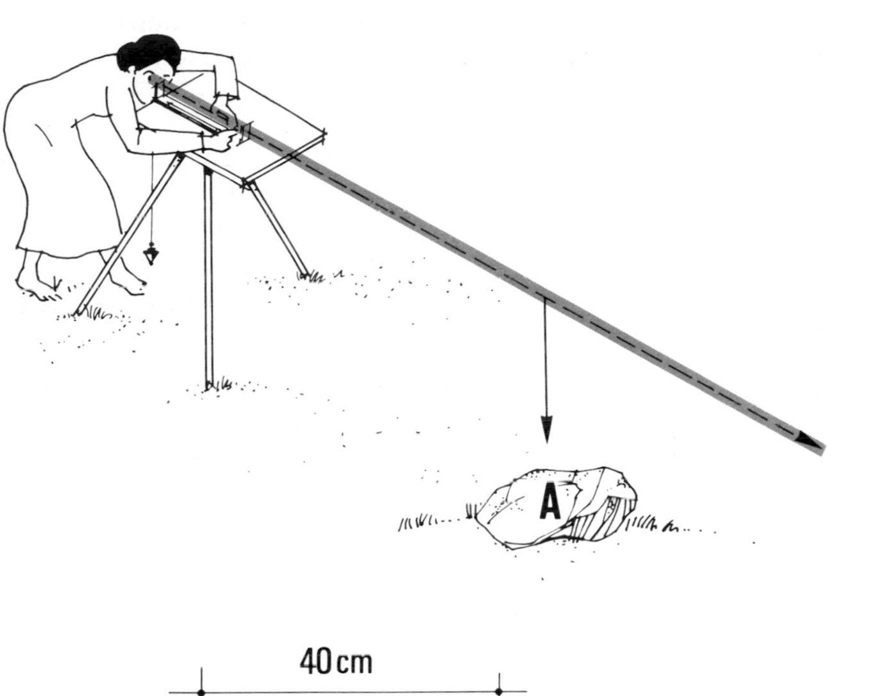

2. First, choose **an appropriate scale** for the map you will draw (see Section 91, steps 4 and 5). Get an estimate of the longest distance you need to map, and decide upon the size of the map you require. If the map is to be fairly large, you can draw it on several sheets of paper, and glue them together.

Example

- You have a plane-table, size 40 x 55 cm.
- You estimate the longest distance to be mapped = 400 m.
- From Table 11, you find that if you use a 1 : 1 000 scale (where 1 cm is equivalent to 10 m), you will need 40 cm to draw this distance on your sheet of paper.
- If this scale is large enough for your purposes, you can use just one sheet of paper.

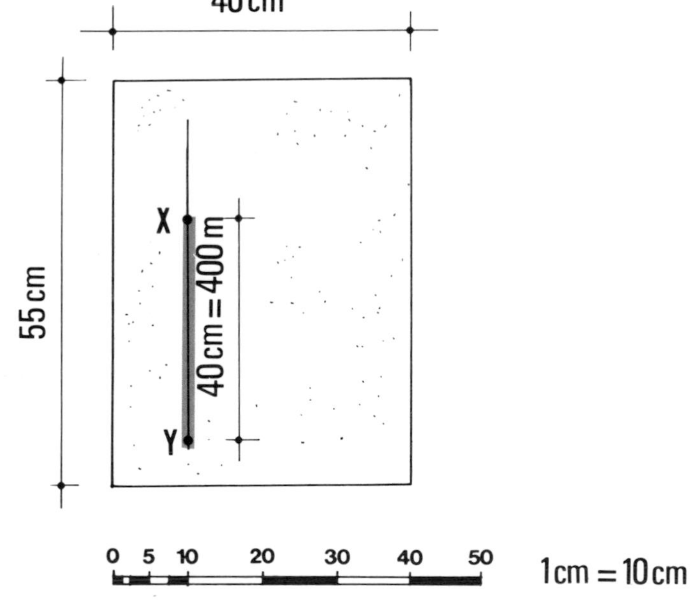

3. Cover the board of your plane-table with paper (see Section 75, steps 34-39). Set up the plane-table (see Section 75, steps 40-44) on or near some **major feature A** of the area you need to map, such as a large rock, a path, a river or a tall tree.

4. Using a well-sharpened pencil with a hard lead, mark a small point and circle on your paper. This is **point a**, the **location of the major feature**, where you have set up your plane-table. Be sure to choose a section of the paper from which you can later map the entire area. For example, if you will be mapping only ahead of point A, begin near the centre of the bottom margin of the plane-table.

Note: you will identify physical features in the field that you need to map with **capital letters**. You will identify the corresponding points that you draw on the plane-table sheet with **lower-case letters**.

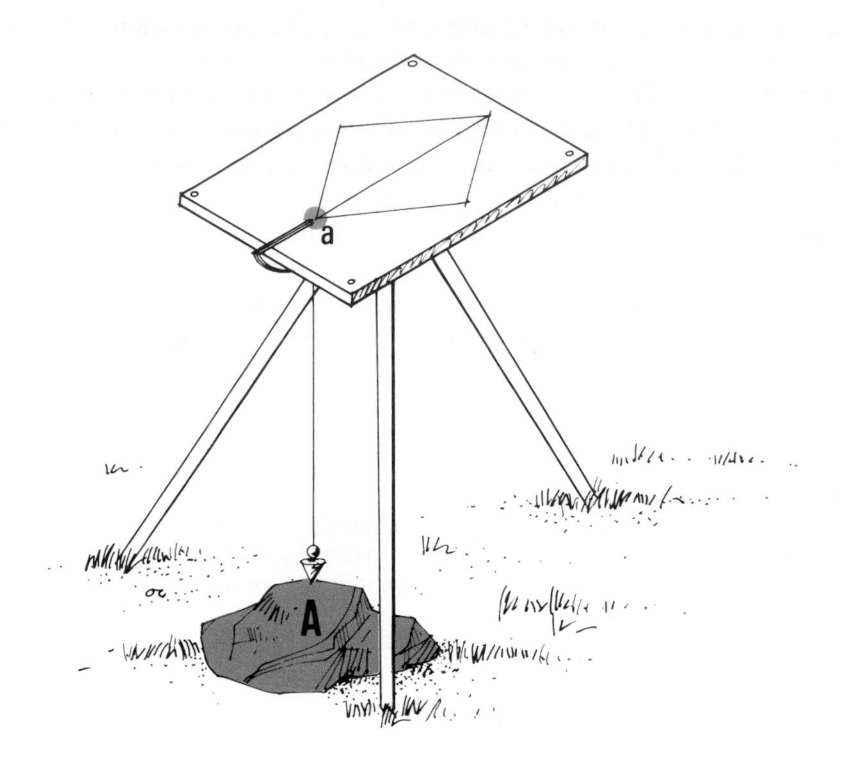

Set up the plane-table at point A

5. Rotate the table so that you will be drawing the map in the **orientation** you need. Using your magnetic compass as a guide, draw arrows showing the magnetic north (see Section 75, steps 45-46).

Note: you should always try to locate the north facing the top of your map. This is a rule which is always applied in professional topographical maps. You may not be able to follow the rule, however, depending on the direction of the longest distance and on the scale you select.

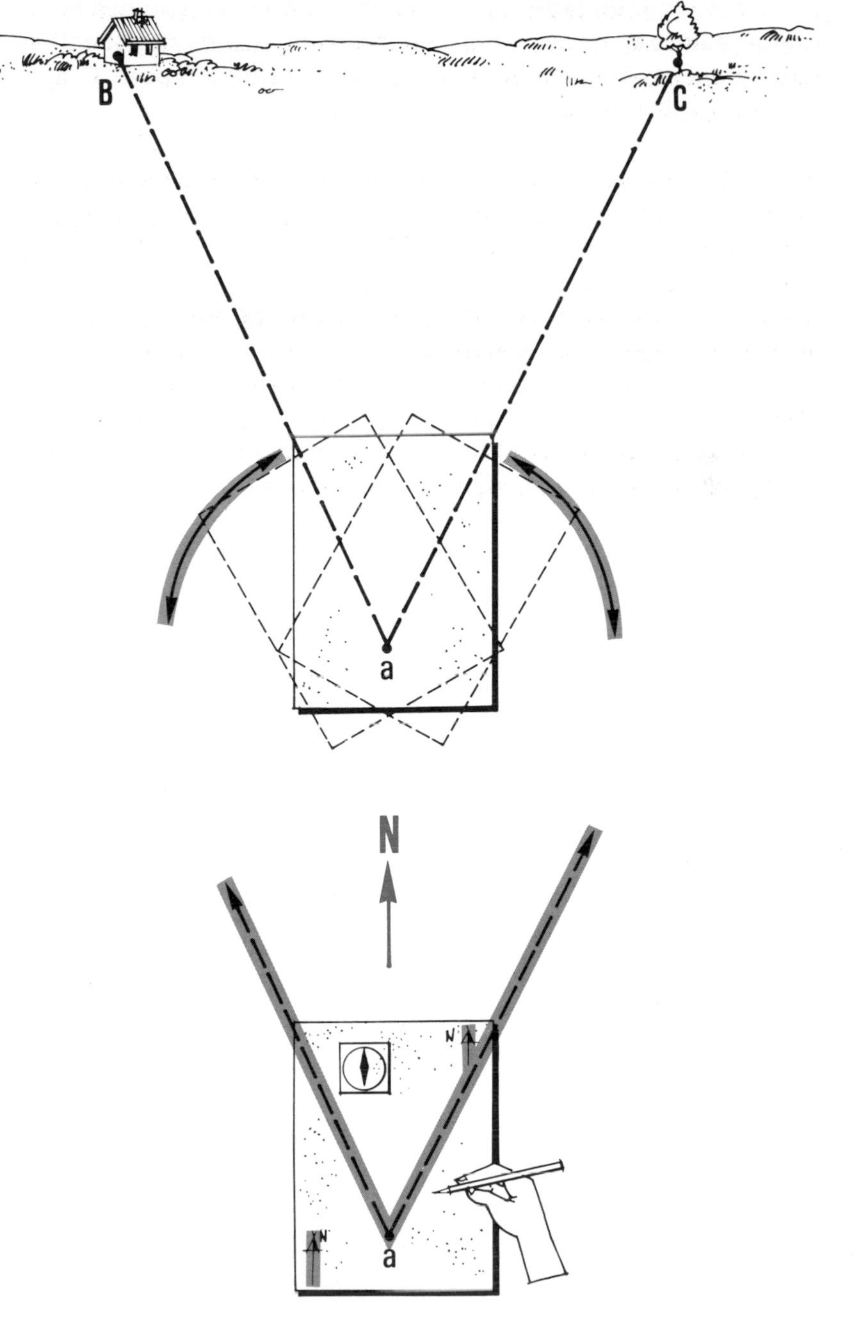

6. Using your **alidade**, sight from the first marked point a to another **major feature B** which you can see from the plane-table location. This could be a small hill, a bend in a path or a ranging pole. Draw a thin line **ax** in this direction.

Note: you can use the alidade much more easily if you place a pin at point a on the board, and then **swivel** the alidade around the pin until you can sight the second point.

7. Measure the horizontal ground **distance** from the plane-table station A to the major feature B. Then mark this distance along line **ax**, starting at point a and scaling it down as line **ab**.

8. Without moving the plane-table from point A, repeat this process for **all other major features C, D, etc.** which you can see, and draw lines **ac, ad**, etc.

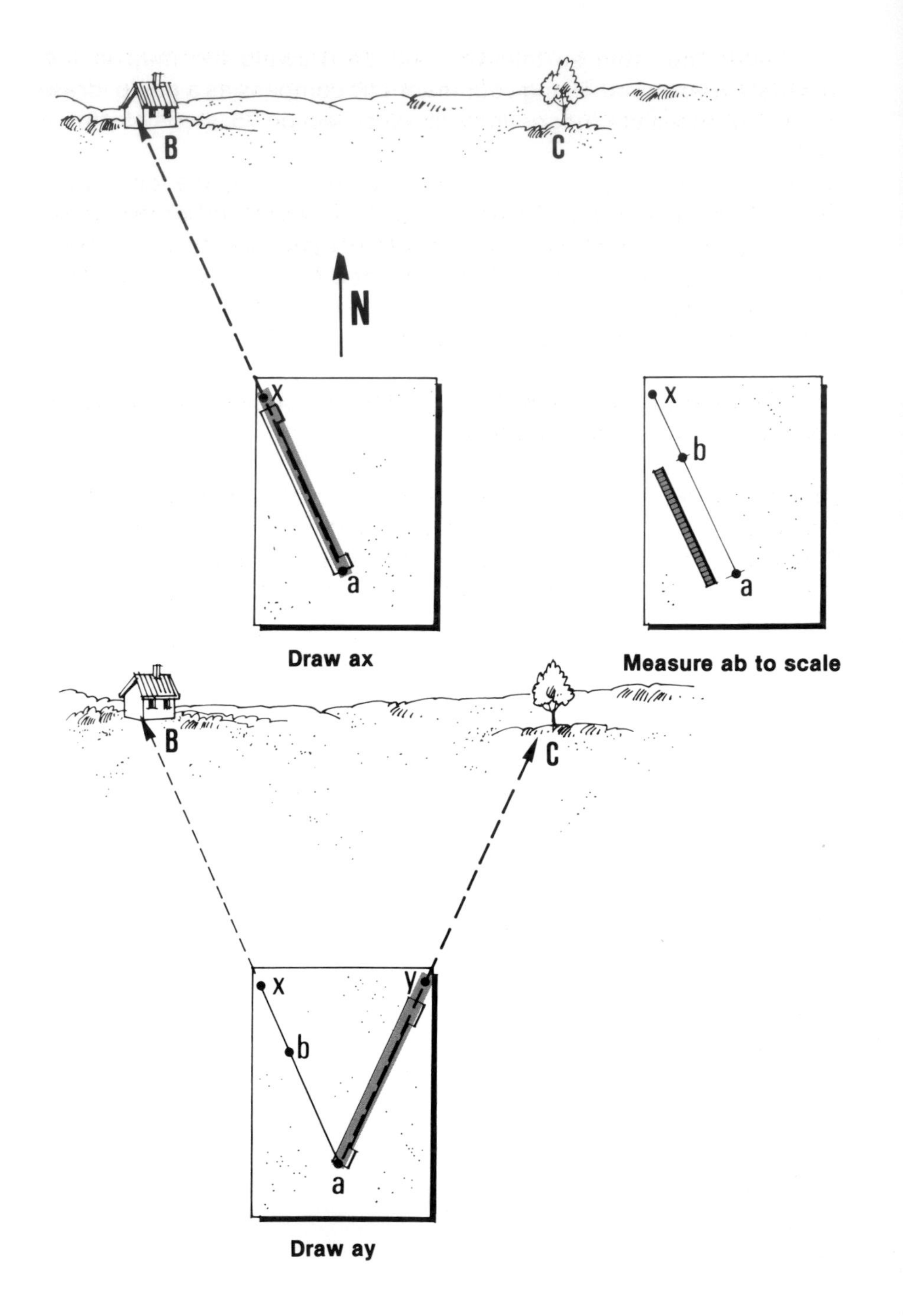

9. Move the plane-table to one of the major features you have just mapped, such as C. Choose a feature from which you can easily map another part of the area, such as the route of a path or the course of a river.

10. Set up the plane-table over this point C. **Reorient the table**. Use the compass and the magnetic north arrows you have already drawn (see step 5 above), or, instead, use the alidade, backsighting along a drawn line which passes through the new station C and a known major feature such as A (see Section 75, step 47).

11. From this new station C, map in the new major features which you can see, as explained above.

12. If necessary, move to other stations to complete the mapping of the entire area. If you need more details in the map, go back to one of the mapped features, reorient the table by backsighting on another mapped feature, and map the details as required.

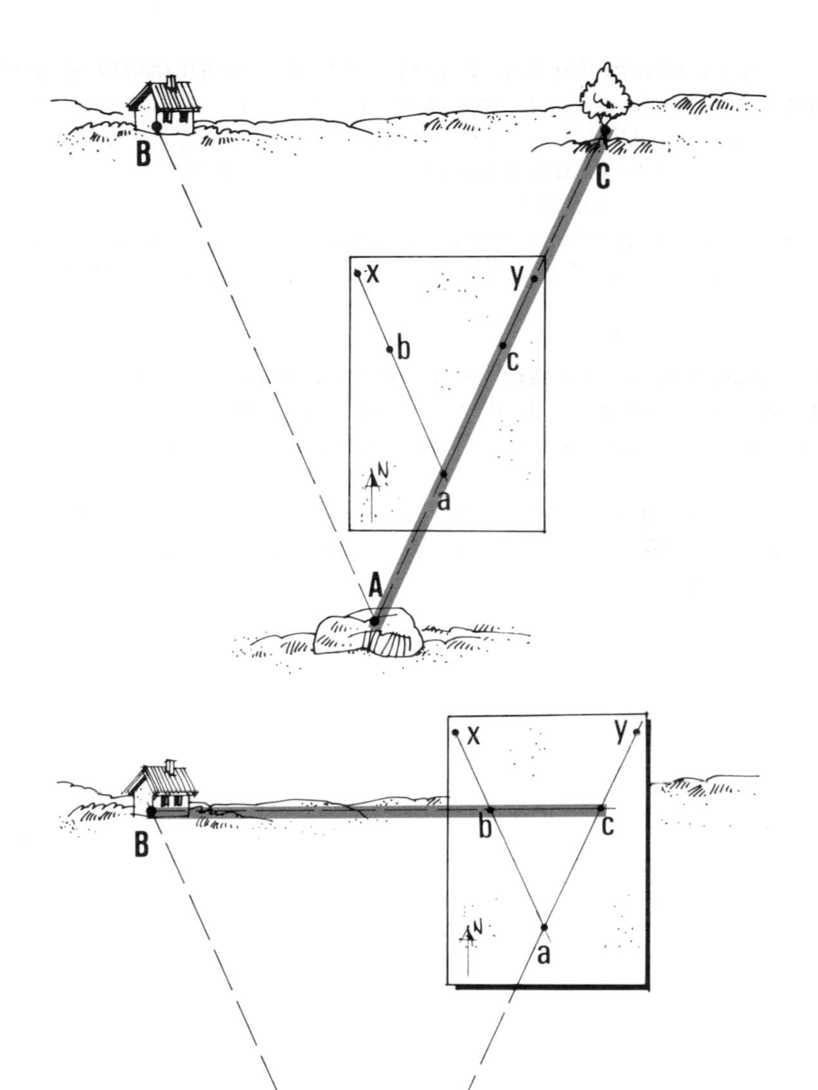

13. You can use the above procedure for plane-tabling in several different situations in the field, such as:

- mapping an **open traverse** (see Section 7l);
- mapping a **closed traverse** (see Section 71);
- mapping by the **radiating method** (see Section 72);
- mapping by the **triangulation method** (see Section 74).

Usually, you will use a combination of some of these surveying methods to map an entire area.

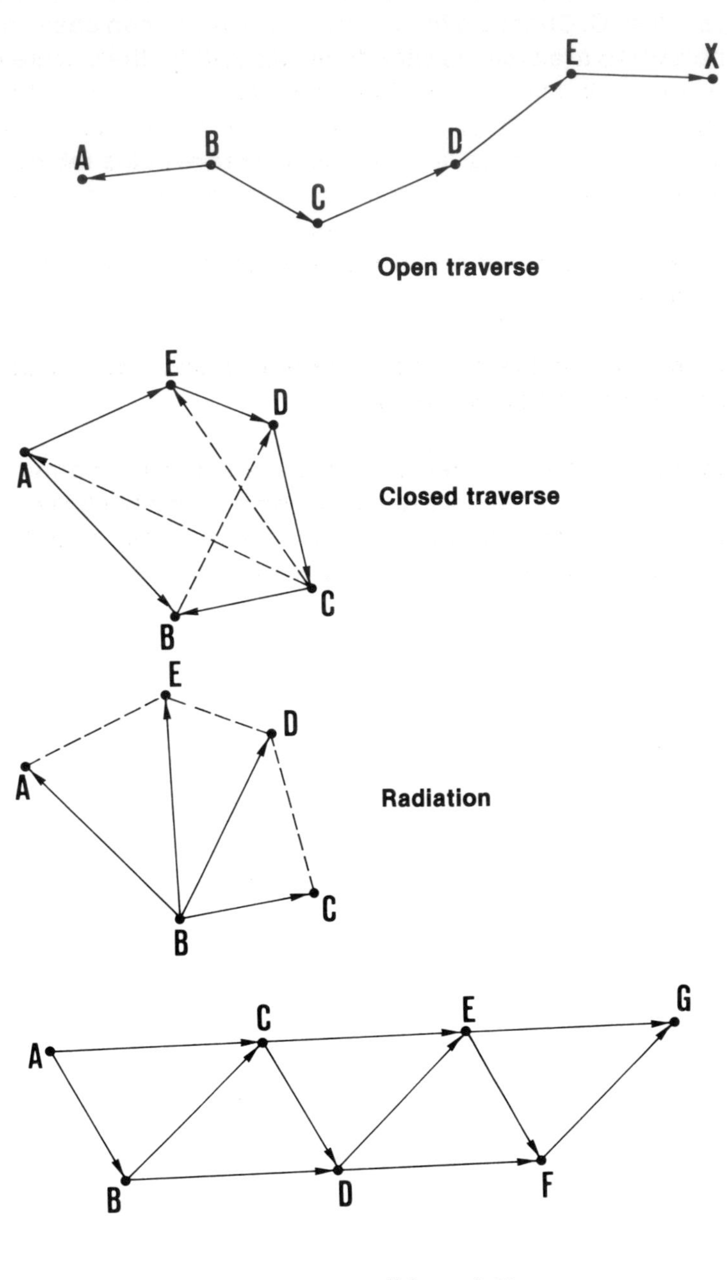

Mapping an open traverse with a plane-table

14. You may need to map an **open traverse** ABCD. To do this, you can, for example, first set up the plane-table at point B, which has a fixed position and from which a line BA of known direction already exists on the ground. Map the location of station B, the direction of BA and the distance BA in turn.

15. Draw the direction of the next station C, measure distance BC, and map point c.

16. Move the plane-table to station C, orient it along CB, and, using the same procedure as above, map point **d**.

Note: if the traverse sections **ba, cb**, etc. on the map are very short, you should mark their directions on the edge of the paper. This will provide longer lines, so that you can line up the alidade along them when you must reorient the plane-table at a new station by backsighting.

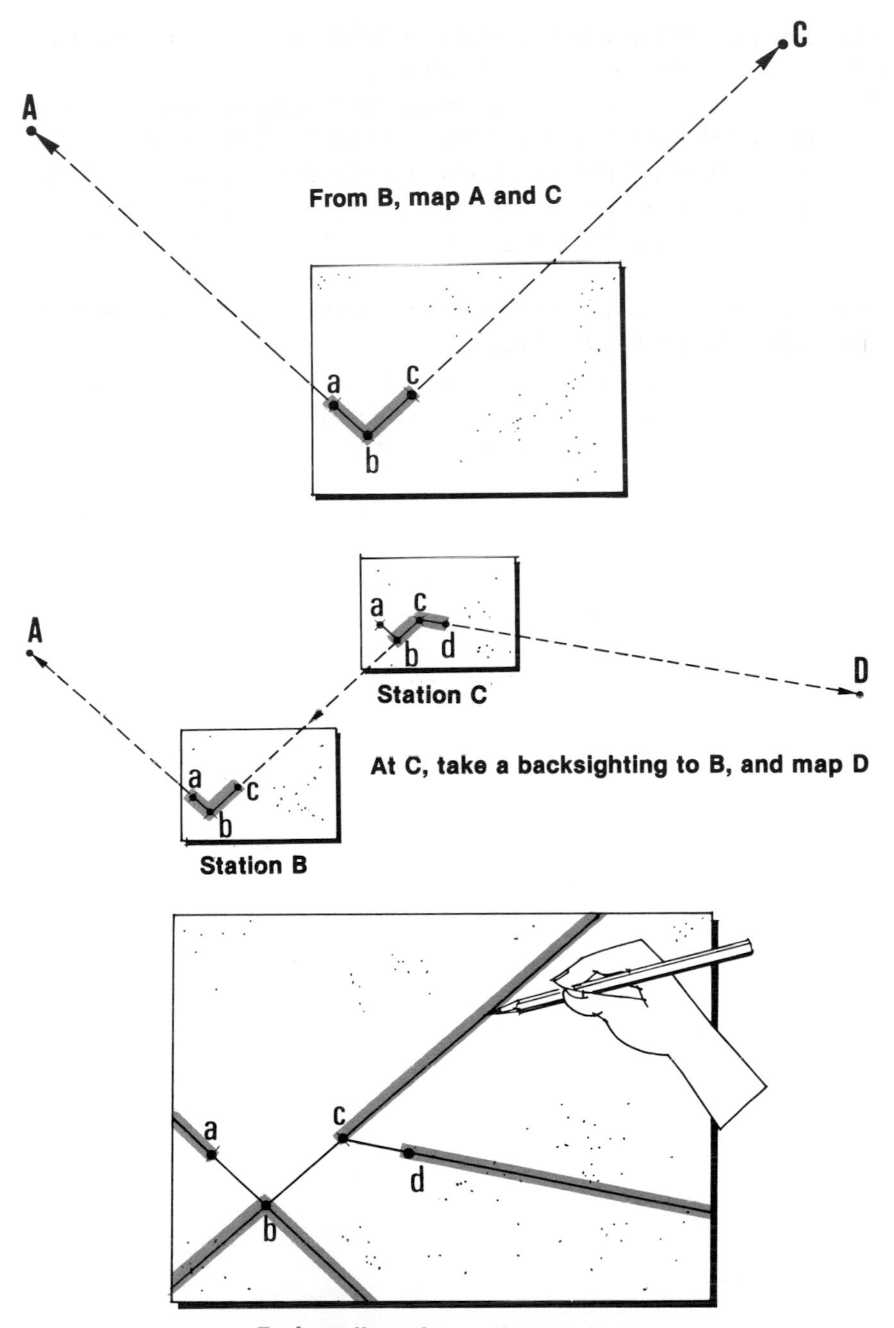

Mapping a closed traverse with a plane-table

17. You need to map a **closed traverse** ABCDEA. First, set up the plane-table at station A and plot this on paper as point **a**; choose a scale and a location on the paper which will allow you to plot the other stations within the limits of the sheet of paper.

18. Using the alidade, take a foresight through point **a** to station B and draw line ax. Measure distances AB and map point **b** on line ax.

19. Move the plane-table to station B, set it up over the point, and orient it by backsighting along line ba on station A. Take a foresight to station C, measure distance BC, and map point **c**.

20. Using this procedure, map the locations of the remaining points on the closed traverse. At the end of the traverse, when you plot the initial station A again, you can see any **error of closure**. If this error is within reasonable limits, correct it, using the graphic method explained in Section 71, step 19.

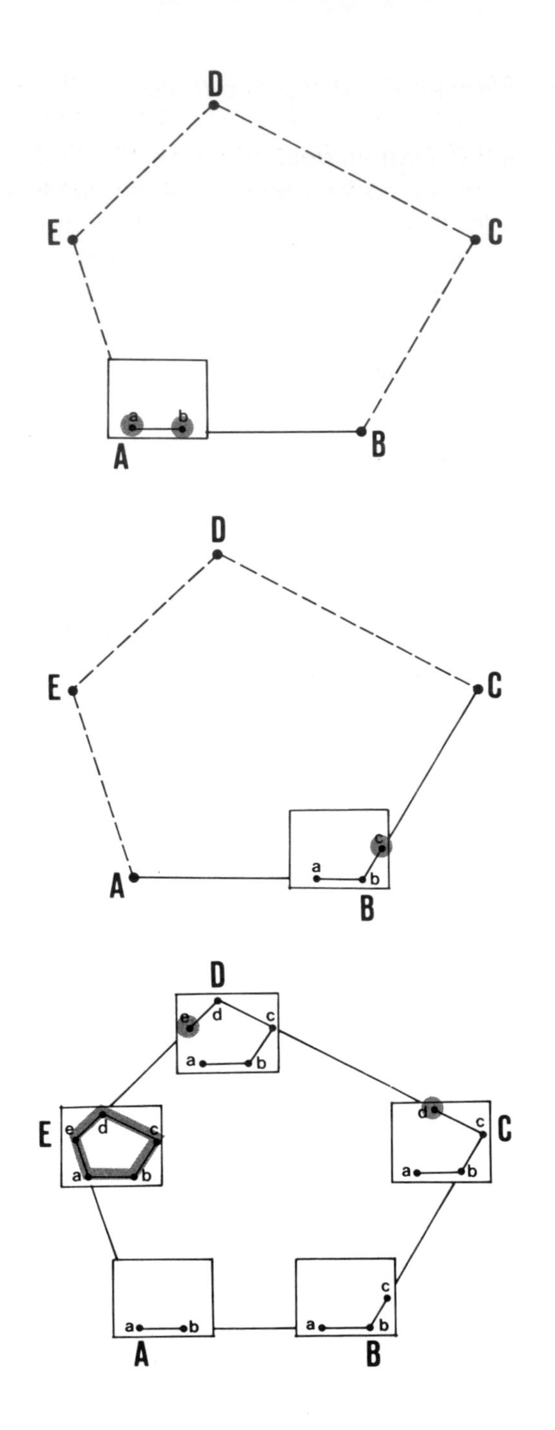

21. From one station on the traverse, you may be able to see two or more of the preceding stations which are not on the same straight line as the station where you are standing (for example, from C to A, from D to B, or from E to B). In this case, check the other parts of the traverse.

Example

From station C, station A is visible. You should check from C the position of point a by backsighting on Station A.

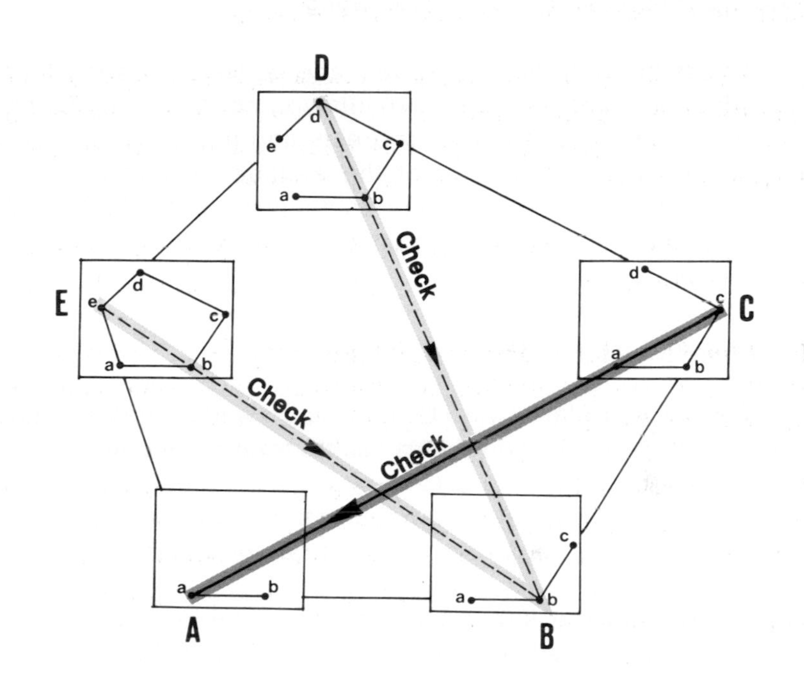

Take additional backsights to check your work

Mapping with a plane-table by radiating

22. To use this method, set up the plane-table at a **central station O**, from which you can see all the points you need to map. Orient the table. On the map, draw lines representing the directions to these ground points; to do this, pivot the alidade around the mapped location of station O. Measure horizontal distances OA, OB, OC, OD and OE, and scale them along each of the drawn lines to map points **a, b, c, d** and **e**.

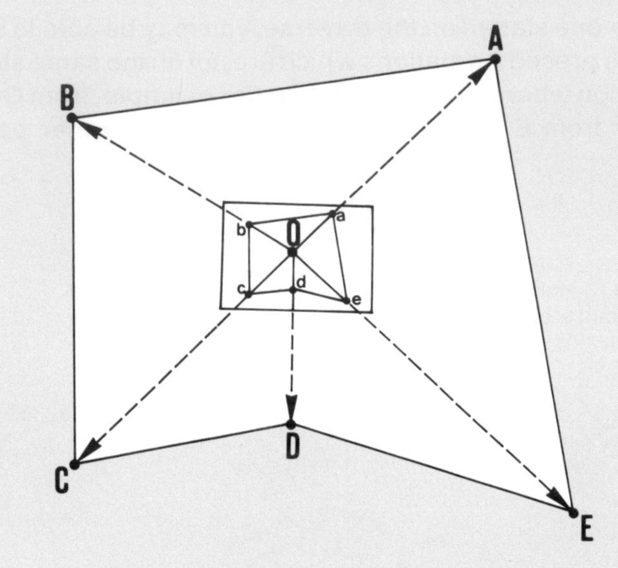

Move the alidade, but not the plane-table

Mapping with a plane-table by triangulation

23. Before you start plane-tabling, you need to find **a base line determined by two known points visible from each other**. This base line can be a known section of an existing traverse, or a line between two points fixed by a previous triangulation. If such a base line is not available, you must accurately determine and measure one.

24. Draw the **base line AB** on the plane-table sheet. Choose a location which will allow you to plot the other features of the map within the limits of the sheet of paper.

25. Set up the plane-table over one of the two end-points of this base line, at point A, for example. Then, with ranging poles, clearly mark the second end-point B, and the third point C that you need to map. You should be able to see point C **from both point A and point B**.

26. Align the alidade along line ab, which represents the base line; orient the plane-table by sighting at the other end-point B of the base along AB.

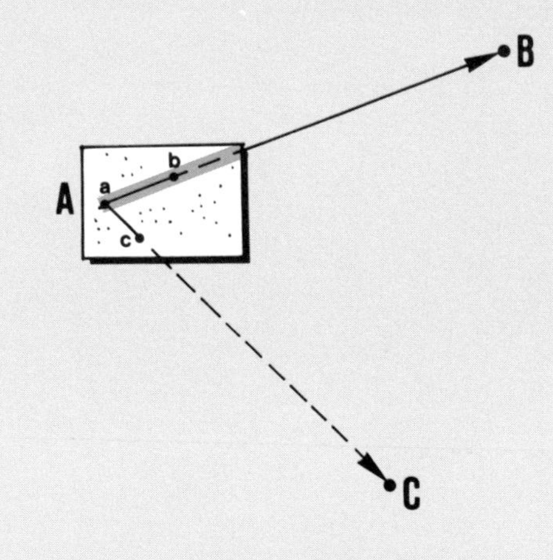

Start from the known base-line

27. Place a pin at point **a** and rotate the alidade around it until you sight point **C**. Draw a thin line from point **a** along the edge of the alidade in the direction of point **C**.

28. Move the plane-table to B. Orient the table with line ba on the map pointing in the direction of ground point A. Place a pin at b and rotate the alidade around it until you sight point C. Draw a thin line from point **a** in the direction of C. **Point c is located** on the map **at the intersection** of line ac (step 27) and line bc.

29. Point C is now known, and you can use it in a similar way to determine other points, taking, for example, BC as a base line to determine D. You can then repeat this mapping process, using each point as it becomes known, as long as each point you need to map is visible from two other known points.

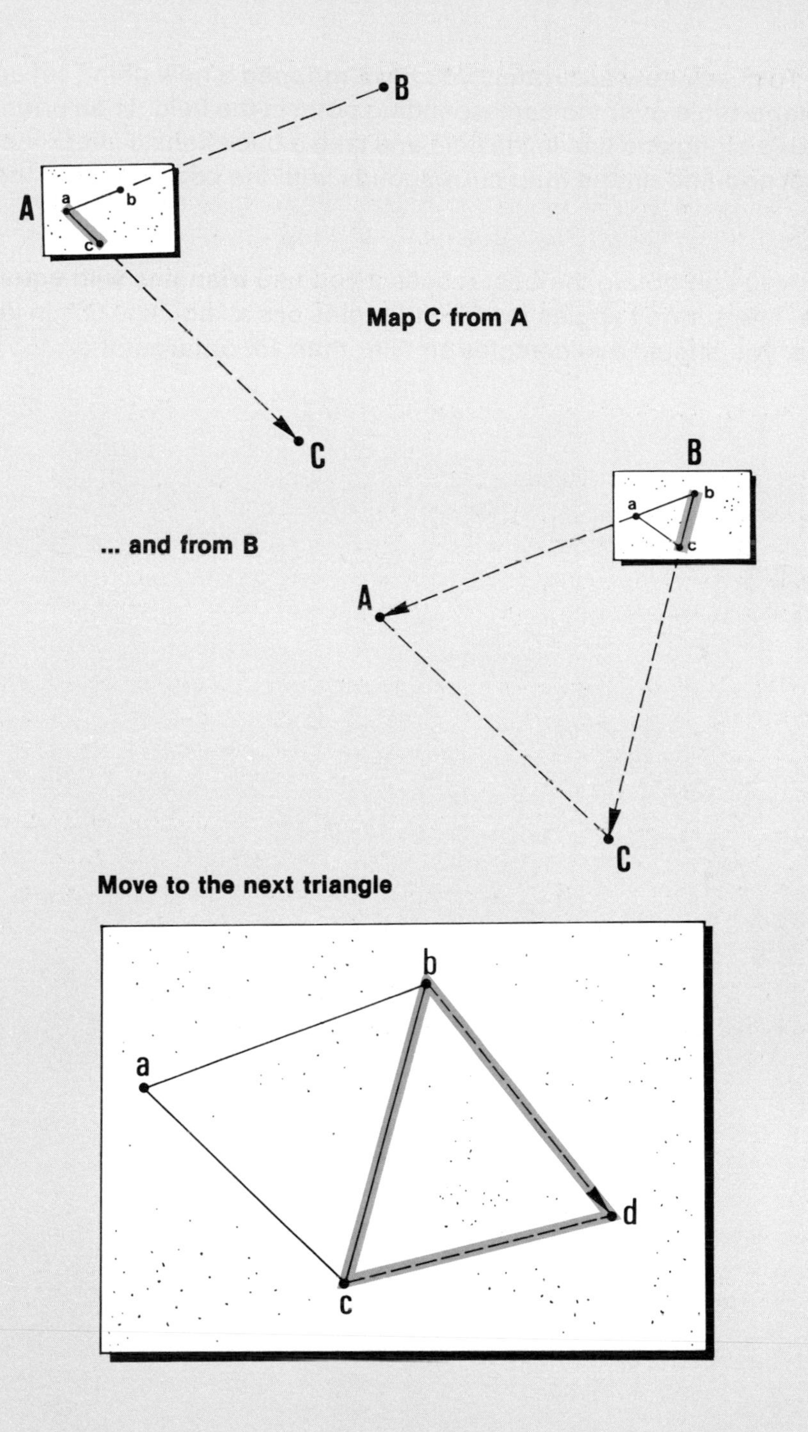

30. To check how accurately you have mapped a new point, set up the plane-table over the corresponding point in the field. Then orient the table along one line in the field and take a backsight to check that the second line on the map corresponds with the correct line in the field.

Note: you can obtain the best results if you use **triangles with equal sides**. The summit angles in these triangles are all equal to 60º. In all cases, you should avoid angles smaller than 15º or larger than 165º.

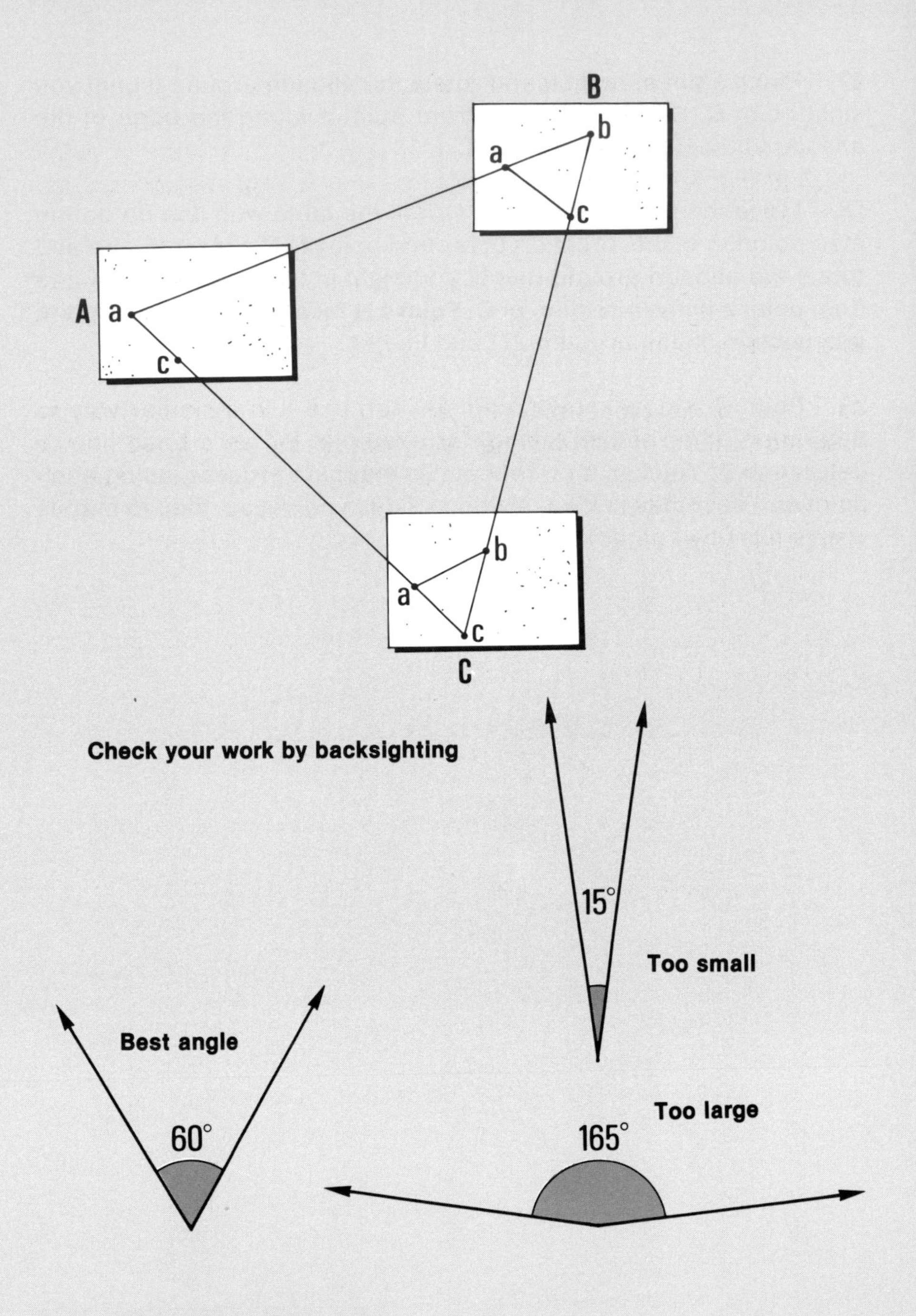

Check your work by backsighting

Choose the proper angle

Mapping with a plane-table by combined methods

31. In most cases, you will map an area with a plane-table by using a **combination of the methods of traversing, radiating and triangulation**.

32. You need to map site ABCDA, which includes such features as a rocky area, a group of houses and a well. Clearly mark points A, B, C and D with ranging poles.

33. Set up the plane-table at corner A of the area. Locate the mapped position of A on the sheet of paper. Be sure to choose a point which will allow you to plot the other features of the map within the limits of the sheet of paper at the drawing scale you have chosen. Orient the sheet by drawing the direction of magnetic north.

34. From station A, you can see the rocks and the houses. By radiating, determine the directions of the rocks and the houses from this station. Then measure and map AB.

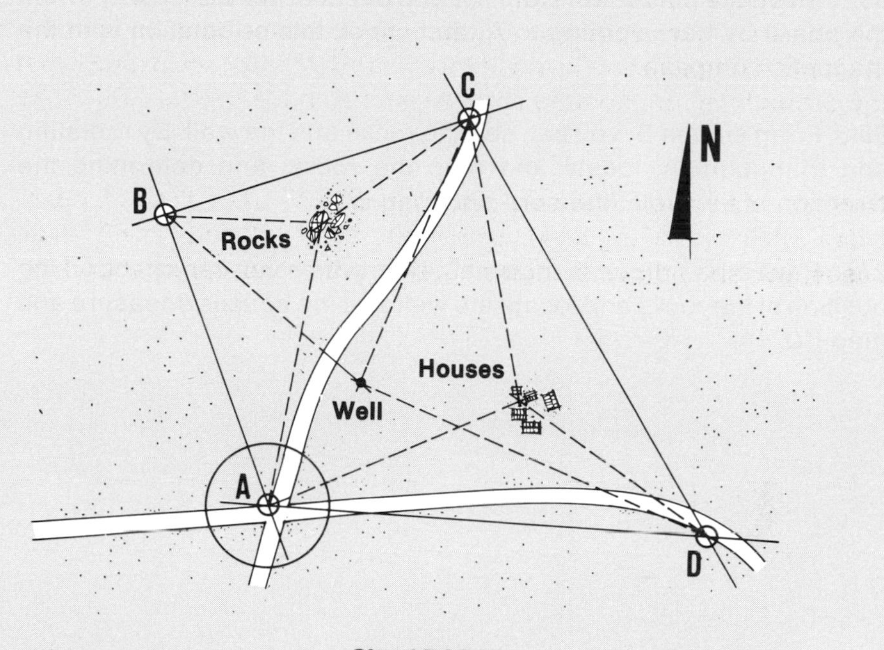

Site ABCDA

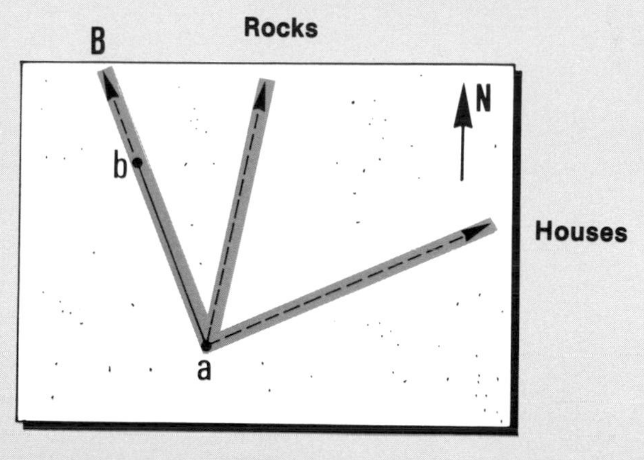

At A

35. Move the plane-table and set it up over corner B. Roughly orient the board by backsighting to A, and check this orientation with the magnetic compass.

36. From station B, you can see the rocks and the well. By radiating and triangulation, locate and map the rocks, and determine the direction of the well. Measure and map BC.

37. Repeat this process at point C, from which you can check on the position of the rocks and locate the well and the houses. Measure and map CD.

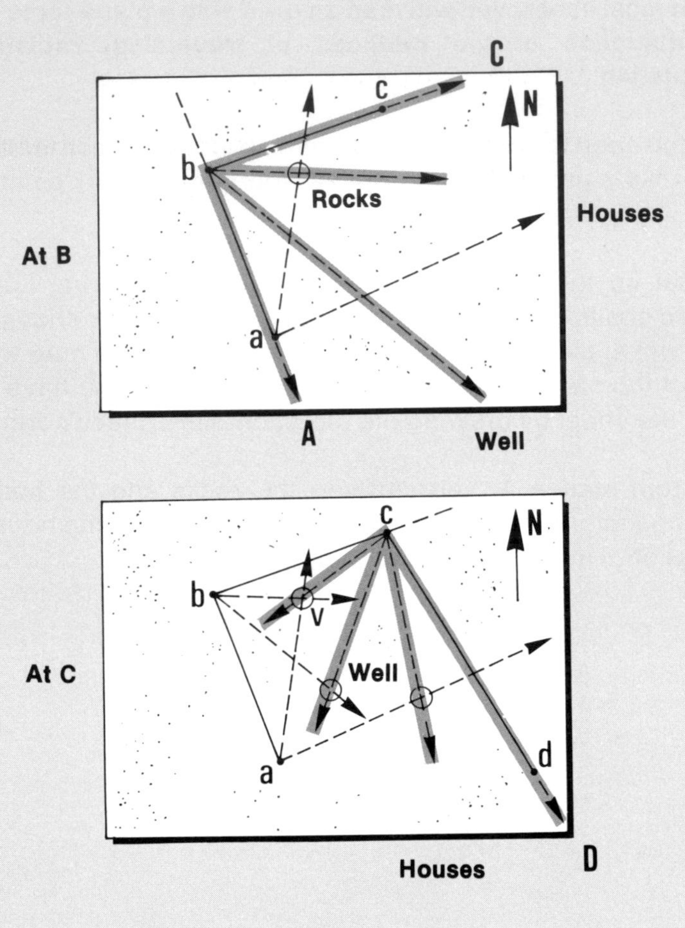

38. Repeat this process at point D, from which you can check on the positions of the houses and the well. Measure and map DA.

39. Check the error of closure of traverse ABCDA and correct it, if possible. If the error is too great, repeat the survey.

40. Finish the map, checking that you have included all the information you need (see Section 90, step 6).

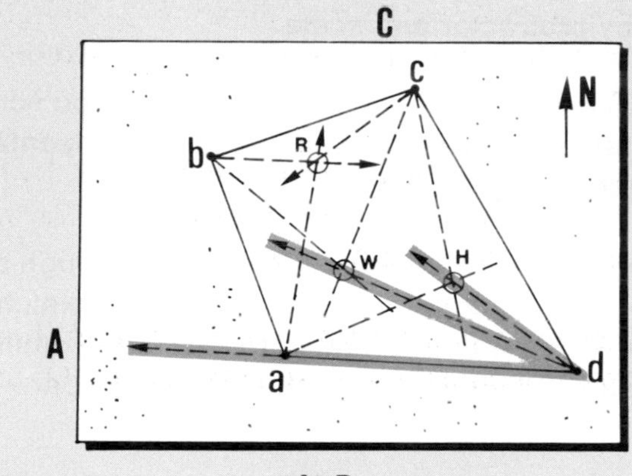

At D

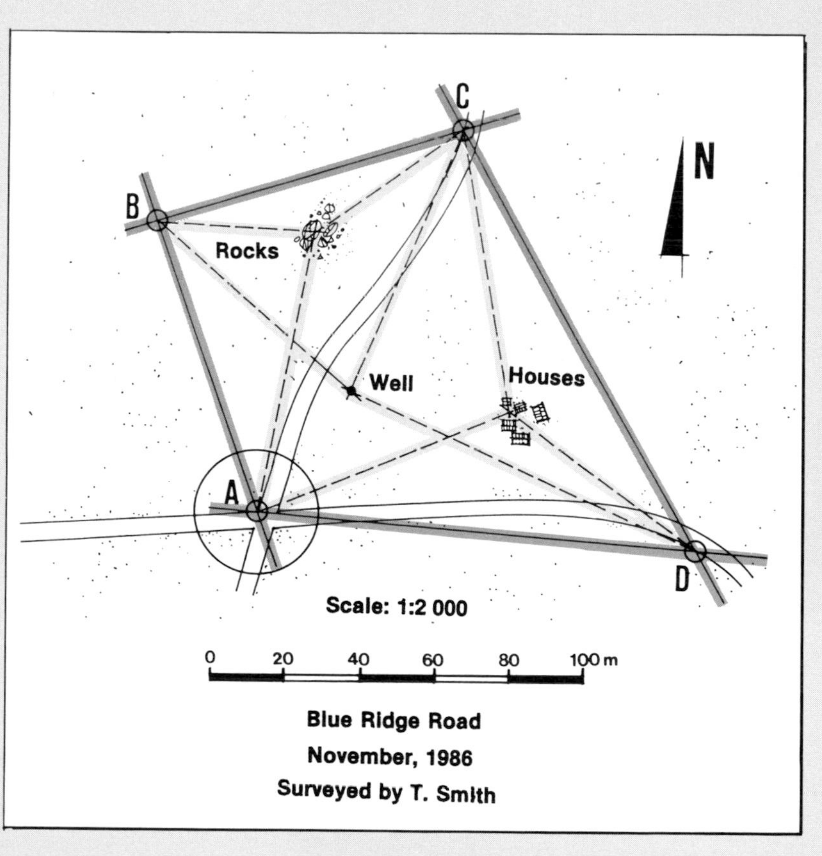

Finished map

93 How to map by protractor and scale

1. When you map in the office, using field records, you will usually plot **horizontal distances** with a ruler/scale, and the **horizontal angles** with a protractor (see Section 33).

2. First, using the scale you have chosen, make a rough sketch of the area to determine its size and shape. From this sketch, decide how large a piece of paper you will need to make everything fit and determine the position of your map on the sheet of paper.

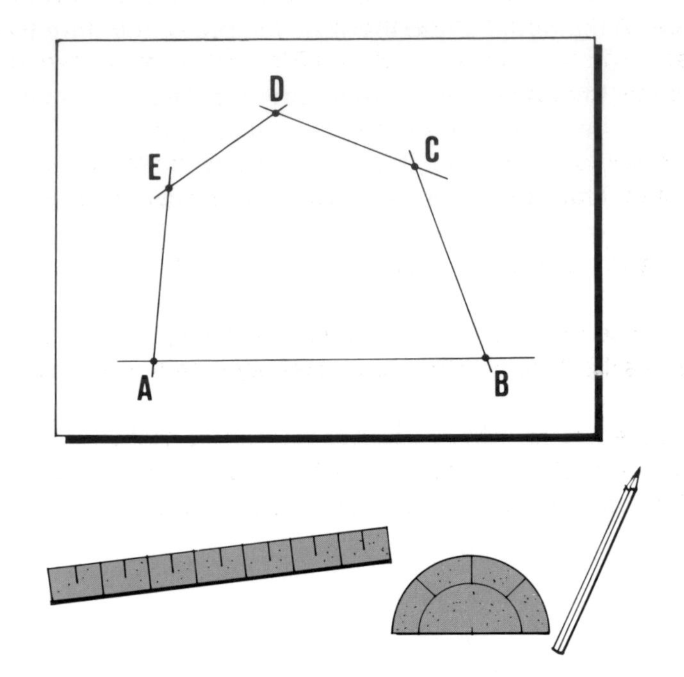

3. Draw the first line in the right place on the paper and determine its length AB, using the selected scale. Using a pencil with a hard lead, accurately mark points A and B on the paper as two dots with a small circle around each.

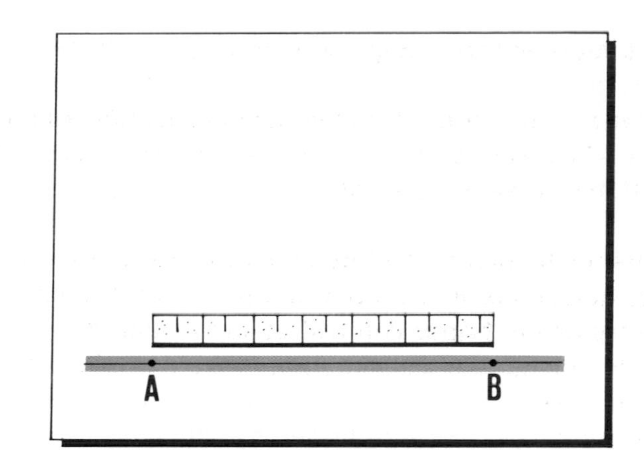

Note: draw the line so that it will **extend beyond the next angle-point B**, a distance greater than the radius R of the protractor.

4. Place the protractor along line AB so that:

- **its centre** is exactly on the second angle-point B; and
- **marks 0° and 180°** line up exactly with line AB.

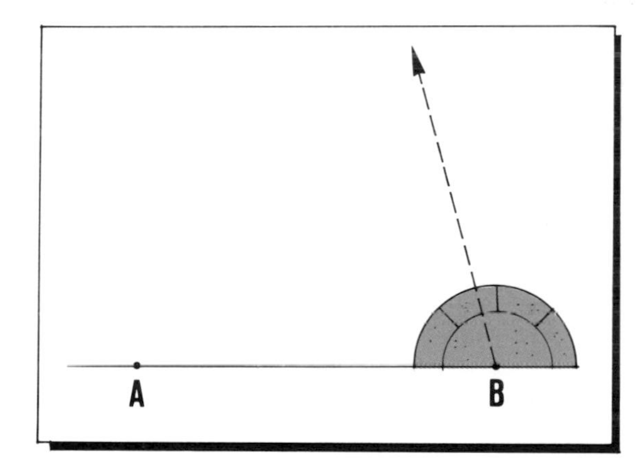

5. Plot the angle, which you have obtained from your field notes, remove the protractor, and draw the second line. Locate and map point C according to the measured distance and scale.

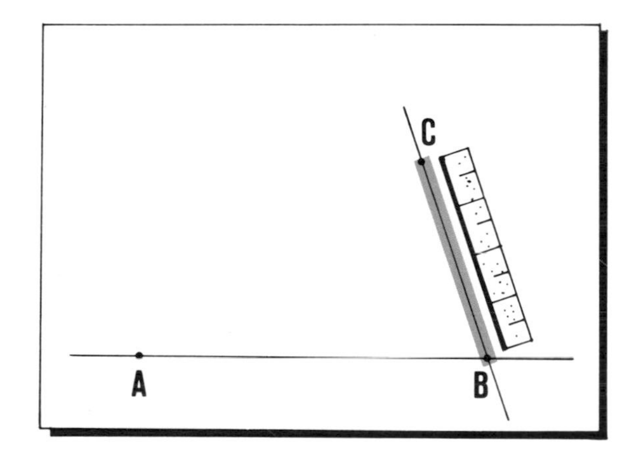

6. Place the protractor along this last line with its centre opposite point C. Lay out the measured angle, and draw the third line. Locate and map point D according to the measured distance and scale.

7. Repeat this process until you have mapped the entire traverse.

8. **Locate the details** on the plan from this traverse line. Plot the positions of buildings, fences, rocky areas, streams, paths, etc., using the scale for distances and the protractor for angles.

9. You can use a method similar to this one to map survey information which you have obtained by radiating, by triangulation, and by offsets.

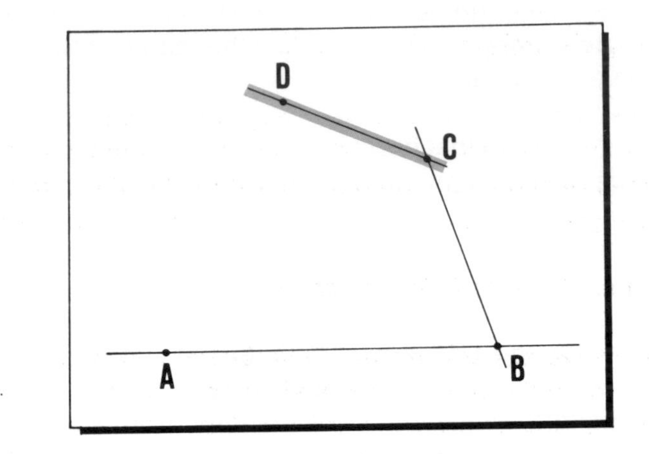

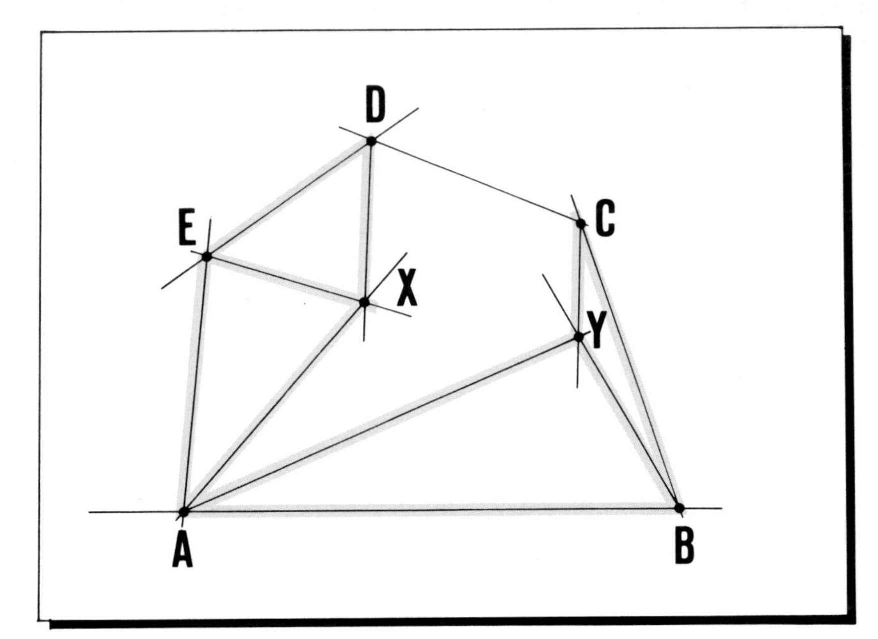

94 How to map contours

What is a contour line?

1. Contour lines are lines drawn **to join points of equal elevation**. On a plan or map, they represent the contours you found and marked in the field (see Section 83). Contour lines show the three-dimensional ground topography of a site on a two-dimensional map or plan.

What are the characteristics of contour lines?

2. As you have already learned (see Section 83, step 7), contours are surveyed on the basis of a selected **contour interval**. Similarly, contour lines are drawn at equal vertical intervals. You should always clearly state the contour interval of the mapped contour lines.

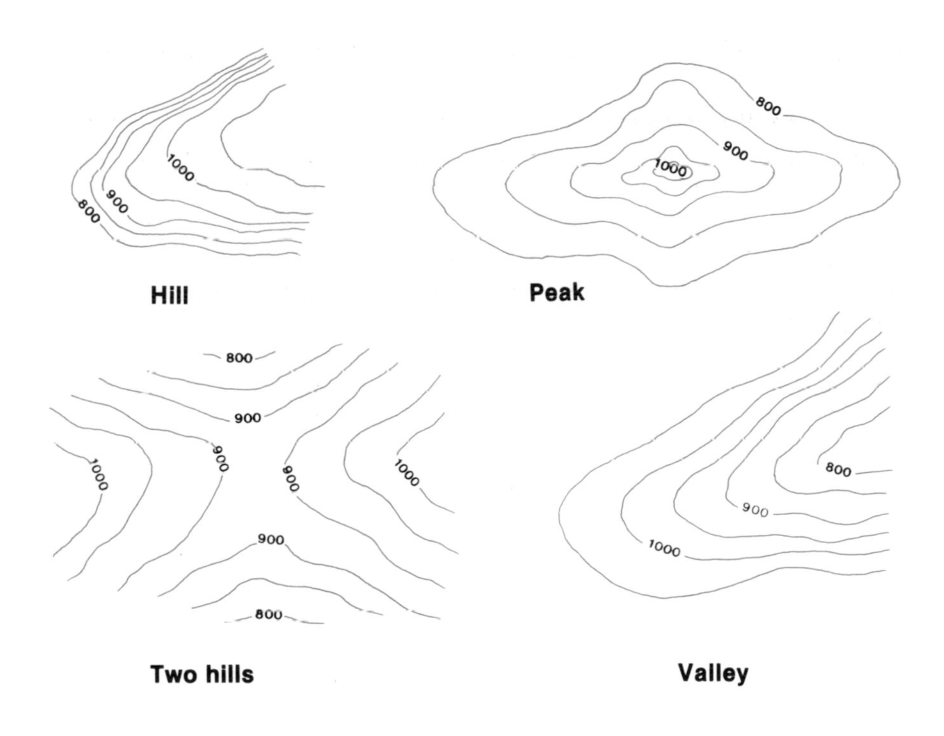

Contour interval = 50 m

3. If you clearly understand the **characteristics of contour lines (CL)**, you will be able to survey, make contour maps and read maps much more easily. The facts to remember are:

- all points on a contour line are at the same elevation;
- contour lines cannot cross each other or divide in any way (such as branching or splitting off);
- contour lines always close on themselves, either within or outside the limits of the map. When they close within the map's limits, they indicate either a **summit** (such as a hill) or a **depression** (such as a valley);
- straight, parallel contour lines indicate horizontal ground;
- evenly spaced contour lines indicate a uniform, or regular, ground slope;
- the closer the contour lines, the steeper the slope (see Note);
- widely spaced contour lines indicate a gentle slope;
- closely spaced contour lines indicate a steep slope;
- the steepest slope is always at right angles to the contour lines;
- contour lines cross ridges perpendicularly;
- contour lines cross river valleys following a U- or V-shaped path.

Note: when two contour lines of equal elevation are near each other, the land between them is often flatter than the general trend of slope but its slope is indeterminate (unknown).

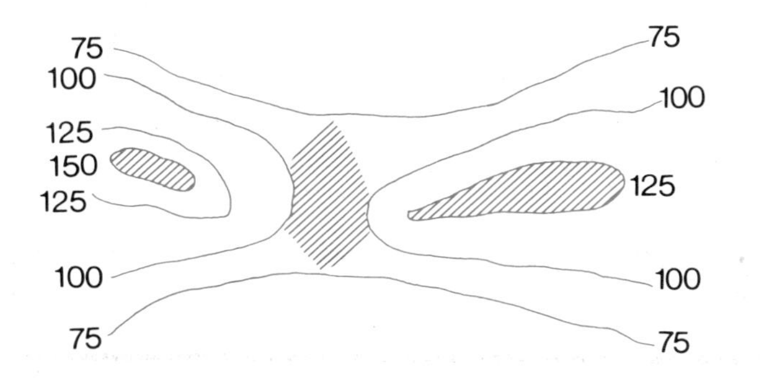

Areas of indeterminate slope

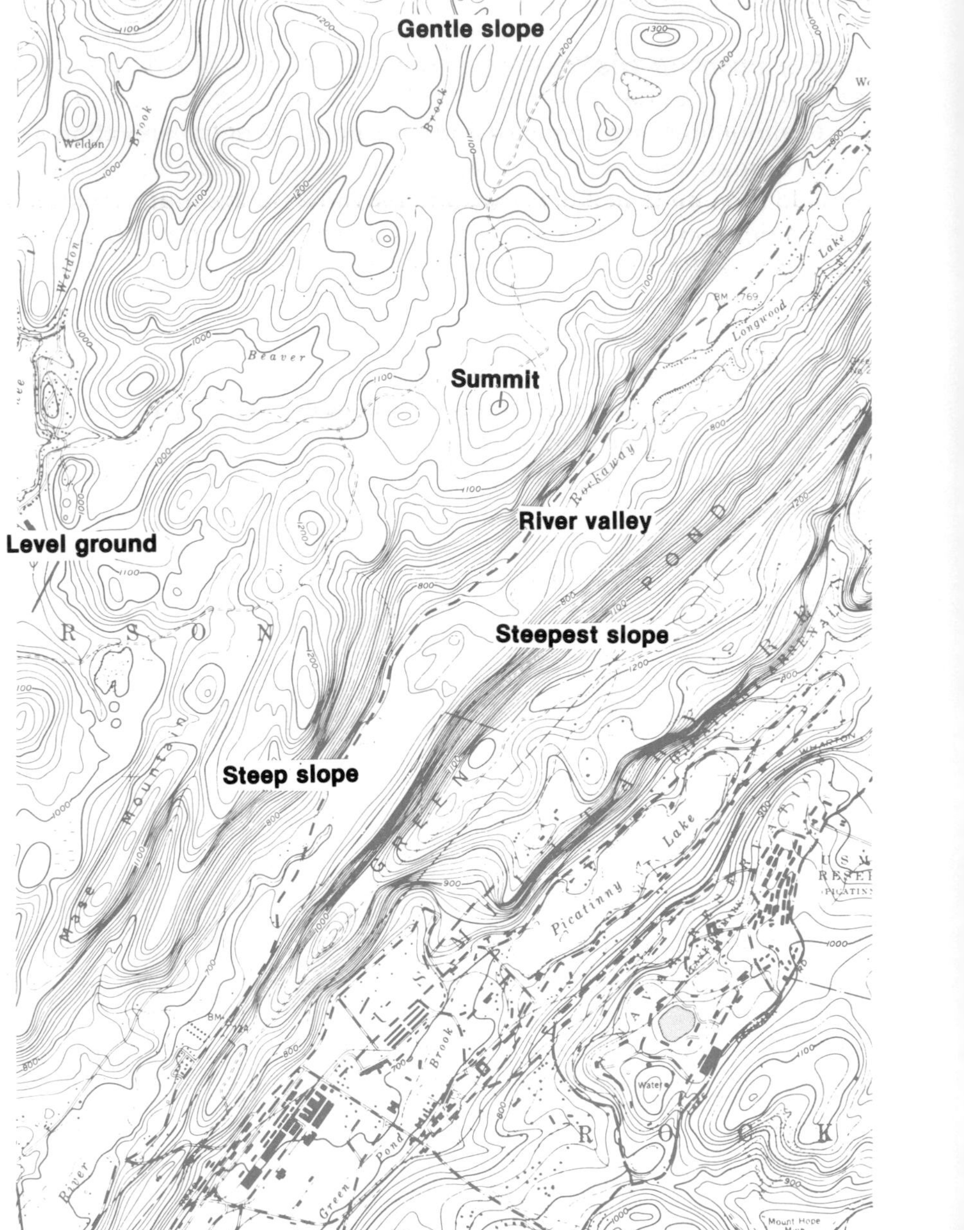

Making a contour map

5. First prepare **a planimetric map** of the area. This is a map showing the boundaries of the land, the surveying stations, the major physical features and all available details (see Sections 92 and 93).

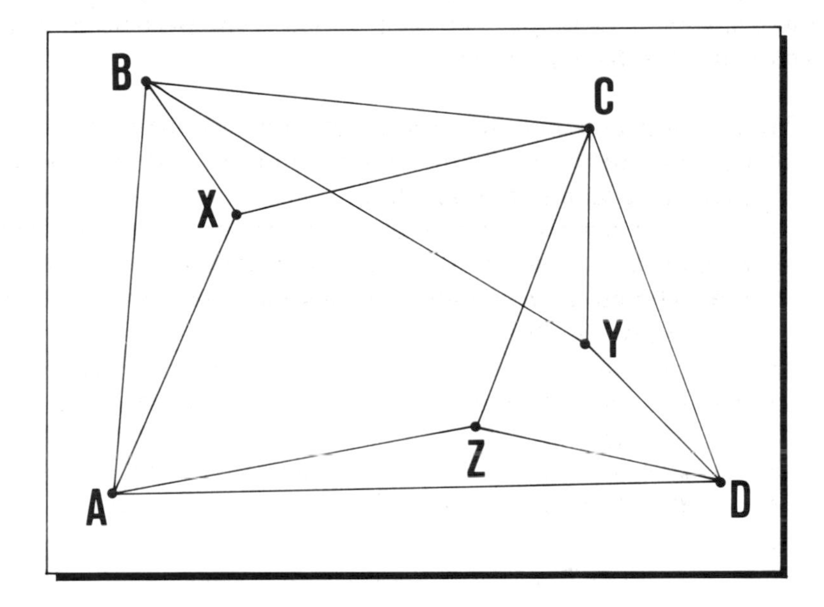

6. Add **the points of known ground elevation** to the map. To locate these points on the map, use a distance scale and, if necessary, a protractor for determining any angles. Write the elevations next to the points.

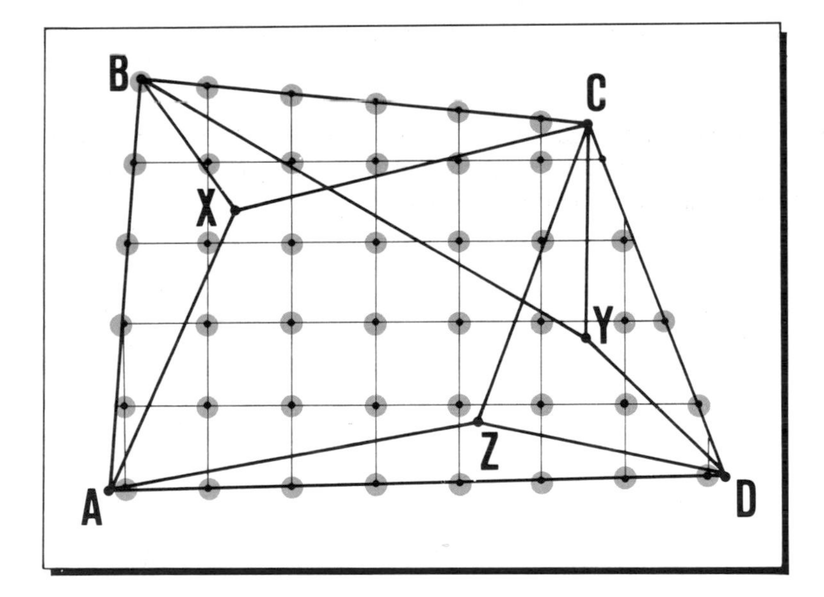

TABLE 12

Contour intervals (metres)

Topography	*Map Scale*		
	Greater than 1 : 1 000	1 : 1 000 to 1 : 10 000	smaller than 1 : 10 000
Flat	0.15 to 0.3	0.3 to 0.6	0.6 to 3
Gently sloping	0.3 to 0.6	0.6 to 1.5	1.5 to 3
Hilly	0.6 to 1.5	1.5 to 3	3 to 6

Choosing the contour interval of contour lines

4. Before drawing the contour lines on a plan or map, you must choose the **contour interval** you will use. The contour interval mainly depends on the accuracy or scale you need for the drawing, and on the topography of the area (see **Table 12**). A smaller contour interval, such as 0.15 m, 0.25 m or 0.5 m, is generally used for flat or gently sloping areas. Remember that most fish-farm sites are located in such areas.

7. Find the points of **lower ground elevation**. Then, according to the **contour interval** you have chosen, determine which elevation represents the first contour line you need to draw.

8. The first contour line will pass **between** ground points with elevations which are lower and higher than the elevation of the contour points. Carefully locate the path of the contour line between these higher and lower points, as you draw. Note that contour lines are usually curved, not straight. You should draw them free-hand, rather than using a ruler to connect the points.

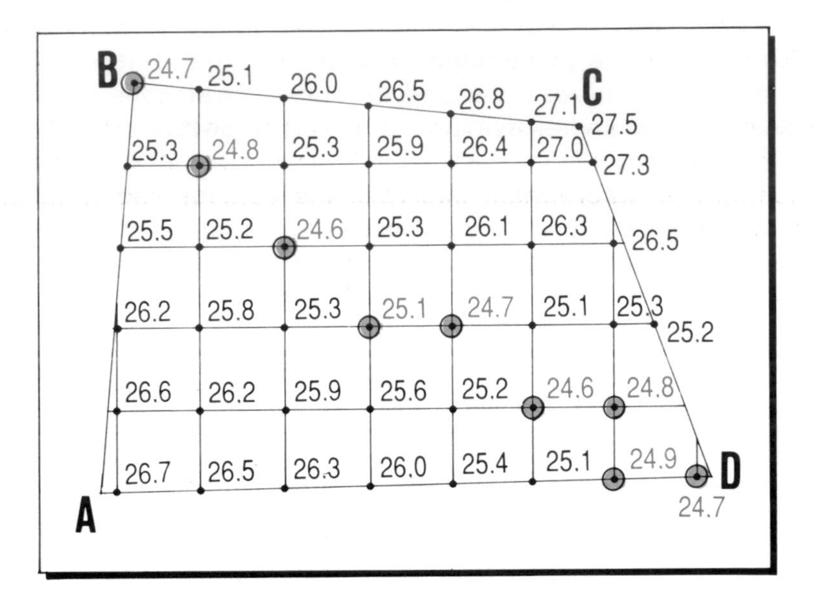

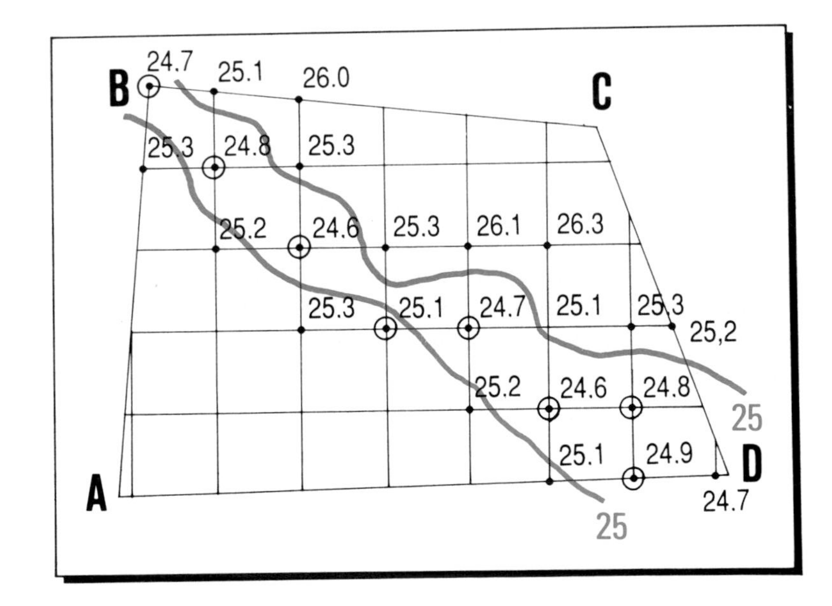

9. Using the same procedure, draw the other contour lines. Show the progressively higher elevations as multiples of the selected contour interval.

Note: contour lines are **only** drawn for elevations which are multiples of the contour interval. Show the elevations of the contours by writing in numbers at appropriate intervals; the contour line is usually broken to leave a space for the number.

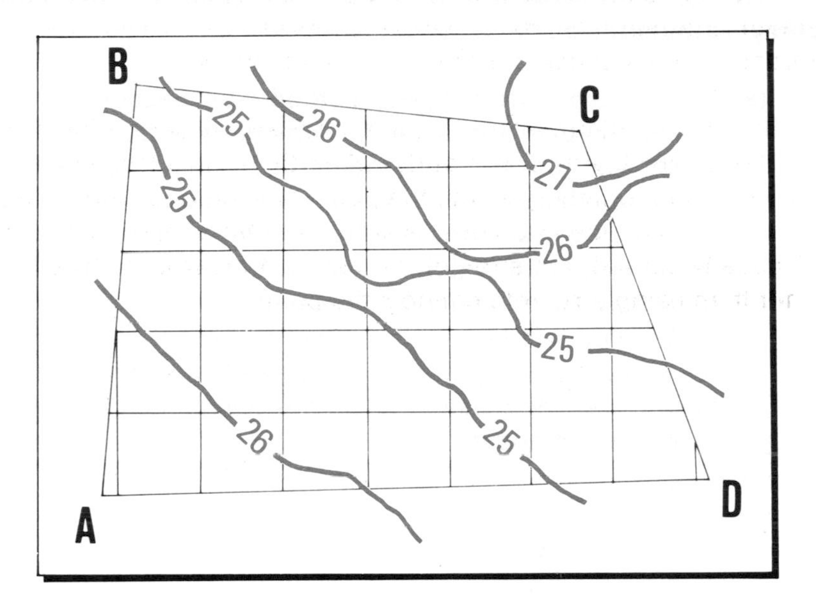

10. This general procedure may vary, depending on **the contour surveying method** you have used in the field.

(a) If you have used **a direct method** (see Section 83, steps 10-29), the plan survey of the contours you have identified gives you all the information you need to map the corresponding contour lines. You will reduce the measured distances to scale, and use **the parallel lines** marked on the ground as a background to the contour lines.

(b) If you have used **an indirect method** (see Section 83, steps 30-33), you will lay out **the pattern of lines** roughly in the drawing, map the points of known elevation and note their elevations. Then, estimate the position of the contour line, as explained above.

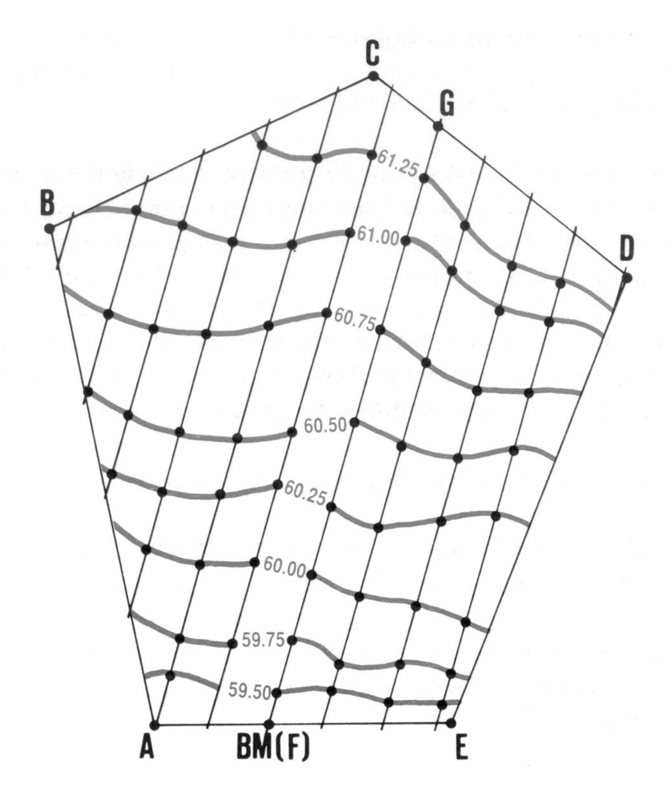

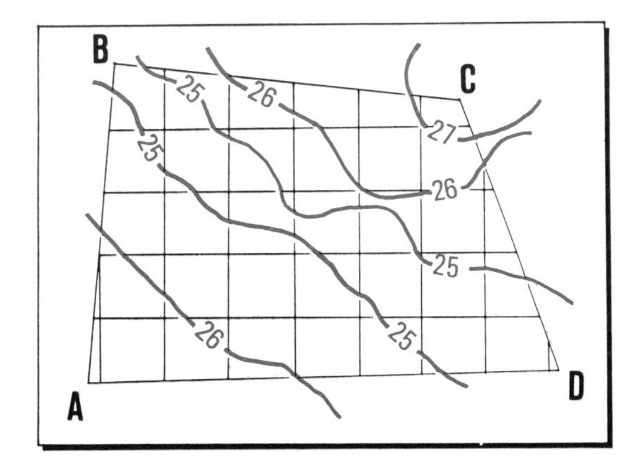

95 How to plot longitudinal profiles

Why are longitudinal profiles plotted?

1. Longitudinal profiles are plotted **to show relative elevations on a plan**. When you design a fish-farm, longitudinal profiles help you to determine the route and the bottom slope of such works as water-supply and drainage canals. They are also useful when you need to estimate the amounts of earth you need to dig out or build up on a site (called the volumes of earthwork), and when you choose sites for the construction of reservoir dams and river barrages (small dams that channel the water into ditches or canals).

Information from which longitudinal profiles can be plotted

2. You plot a longitudinal profile as **a continuous line drawn through points of known elevations**. The information you use for this can be:

- ground elevations, which are separated by known distances, along several lines (see Section 82); or
- a contour map (see Section 94).

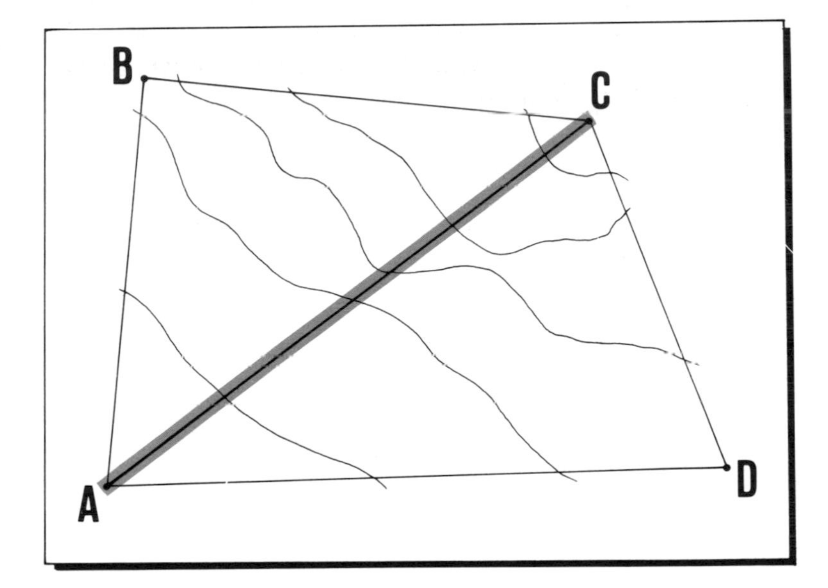

Scales to be used for longitudinal profiles

3. You need two different scales to be able to plot longitudinal profiles:

- **a horizontal scale**, which reduces horizontal ground distances; or
- **a vertical scale**, which reduces vertical elevations.

Both scales should use **the same unit of length**. This is usually the metre.

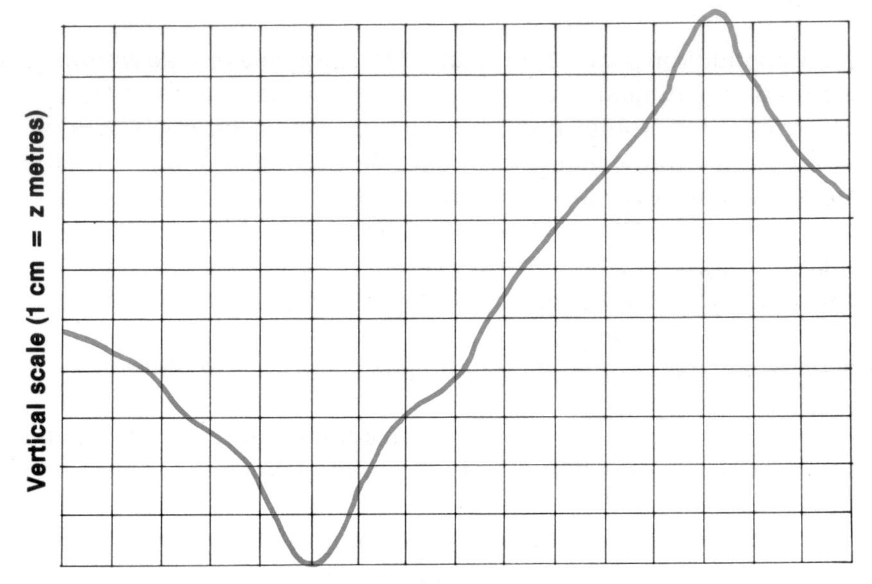

4. **The horizontal scale** of the profile should preferably be the same as the scale of the plan or map.

Example

If the scale of the plan is 1 cm per metre, the horizontal scale of the longitudinal profile should also be 1 cm per m.

5. In most aquaculture surveys, the differences in elevation are very small in comparison to the horizontal distances. When you plot longitudinal profiles for such a survey, you will therefore need to make the differences in elevation seem larger. You can use a vertical scale which is from 10 to 100 times larger than the horizontal scale.

Example

Horizontal scale	*Vertical scale*
1 cm per 25 m	1 cm per 2.5 m
1 cm per 10 m	1 cm per 0.25 m

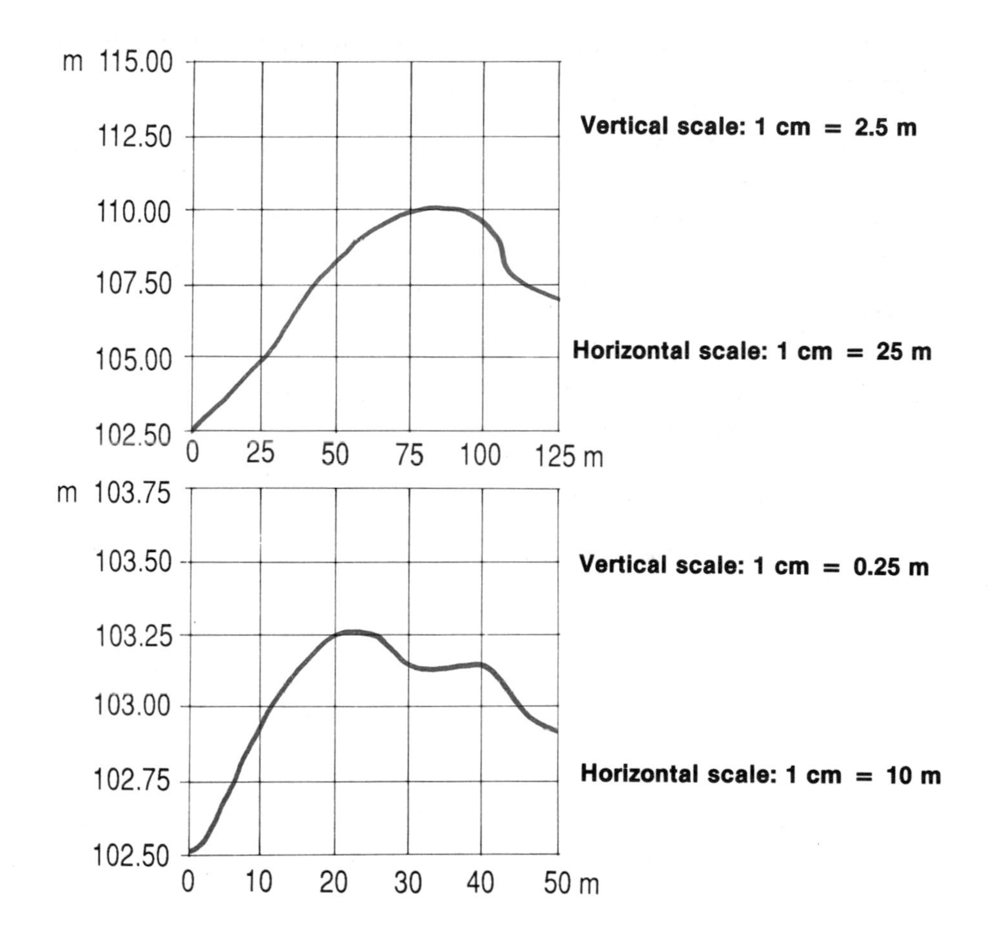

Plotting profiles from contour maps

6. Get some sheets of **square-ruled millimetric paper**. Or, use the page provided at the end of this manual, placing it under a sheet of transparent tracing paper on which you will plot your profiles.

7. On the **contour map**, draw line AB along which you need to determine the longitudinal profile. Study the range of the elevations you will plot, choose the vertical scale, and decide where to start your drawing so that it will fit within the limits of the sheet of paper.

Example

Contour map with contour interval = 2 m;
contour lines from 484 m to 506 m;
horizontal scale 1 cm = 20 m;
vertical scale 1 cm = 0.25 m.

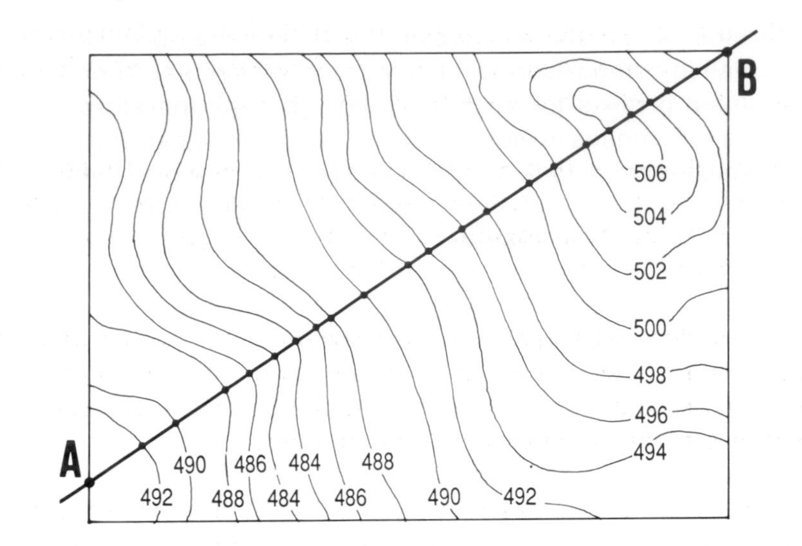

Contour interval 2 m

Horizontal scale: 1 cm = 20 m

Total vertical distance 506 m — 484 m = 22 m

Vertical scale: 1 cm = 2 m

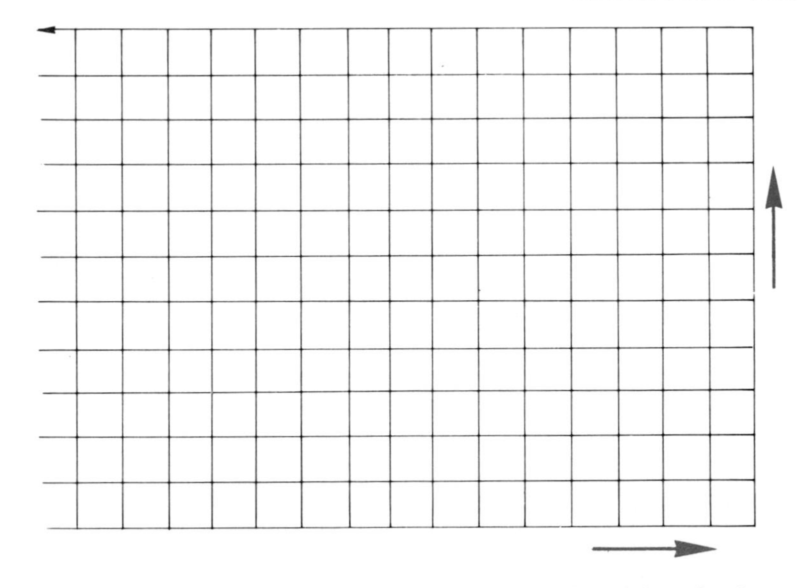

Horizontal scale: 1 cm = 20 m

8. Cut **a strip of paper** a little longer than the longitudinal profile AB you need to draw and about 2 cm wide. Place this paper strip on the contour map with one edge exactly on line AB.

9. Mark points A and B with thin vertical lines to indicate the end-points of the longitudinal profile. In a similar way, mark **the position of each of the contour lines** along the edge of the strip. Note the elevations of the main contour lines next to their mark.

10. Place **the paper strip on the drawing sheet**. Its marked edge should line up with the horizontal line representing **the lowest elevation** present (484 m) in the longitudinal profile. Align point A on the strip with the starting point of the drawing.

11. Transfer all the pencilled marks from the paper strip to the drawing and note the main elevations next to their marks.

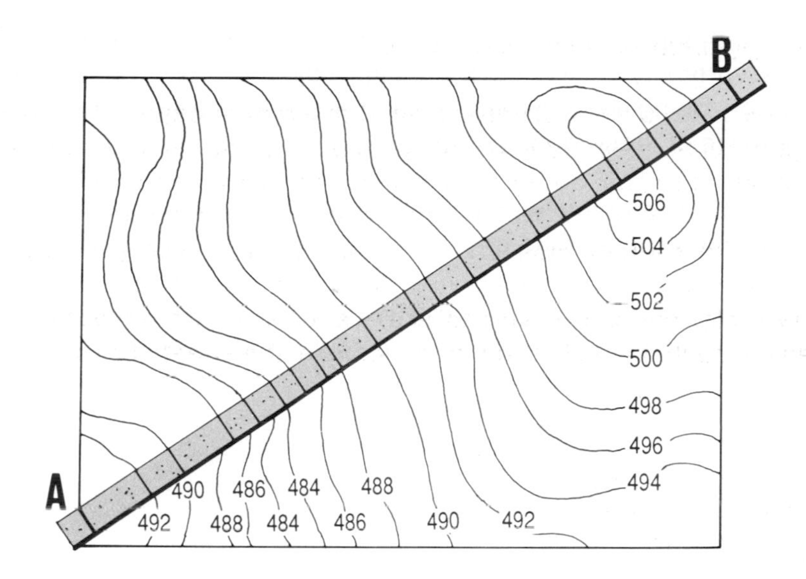

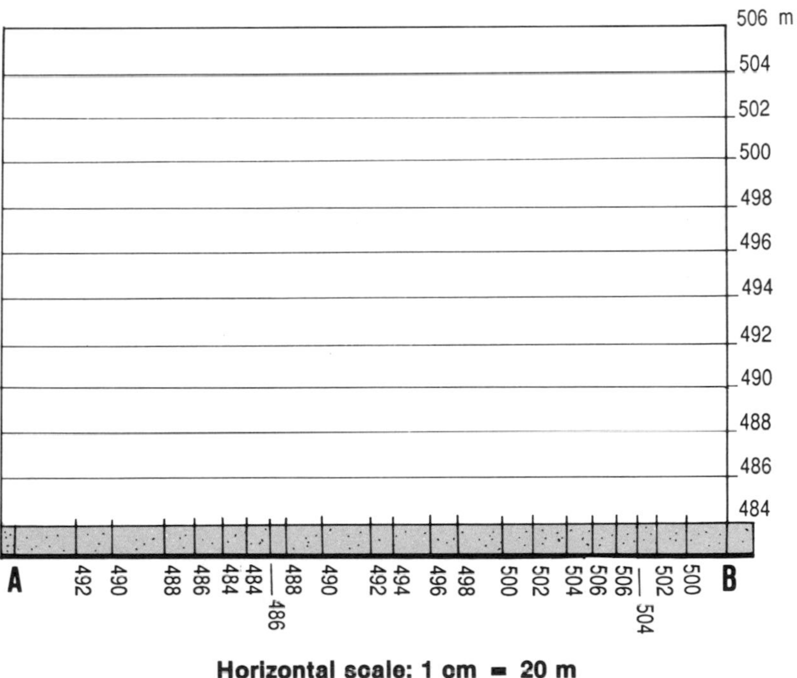

12. Using the vertical scale as a reference, **transfer each of these marks** vertically up to the horizontal line that corresponds to its elevation. Using a sharp pencil with a hard lead, make a small circled dot at each of these points on the lines.

13. Join these points with **a continuous line**, which represents the longitudinal profile of the ground along selected line AB.

Note: you can only apply this method if the horizontal scale of your drawing is the same as the distance scale of the contour map.

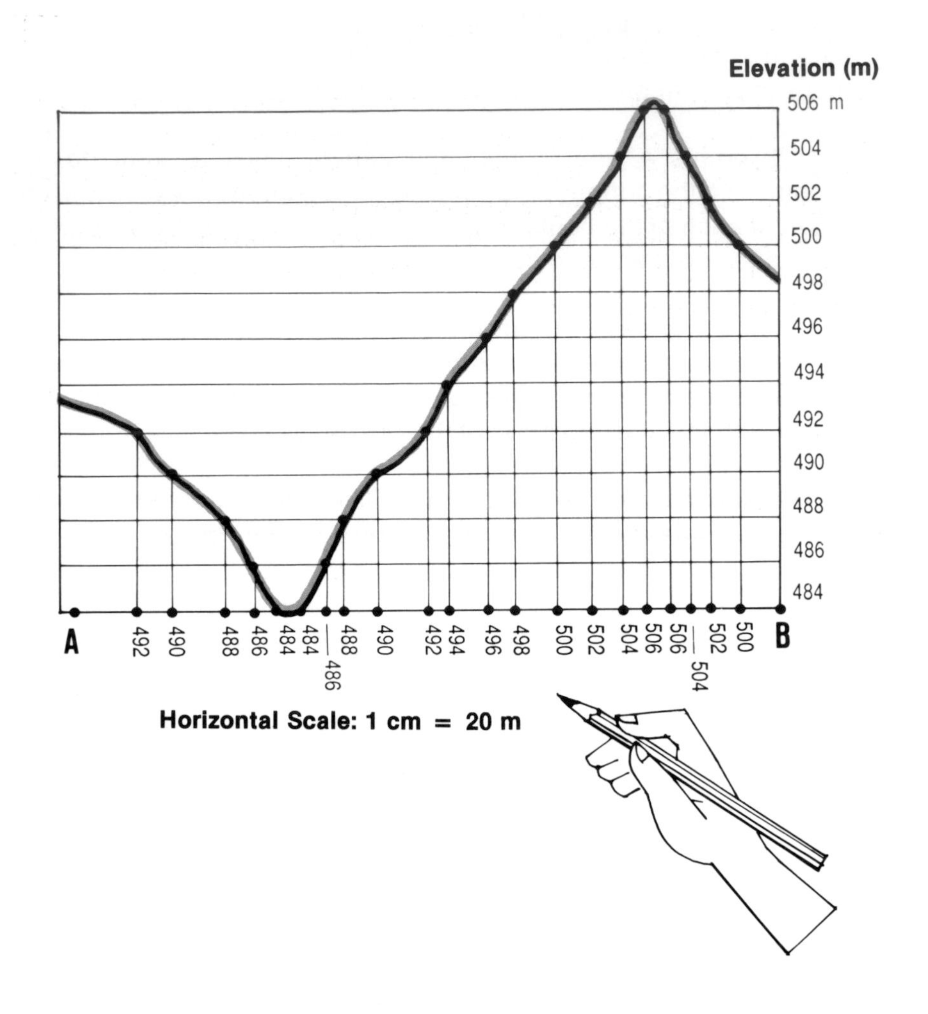

14. You can use measurements of distances and elevations from a field survey to plot profiles. Along the horizontal axis, first plot the positions of the survey stations which you have located, for example at **regular intervals along a centre-line** (see Section 82) using the horizontal scale (here 1 cm = 10 m) as a basis. Next to each of these points, mark its distance from the starting point of the profile, the cumulative distance* (in m).

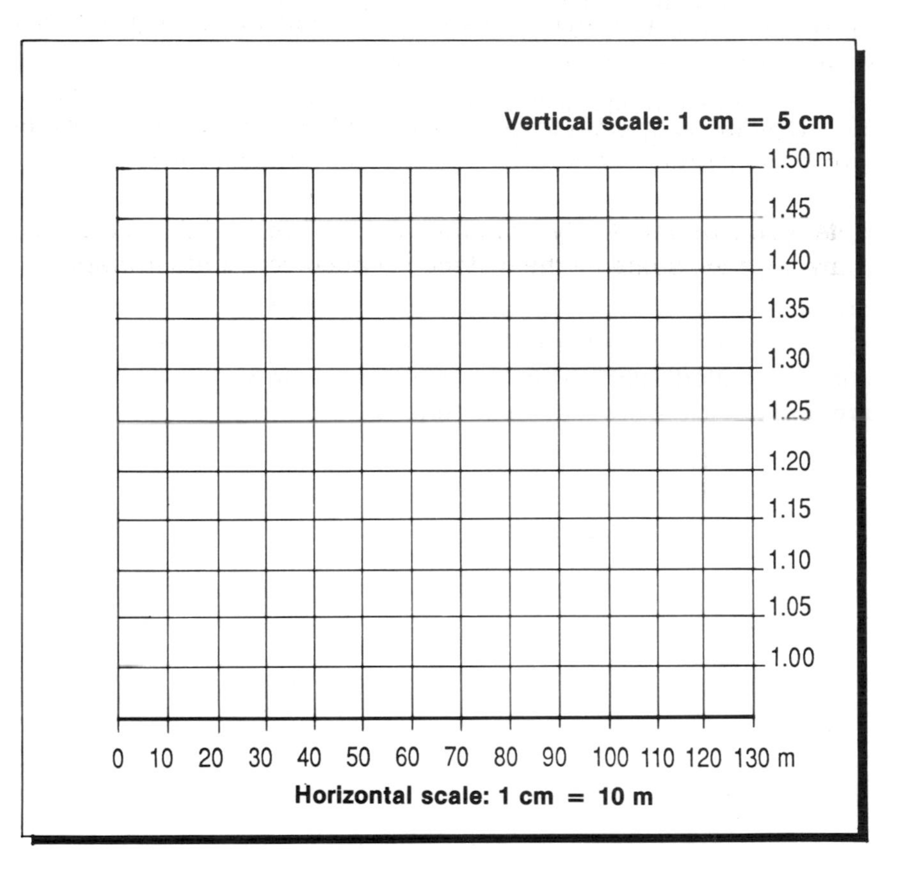

15. For each of these points, **plot the elevations** on vertical lines, using the vertical scale (1 cm = 5 cm) and the two extreme elevations (1.34 m and 1.06 m) as bases.

16. Join these points with **a continuous line**, which represents the profile of the ground along the centre-line.

17. Add more information, such as the elevations of the bench-mark (BM) and of any turning point (TP). If you also plot **the proposed canal slope** (0.15 cm/m = 7.5 cm/ 50 m), you can use the drawing to easily locate areas where you need to raise the land to a required level (called a **fill***), or places where you need to dig a channel (called a **cut***). Then you can use the drawing to estimate the amount of earthwork these will require.

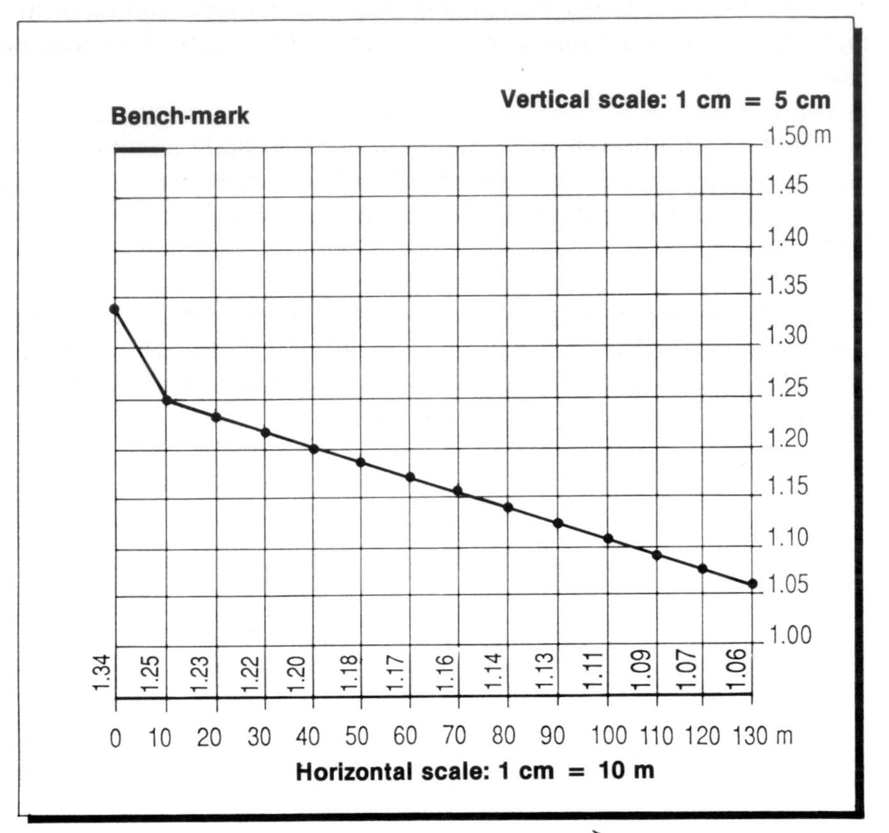

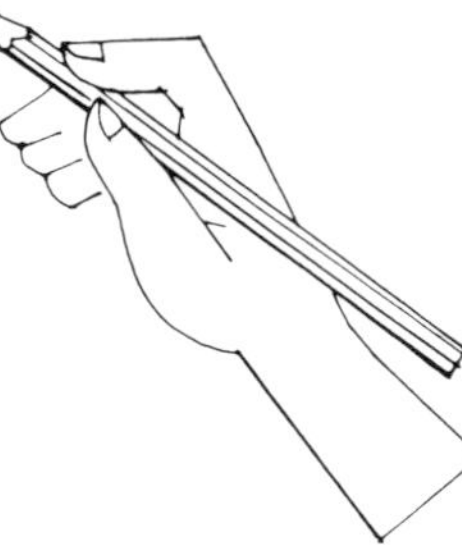

96 How to plot cross-section profiles

1. You can plot cross-section profiles either from contour maps (see Section 94) or from levelling-survey information (see Section 82).

2. A good example of when to use a cross-section profile plotted from a contour map is for **a study of a river valley** when you want to create a water reservoir, or build a small barrage that will raise the water level and fill the fish-ponds by gravity.

3. If you use the information from a levelling survey, you can plot cross-section profiles to calculate volumes of earthwork when you are building water canals and fish-ponds, for example (see next manuals on Constructions, in this series).

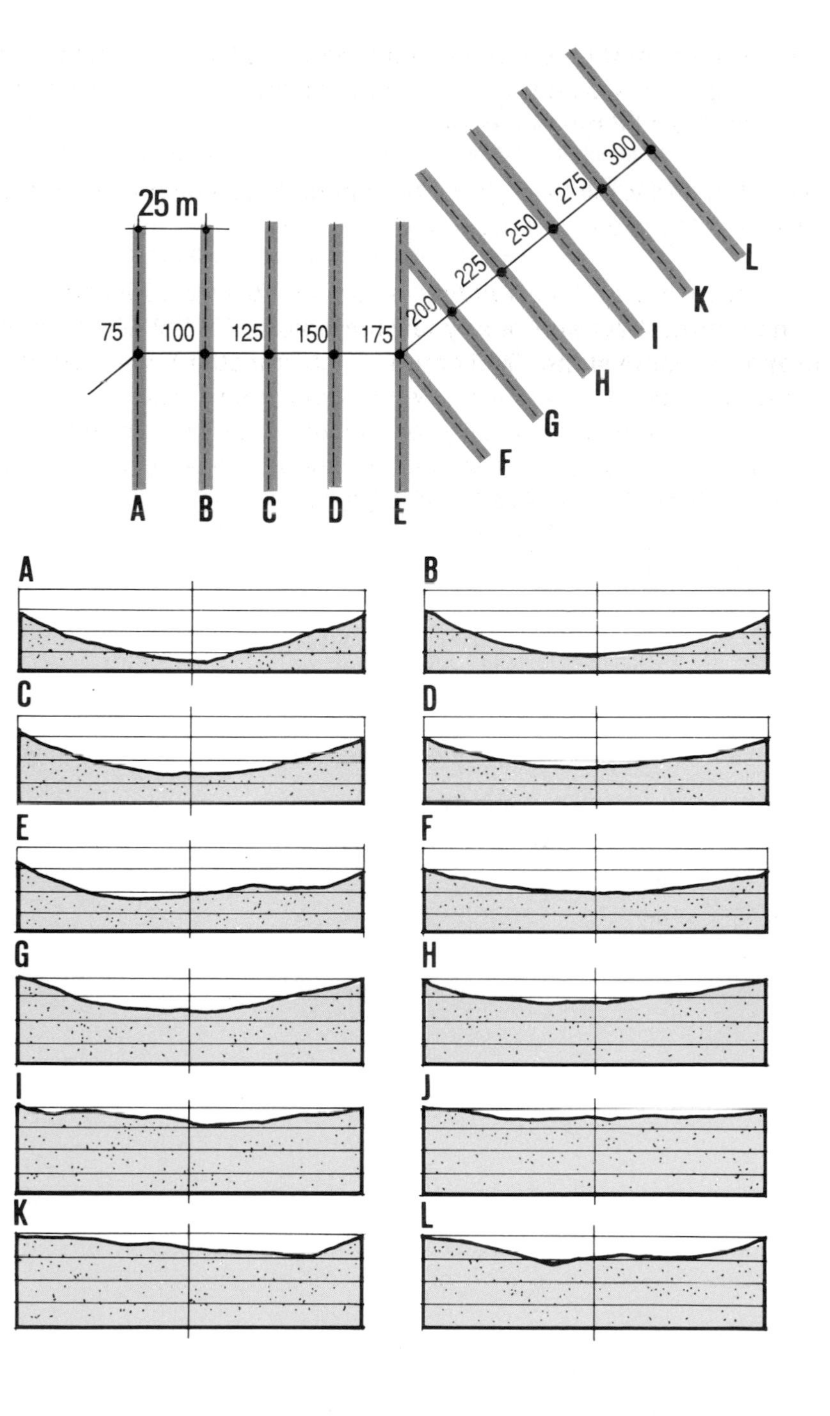

Plotting cross-section profiles from contour maps

4. On the contour map, draw the lines along which you will study the profiles. These lines should be **perpendicular to a longitudinal profile** (see Section 82, step 15).

5. Get several sheets of **square-ruled millimetric paper**, or use the page provided at the end of this manual under transparent tracing paper. Plot the cross-section profiles with the help of a marked paper strip (as described in Section 95, steps 8-13).

6. Remember that:

- **the horizontal scale** of the drawing should be the same as the distance scale of the contour map; and
- **the vertical scale** of the drawing should be from 10 to 20 times larger than the horizontal scale.

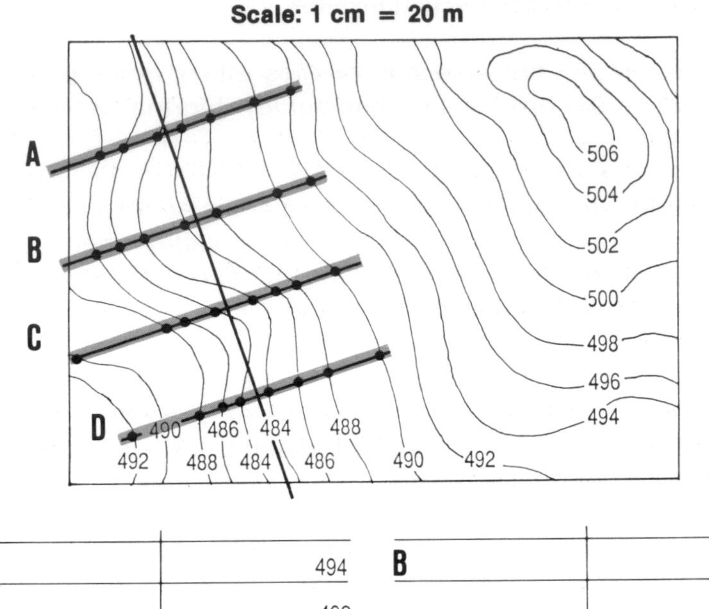

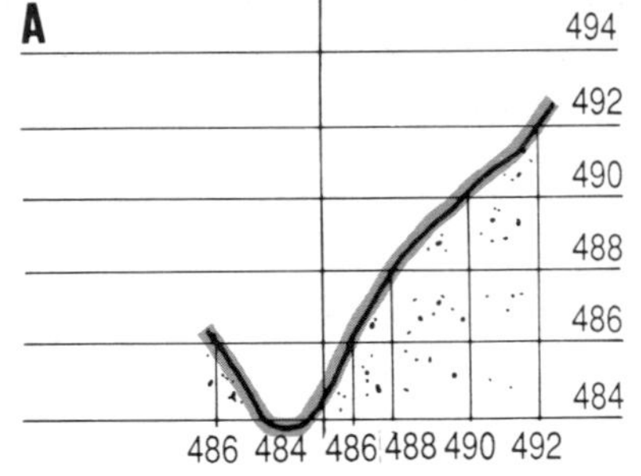

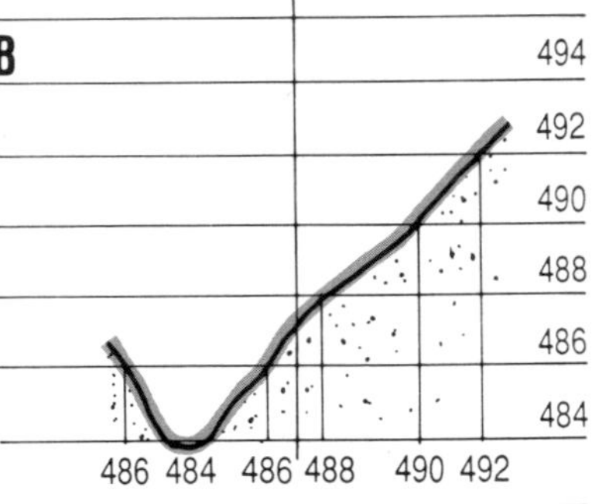

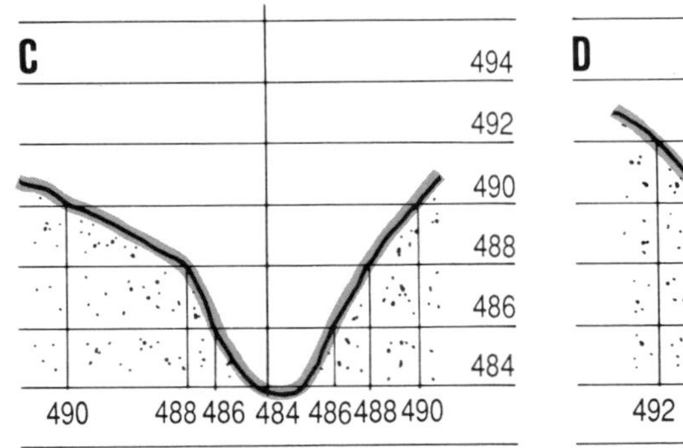

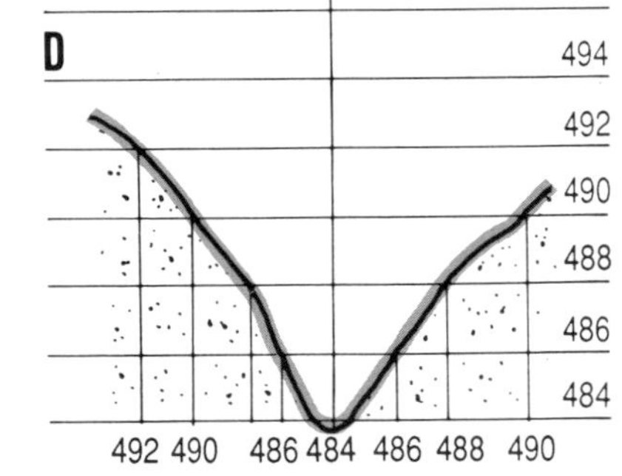

Cross-sections of a valley

Plotting cross-section profiles for earthwork estimates

7. To estimate how much earthwork you need to do, you can usually plot cross-sections to a scale of either **1 cm per metre** or **1 cm per 0.5 m**. Use the larger scale when the amount of a cut or fill is small. **Horizontal scales and vertical scales** should be identical, so that you can obtain a true surface area from the scaled dimensions.

8. You can plot best on square-ruled millimetric paper. You can also use the last page of this book, placed under a sheet of transparent tracing paper.

9. Draw a **vertical centre-line (LL)** representing the centre-line of the cross-section profile. LL should follow one of the heavier lines of the squared-ruled paper.

10. On both sides of this centre-line, draw **the ground profile** EFD on the basis of your levelling data, using the horizontal scale for distances and the vertical scale for elevations.

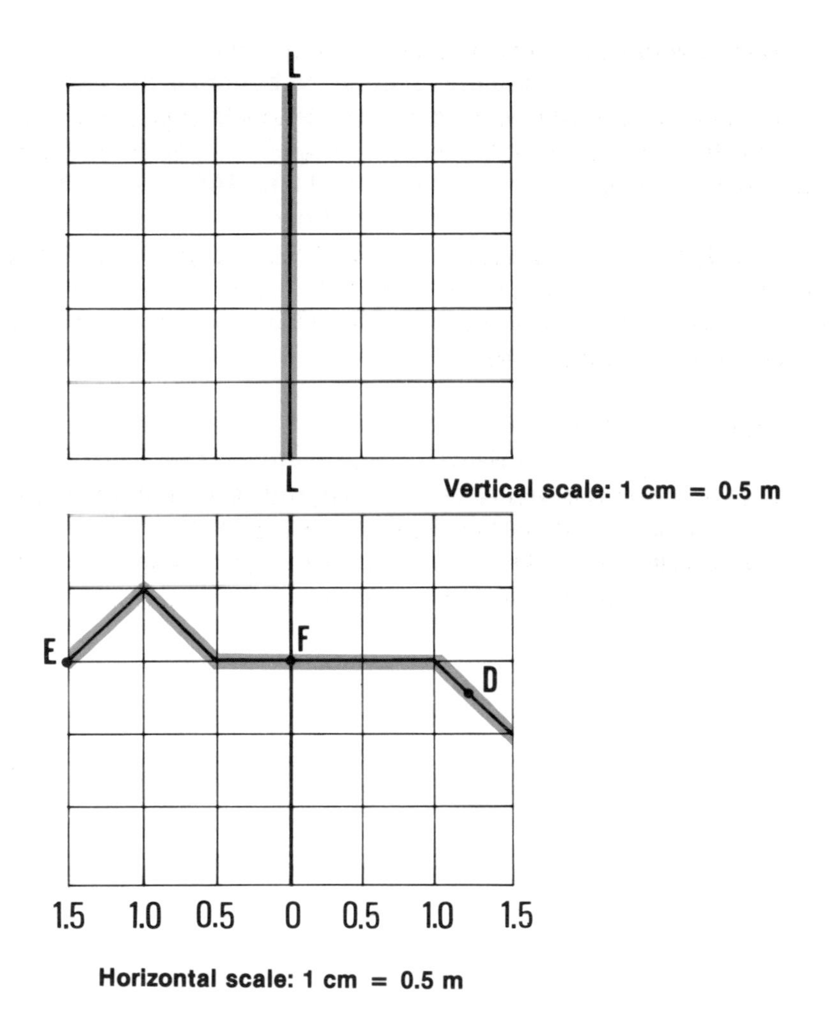

11. From your longitudinal profile, locate **point A** on line LL. In the example, it represents **the elevation of the bottom of the canal** at this particular levelling station (see Section 95, step 17).

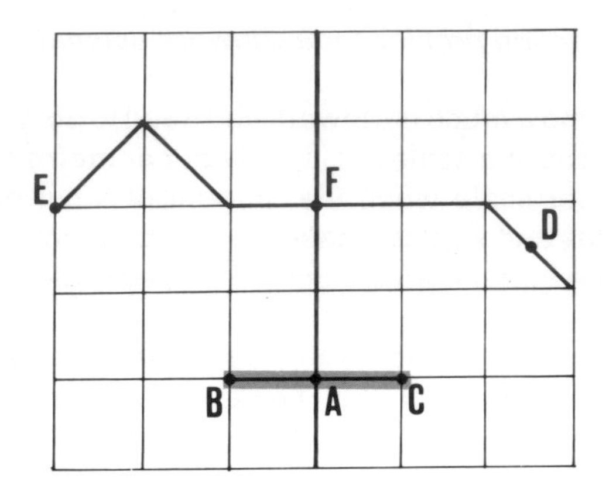

12. Through point A, draw a **horizontal line BAC** to show **the canal bottom**. Make sure that AB = AC, and each is half the width of the canal bottom.

13. Through B and C, draw lines BE and CD representing **the sides of the canal** (for example, with a slope of 1.5:1). These two lines intersect the ground surface at points E and D.

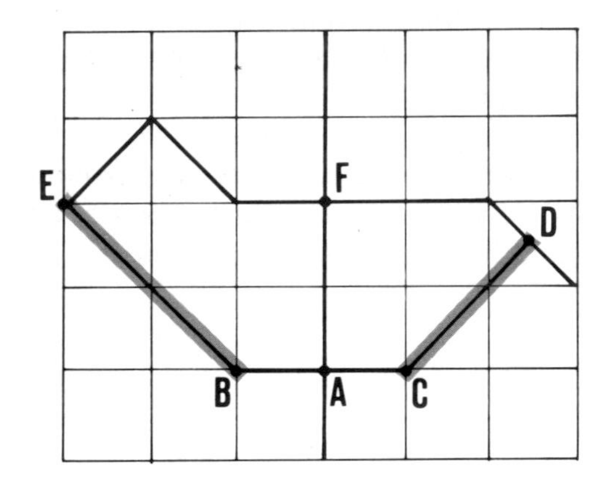

14. The cross-section EBACDFE represents a vertical section of the earth. You can then easily calculate **the area of this cross-section** (see, for example, Section 103). Using this area as a basis, you can estimate the volume of earth you need to remove from this location along the centre-line of the canal.

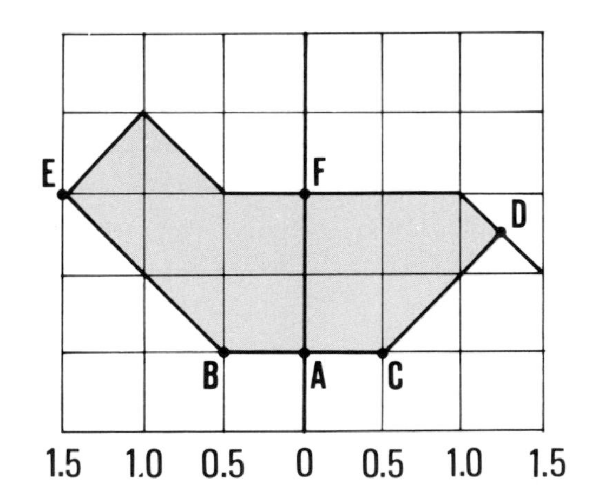

10 MEASUREMENT OF AREAS

101 Introduction

1. One of the main purposes of your topographical survey may be to determine **the area of a tract of land** where you want to build a fish-farm. From existing topographical maps, you may need to calculate the area of a watershed or of a future reservoir (see **Water**, Volume 4 in this series).

Note: in land surveying, you should regard land areas as **horizontal surfaces**, not as the actual area of the ground surface. You always measure **horizontal distances**.

2. You will often need to know the areas of **cross-section profiles** (see Section 96) to calculate the amount of earthwork you need to do.

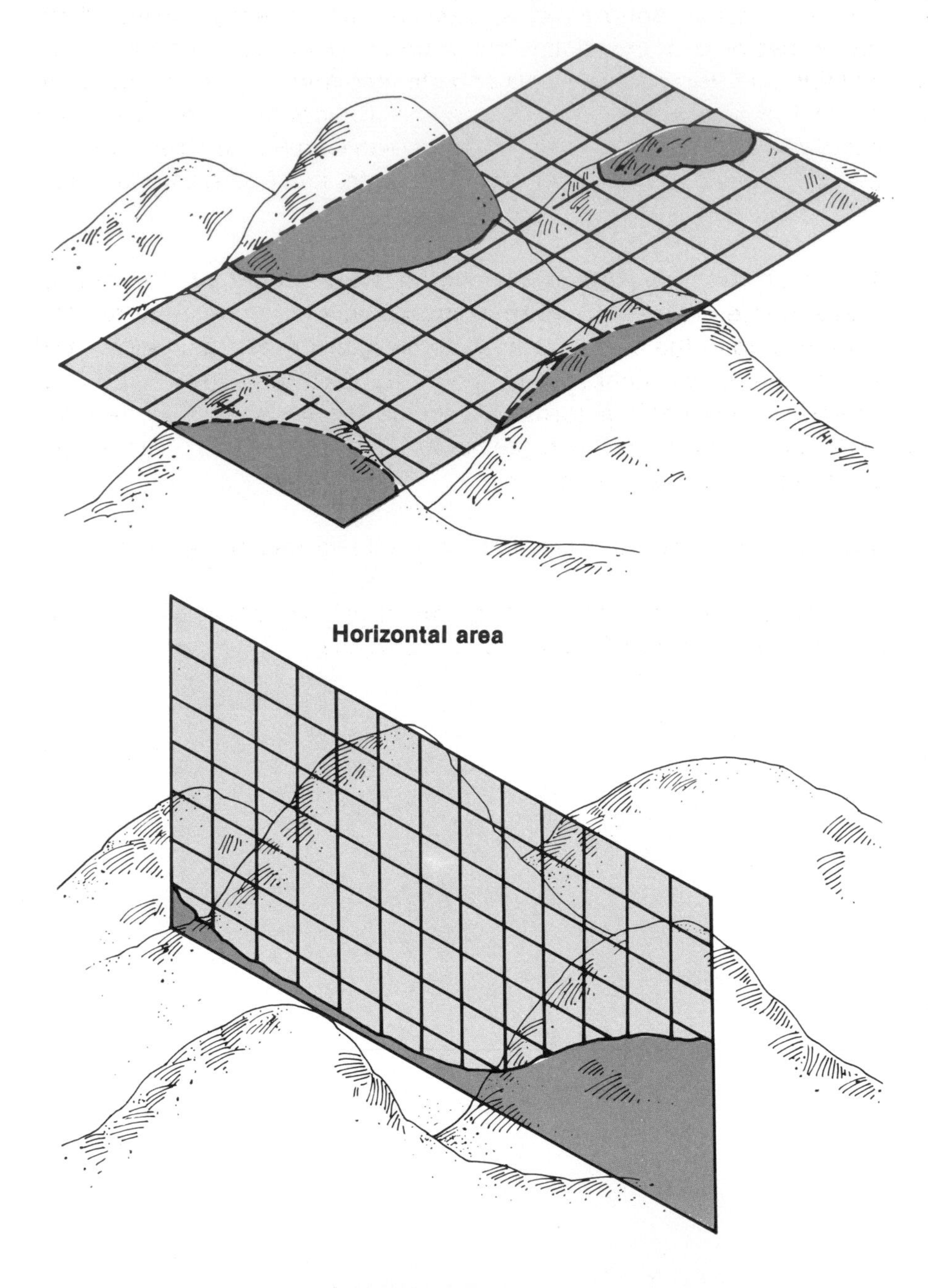

3. You may determine areas either directly from **field measurements**, or indirectly, from **a plan or map**. In the first case, you will find all the measurements of distances and angles you need by surveying, and you will calculate the areas from them. In the second case, you will draw a plan or map first (see Chapter 9). Then you will get the dimensions you need from the scale, and determine the area on that basis.

4. There are several simple methods available for measuring areas. Some of these are **graphic methods**, where you compare the plan or map of the area you need to measure to a drawn pattern of known unit sizes. Others are **geometric methods**, where you use simple mathematical formulas to calculate areas of regular geometrical figures, such as triangles, **trapeziums***, or areas bounded by an irregular curve.

Note: a trapezium is a four-sided polygon with two parallel sides.

5. The simple methods will be described in detail in the next sections. They are also summarized in **Table 13**.

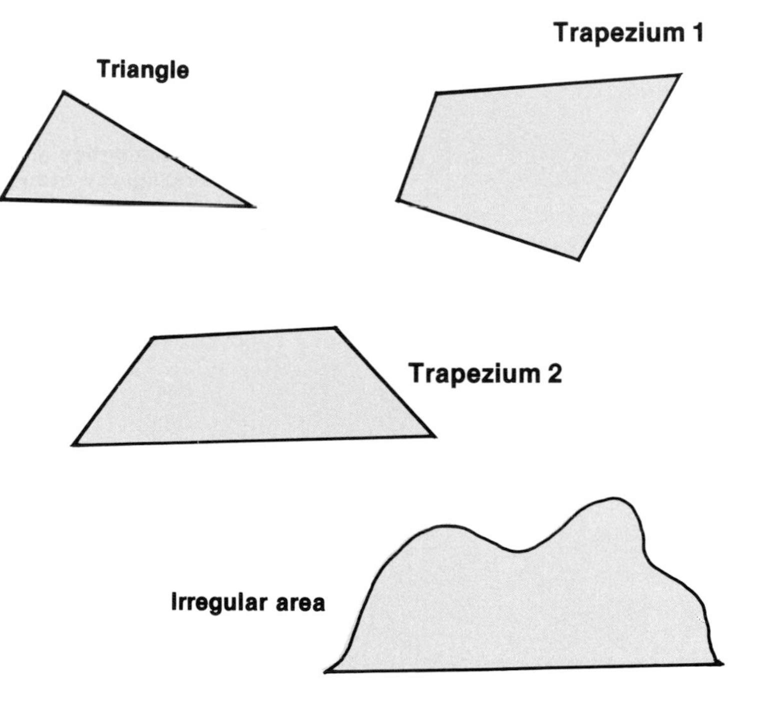

TABLE 13

Simple area measurement methods

Section	*Method*	*Remarks*
102	Strips	Graphic method giving rough estimate
103	Square-grid	Graphic method giving good to very good estimates
104	Subdivision into regular geometric figures, triangles, trapeziums	Geometric method giving good to very good estimates
105	Trapezoidal rule	Geometric method giving good to very good estimates Suitable for curved boundary

102 How to use the strips method for measuring areas

1. Get a piece of **transparent paper**, such as tracing paper or light-weight square-ruled millimetric paper. Its size will depend on the size of the mapped area you need to measure.

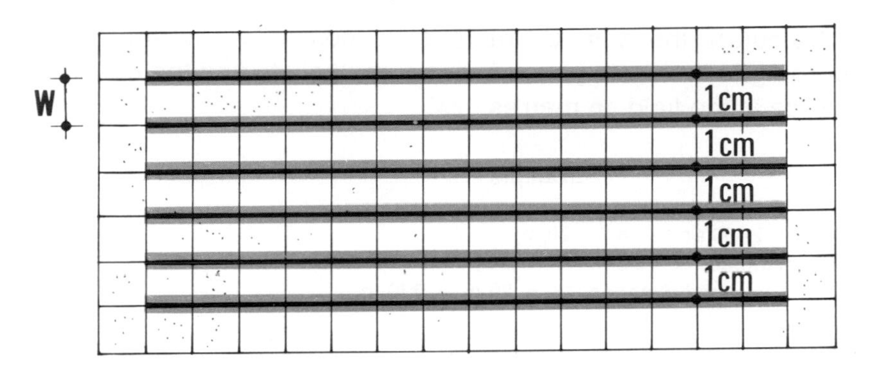

W = 1 cm

2. On this paper, draw **a series of strips**, by drawing a series of parallel lines at a regular, fixed interval. Choose this **strip width W** to represent a certain number of metres. You can follow the scale of the plan or map to do this (see Section 91).

Example

Scale 1:2000; strip width W = 1 cm = 20 m.
Scale 1:50000; strip width W = 1 cm = 500 m.

Note: the smaller the strip width, the more accurate your estimate of the land area will be.

3. Place the sheet of transparent paper over the plan or map of the area you need to measure, and attach it securely with drawing pins or transparent tape.

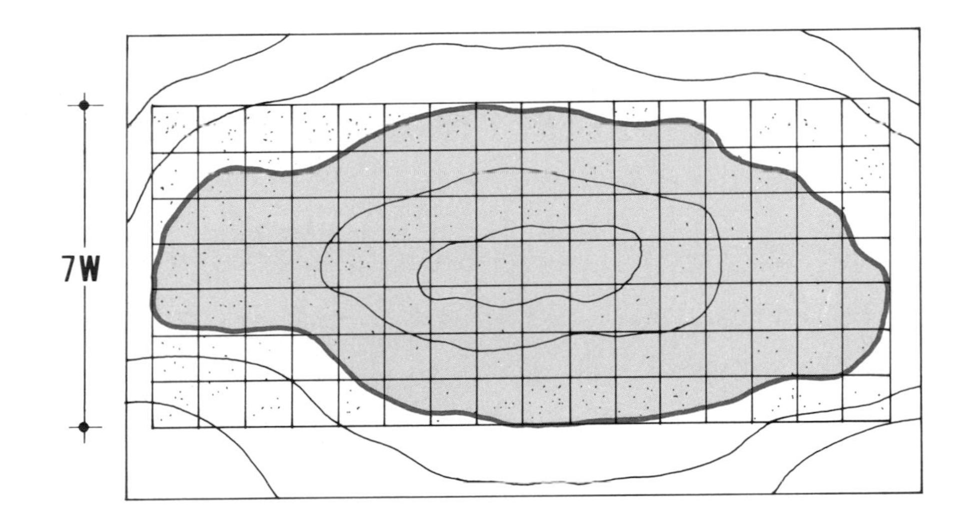

Scale: 1: 2.000

4. For each strip, measure the distance AB in centimetres **along a central line** between the boundaries of the area shown on the map.

5. Calculate the sum of these distances in centimetres. Then, according to the scale you are using, multiply to find the equivalent distance in the field, in metres.

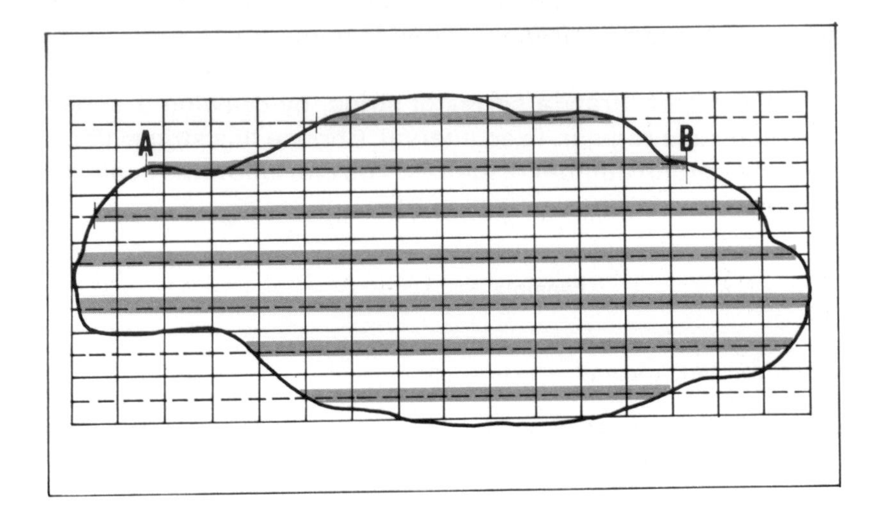

Example

Scale is 1:2000 and 1 cm = 20 m.
Sum of distances = 16 cm.
Equivalent ground distance: 16 × 20 m = 320 m.

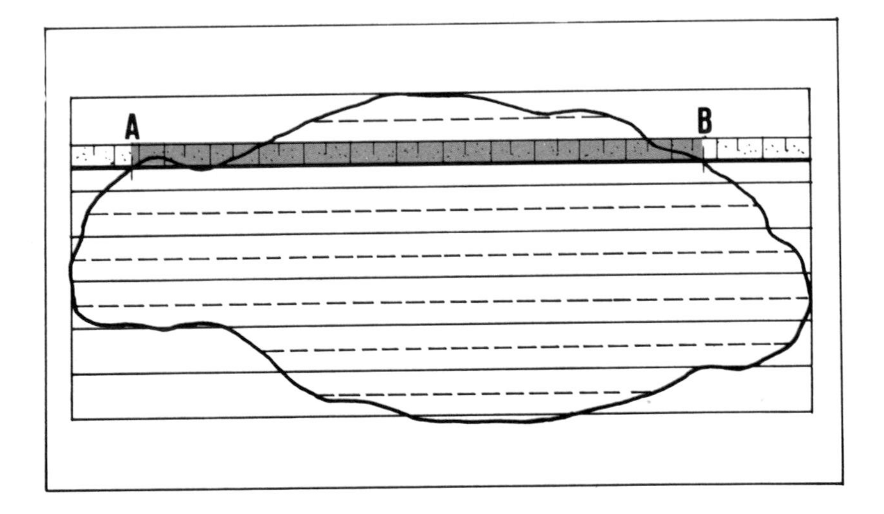

6. Multiply this sum of real distances (in metres) by the equivalent width of the strip W (in metres) to obtain a **rough estimate** of the total area in **square metres** (abbreviated as m²).

Example

Sum of equivalent distances is 320 m.
Strip width (1 cm) is equivalent to 20 m.
Land area: 320 m × 20 m = 6 400 m² or 0.64 ha

Note: 10 000 m² = 1 hectare (ha)

7. Repeat this procedure at least once to check on your calculations.

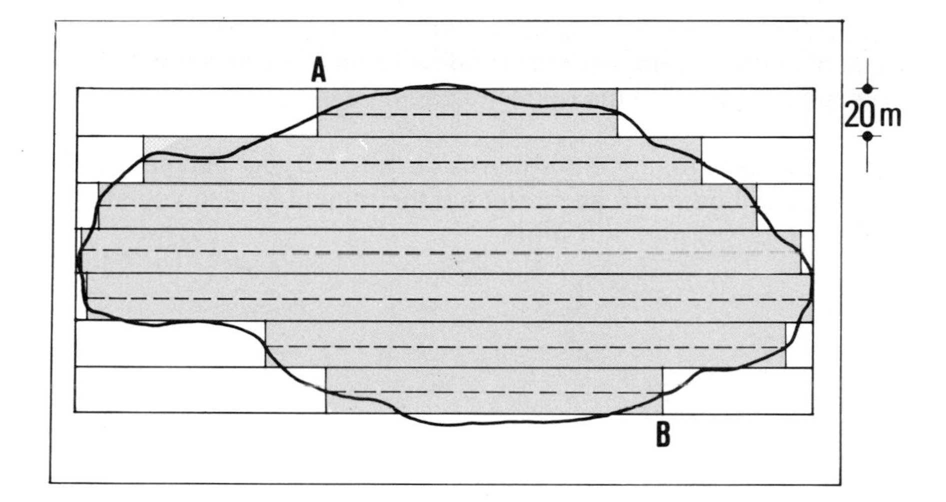

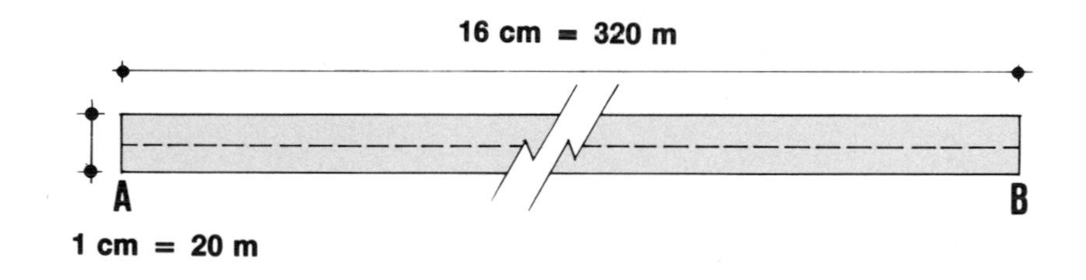

103 How to use the square-grid method for measuring areas

1. Get a piece of transparent square-ruled paper, or draw a square grid on transparent tracing paper yourself. To do this, trace **a grid made of 2 mm × 2 mm squares inside a 10 cm × 10 cm square**, using the example given on the opposite page.

Note: if you use smaller unit squares on the grid, your estimate of the land area will be more accurate; but the minimum size you should use is 1 mm × 1 mm = 1 mm^2.

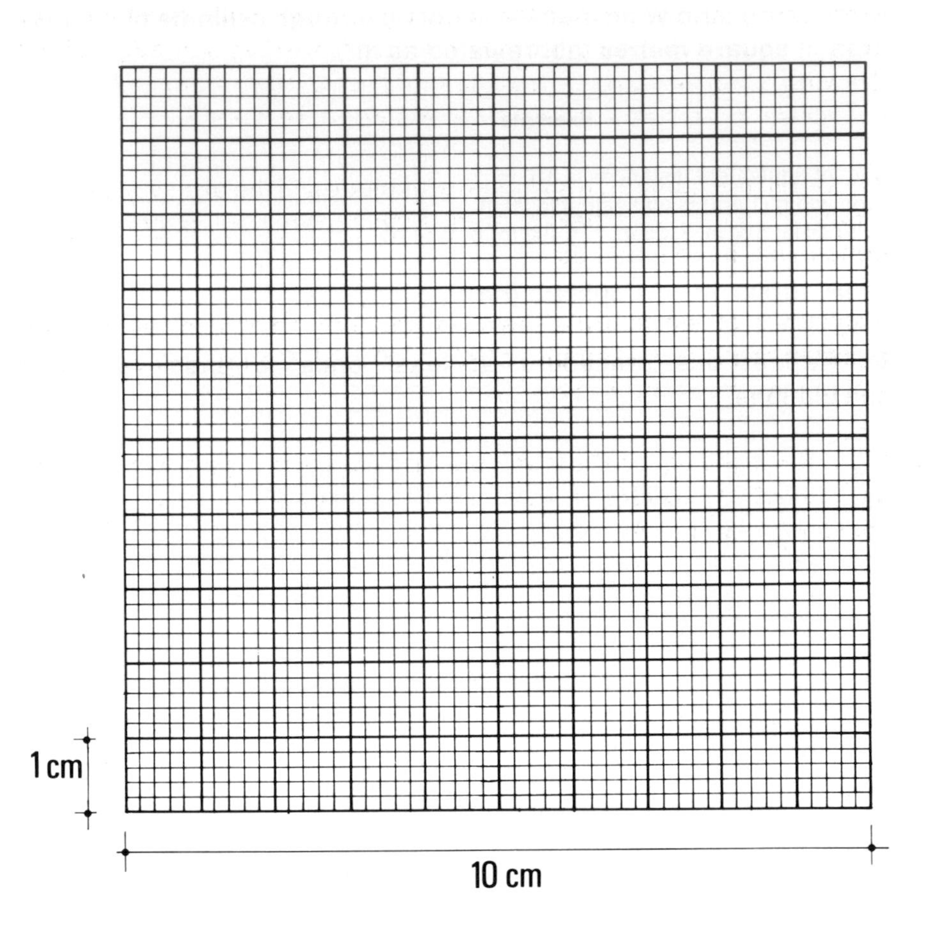

2. Place this transparent grid over the drawing of the area you need to measure, and attach it to the drawing securely with thumbtacks or tape. If your grid is smaller than this area, start at one edge of the drawing. Clearly mark the outline of the grid, then move to the next section and proceed in this way over the entire area.

3. Count the number of **full squares** included in the area you need to measure. To avoid mistakes, mark each square you count with your pencil, making a small dot.

Note: towards the centre of the area, you may be able to count **larger squares** made, for example, of 10 × 10 = 100 small squares. This will make your work easier.

4. Look at the squares **around the edge** of the drawing. **If more than one-half of any square** is within the drawing, count and mark it as a full square. Ignore the rest.

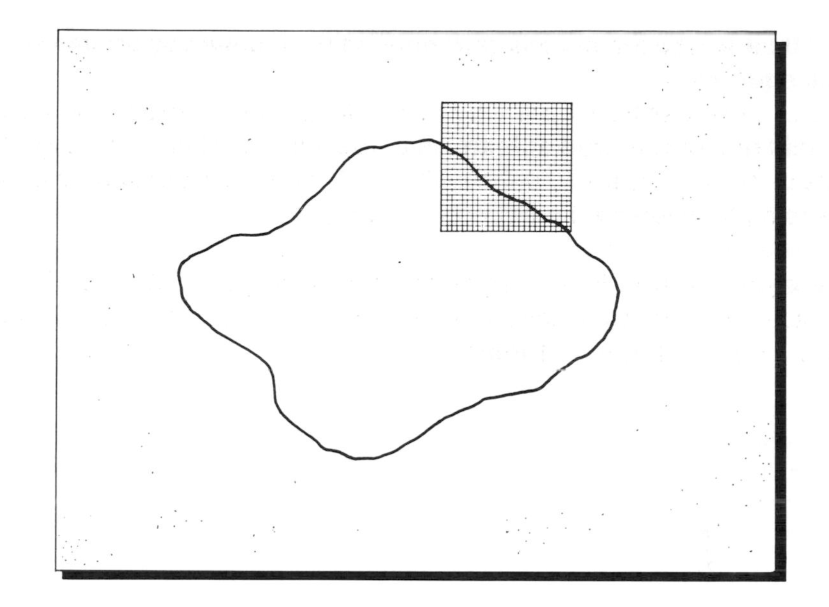

Half or more squares

5. Add these two sums (steps 3 and 4), to obtain the **total number T of full squares**.

6. Add the sums again at least once to check them.

7. Using **the distance scale** of the drawing, calculate the **equivalent unit area** for your grid. This is the equivalent area of one of its small squares.

Example

- Scale 1:2000 or 1 cm = 20 m or **1 mm = 2 m**
- Grid square size is 2 mm × 2 mm
- Equivalent unit area of grid = 4 m × 4 m = **16 m²**

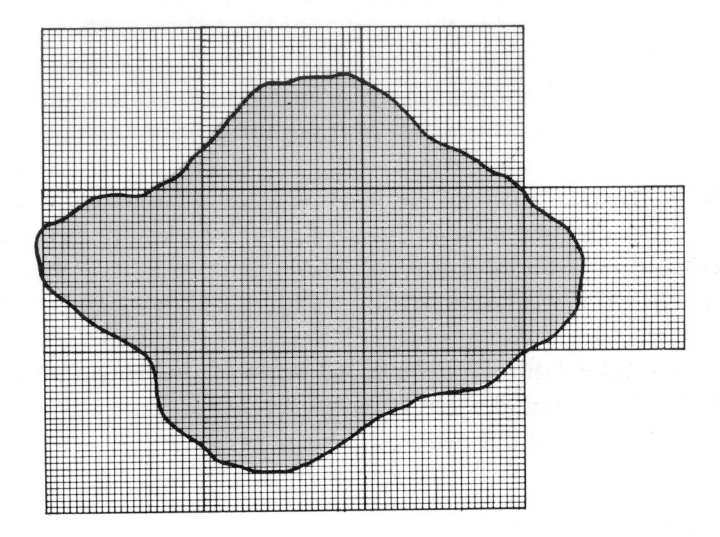

Scale:

1 cm = 20 m

1 mm = 2 m

2 mm = 4 m

2 mm x 2 mm = 4 m x 4 m = 16 m²

8. Multiply the equivalent unit area by the total number T of full squares to obtain a fairly good estimate of the measured area.

Example

- Total count of full squares T = 256
- Equivalent unit area = 16m²
- Total area = 256 × 16 m² = 4096 m²

Note: when you work with **large-scale plans** such as cross-sections, you can improve the accuracy of your area estimate by modifying step 5, above. To do this, look at all the squares around **the edge of the drawing** which are crossed by a drawing line. Then, estimate by sight the **decimal part of the whole square** that you need to include in the total count (the decimal part is a fraction of the square, expressed as a decimal, such as 0.5, which equals 5/10).

Example

Square A = 0.5; B = 0.1; C = 0.9.

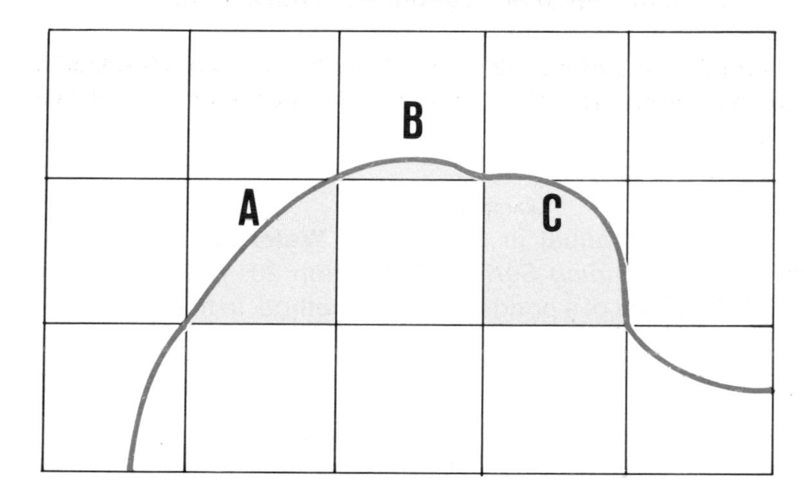

104 How to subdivide the area into regular geometrical figures

1. When you need to measure areas directly in the field, divide the tract of land into **regular geometrical figures,** such as triangles, rectangles or trapeziums. Then take all the necessary measurements, and calculate the areas according to **mathematical formulas** (see **Annex 1**). If a plan or map of the area is available, you can draw these geometrical figures on it, and find their dimensions by using the reduction scale.

2. In the first manual in this series, **Water for Freshwater Fish Culture**, *FAO Training Series* (4), Section 20, you learned how to calculate the area of a pond using this method. In the following steps, you will learn how to apply it under more difficult circumstances.

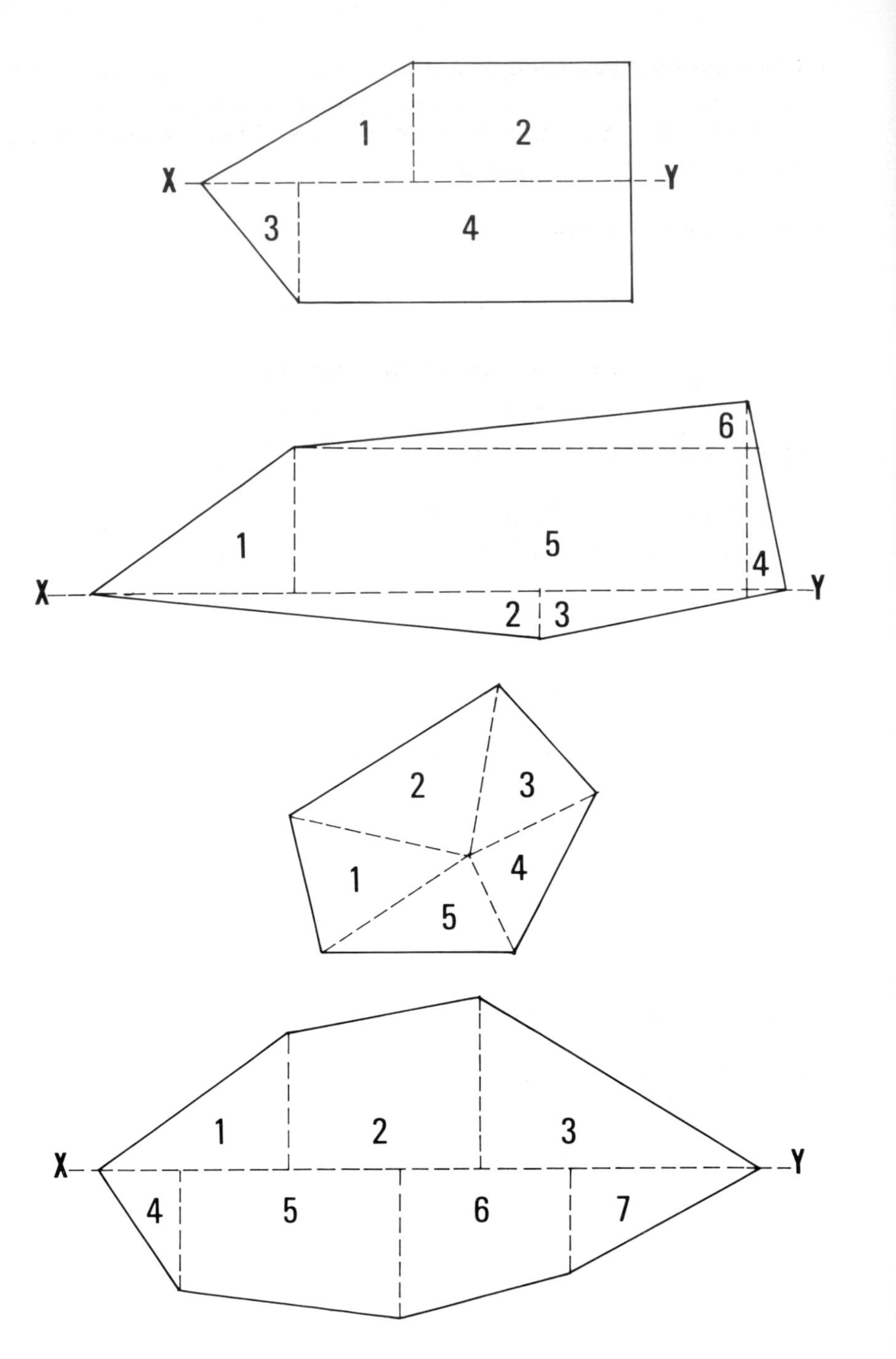

3. You can easily calculate the area of **any triangle** when you know the dimensions of:

all **three sides** a, b and c

$$\textbf{Area} = \sqrt{s(s-a)(s-b)(s-c)}$$

where **s** = (a + b + c) ÷ 2;

Two sides (b, c) and **the angle** BAC between them (called the included angle), using **Table 14** to obtain the **sine value** of the angle

$$\textbf{Area} = (bc \sin BAC) \div 2$$

obtaining **sin BAC** from Table 14.

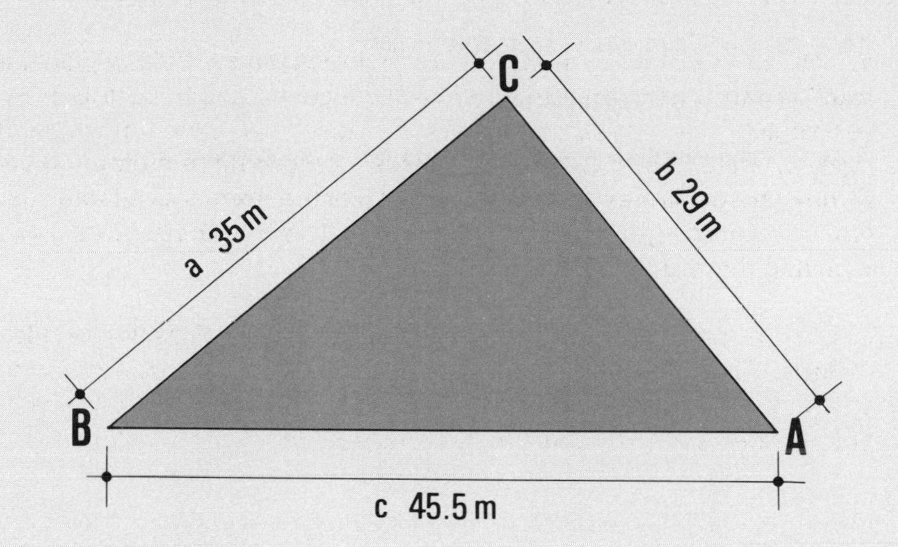

$$\textbf{Area} = \sqrt{s\,(s - a)(s - b)(s - c)}$$

Example

If **a** = 35 m; **b** = 29 m; and **c** = 45.5 m

Then **s** = (35 m + 29 m + 45.5 m) ÷ 2 = 54.75 m

Area^2 = 54.75 m (54.75 m – 35 m) (54.75 m – 29 m)(54.75 m – 45.5 m)
= 54.75 m × 19.75 m × 25.75 m × 9.25 m = 257 555 m^4

Area = $\sqrt{257\,555\ m^4}$ = 507 m^2

Example

If **b** = 29 m; **c** = 45.5 m; and angle **BAC** = 50º

Then sin BAC = 0.7660 (Table 14)

Area = (29m × 45.5m × 0.7660) ÷ 2 = 1010.737 ÷ 2 = 505.3685 m²

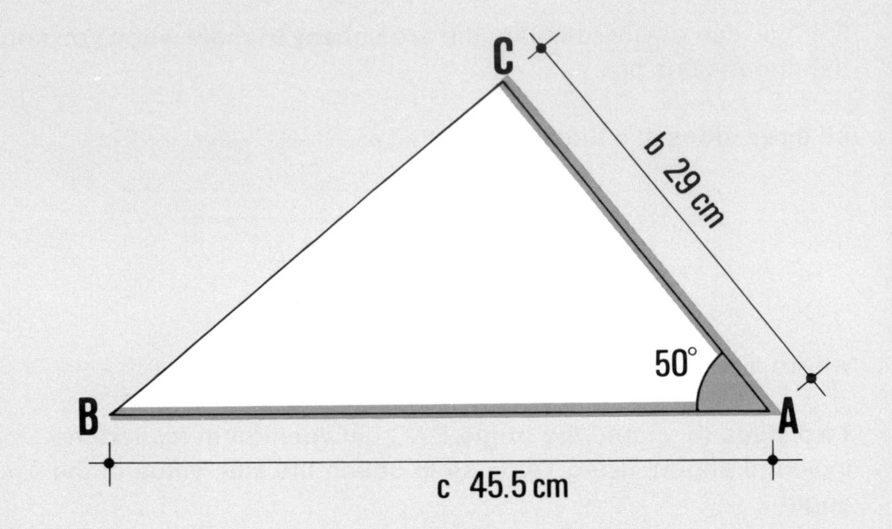

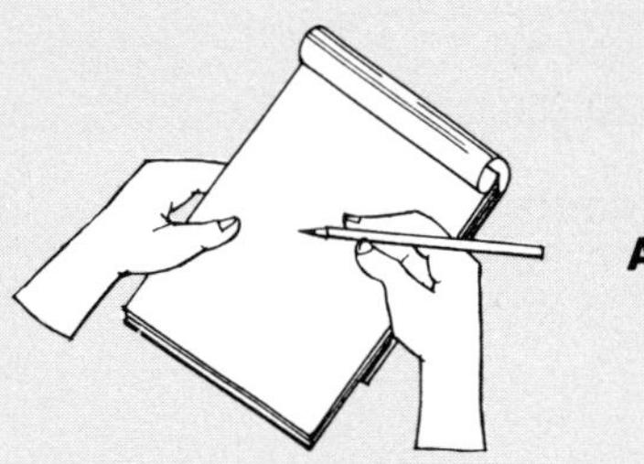

Area = (bc sin BAC)÷2

TABLE 14

Sine values of angles

Degree	*Sine*	*Degree*	*Sine*	*Degree*	*Sine*
1	0.0175	31	0.5150	61	0.8746
2	0.0349	32	0.5299	62	0.8829
3	0.0523	33	0.5446	63	0.8910
4	0.0698	34	0.5592	64	0.8988
5	0.0872	35	0.5736	65	0.9063
6	0.1045	36	0.5878	66	0.9135
7	0.1219	37	0.6018	67	0.9205
8	0.1392	38	0.6157	68	0.9272
9	0.1564	39	0.6293	69	0.9336
10	0.1736	40	0.6428	70	0.9397
11	0.1908	41	0.6561	71	0.9455
12	0.2079	42	0.6691	72	0.9511
13	0.2250	43	0.6820	73	0.9563
14	0.2419	44	0.6947	74	0.9613
15	0.2588	45	0.7071	75	0.9659
16	0.2756	46	0.7193	76	0.9703
17	0.2924	47	0.7314	77	0.9744
18	0.3090	48	0.7431	78	0.9781
19	0.3256	49	0.7547	79	0.9816
20	0.3420	50	0.7660	80	0.9848
21	0.3584	51	0.7771	81	0.9877
22	0.3746	52	0.7880	82	0.9903
23	0.3907	53	0.7986	83	0.9925
24	0.4067	54	0.8090	84	0.9945
25	0.4226	55	0.8192	85	0.9962
26	0.4384	56	0.8290	86	0.9976
27	0.4540	57	0.8387	87	0.9986
28	0.4695	58	0.8480	88	0.9994
29	0.4848	59	0.8572	89	0.9998
30	0.5000	60	0.8660		

4. Subdivide the tract of land into triangles. For a **four-sided area**, you can do this in two ways.

- You can join **two opposite angles** with a straight line BD. Measure the length of BD to find the length of the three sides of each of the two triangles, then calculate their areas (see step 3, above). The sum of the two triangular areas is the total area.
- You can proceed **by radiating** from central station O. Measure consecutive angles AOB, BOC, COD and DOA. Then measure distances OA, OB, OC and OD from O to each corner of the site and calculate the area of each triangle (see step 3, above). The sum of the four triangular areas is the total area.

5. On a land tract with more than four sides, you can subdivide its area into triangles:

- by radiating from a central station 0 (see step 4, above); or
- by radiating from a lateral station, such as A.

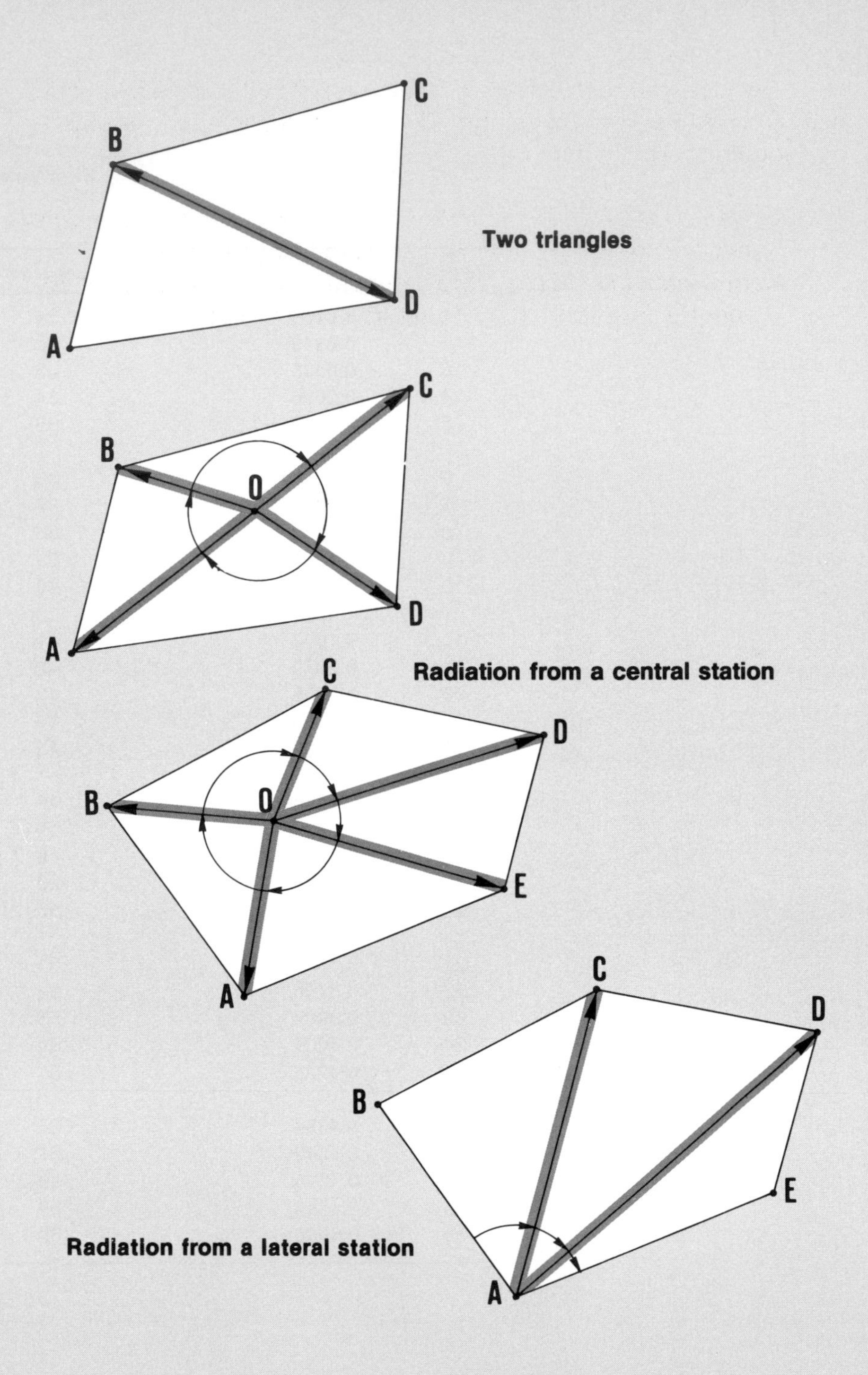

Two triangles

Radiation from a central station

Radiation from a lateral station

6. Check on your calculations. If you have found the area by using two opposite angles, use the first procedure. If you have proceeded by radiating, use the second.

- Repeat the measurement of the total area by using **the other two triangles** ABC and ACD, formed by straight line AC.
- Alternatively repeat the measurements of angles and lengths from either the same station or a different one.

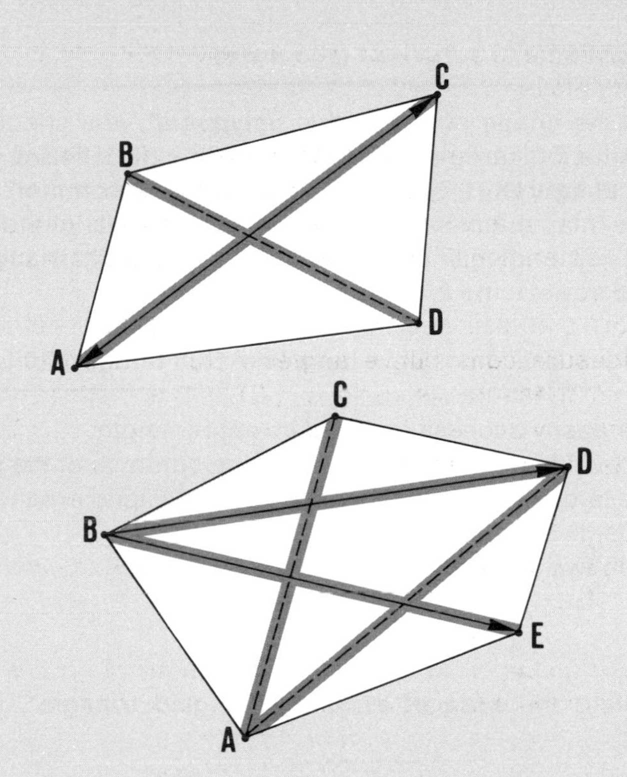

Using a base line to subdivide land areas

7. When the shape of the land is **polygonal***, you should usually subdivide the total area you need to measure **into a series of regular geometrical figures** (1-7 in the example) **from a common base line AD**. You will lay out offsets from the other summits of the **polygon*** which are perpendicular to this base line to form **right triangles** 1, 3, 4 and 7, and **trapeziums** 2, 5 and 6.

8. When you are choosing **a base line**, remember that it should:

- be easily accessible along its entire length;
- provide good sights to most of the summits of the polygon;
- be laid out along the longest side of the land area to keep the offsets as short as possible;
- join two polygon summits.

9. Calculate **the area of each right-angled triangle***, using the formula:

Area = (base × height) ÷ 2

10. Calculate **the area of each trapezium**, using the formula:

Area = height × (base 1 + base 2) ÷ 2

where:

- **Base 1** is parallel to **base 2**;
- Height is the **perpendicular distance** from base 1 to base 2.

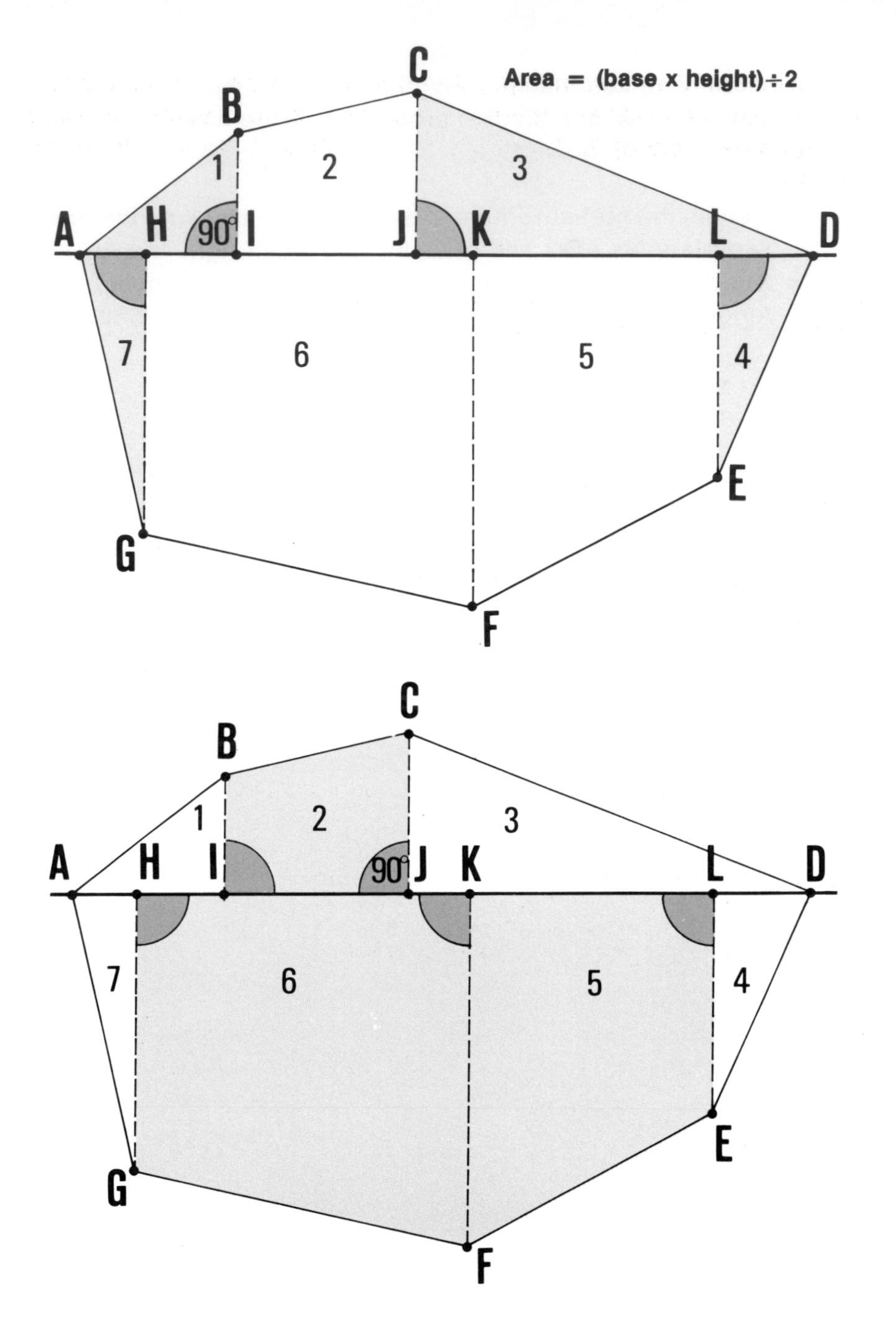

11. Add together all these partial areas to find the total land area. You should use a **table** to enter all the basic dimensions for both **right triangles** (one base) and **trapeziums** (two bases), as shown in the example.

Example

- Along base line AD, measure from point A **cumulative distances** to points H, I, J, K, L, and D, as follows:

Base line (in m)

A	H	I	J	K	L	D
0	2.80	6.50	14.15	16.75	27.25	31.25

- From these measurements, obtain **partial distances** for each section AH, HI, IJ, JK, KL and LD as follows:

Base line (in m)

AH	HI	IJ	JK	KL	LD
2.80	3.70	7.65	2.60	10.50	4.00

- **Measure offsets** HG, IB, ... LE from the base line to each polygon summit:

  HG = 11.80 m; IB = 5.20 m; ... LE = 9.65 m

- Enter these data in the following **table**, and obtain **partial areas** of each lot 1, 2, 3, 4, 5, 6 and 7; the sum is the **total area**.

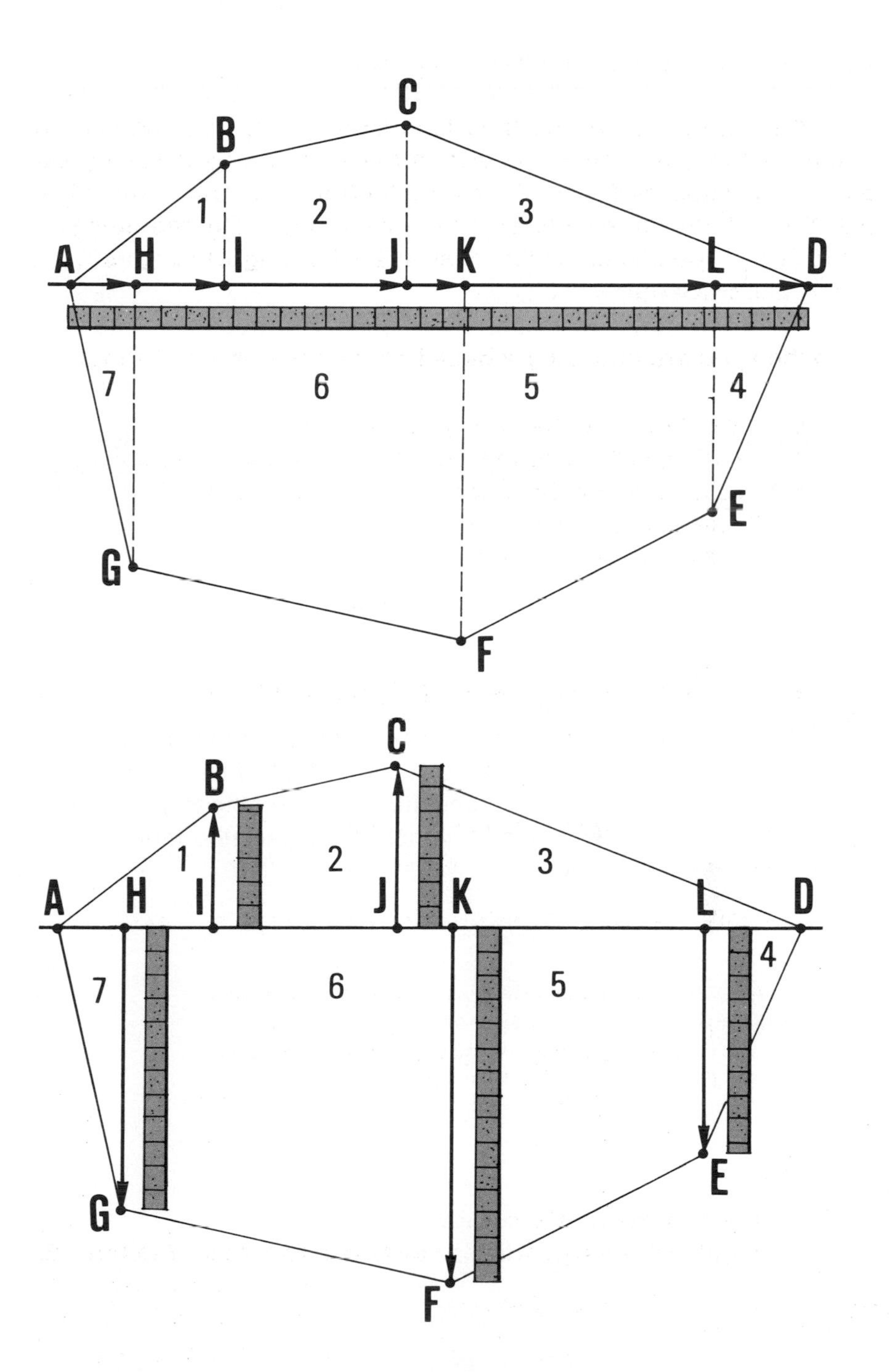

Lot No [1]	Height (m)	Base (m) 1	Base (m) 2	(B¹ + B²) ÷ 2 (m)	Area (m²)
1 TR	5.20	6.50	–	3.25	16.90
2 TP	7.65	5.20	6.20	5.70	43.61
3 TR	6.20	17.10	–	8.55	53.01
4 TR	9.65	4.00	–	2.00	19.30
5 TP	10.50	9.65	14.80	12.22	128.31
6 TP	13.95	14.80	11.80	13.30	185.54
7 TR	11.80	2.80	–	1.40	16.52
Total area					**463.19**

[1] **TR** = right-angled triangle; **TP** = trapezium

Subdividing land areas without base lines

12. When the shape of the land is more complicated than the ones you have just learned to measure, you will have to use more than one base line, and subdivide the area **into triangles**, and **trapeziums** of various shapes. Usually there will be no existing right angle for you to work with and you will have to calculate the area of the trapeziums by taking additional measurements, which will determine their heights along perpendicular lines.

Example

Land tract ABCDEFGHIA along a river is subdivided into five lots 1-5 representing three triangles (1,2,5) and two trapeziums (3 with BE parallel to CD, and 4 with EI parallel to FH). The land boundary forms a closed **polygon**, which has been surveyed as shown.

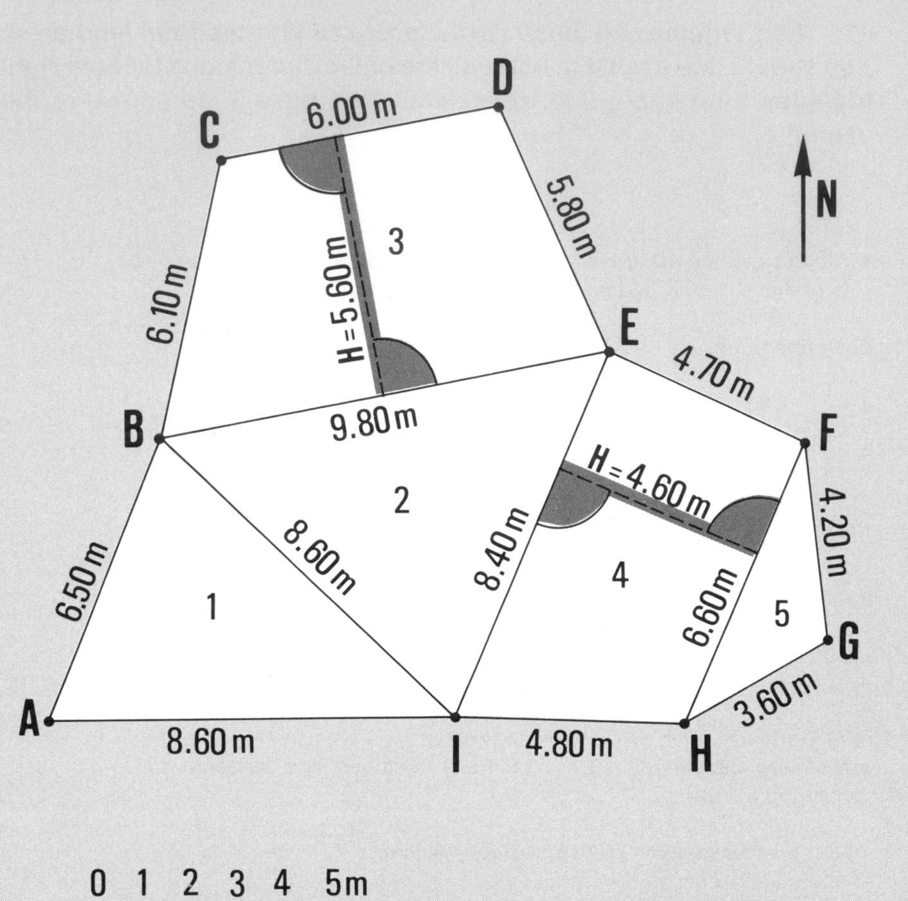

13. Calculate **the areas of triangles** 1, 2 and 5, using the lengths of their three sides and the following formulas:

$\mathbf{s = (a + b + c) \div 2}$
$\mathbf{area = \sqrt{s(s-a)(s-b)(s-c)}}$

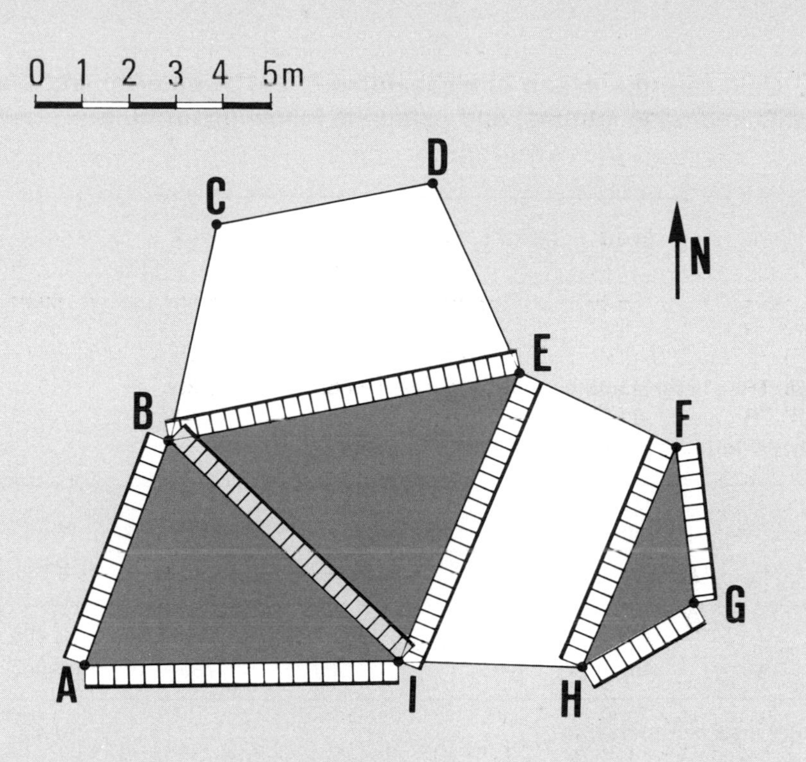

Example

Take measurements of the sides of the triangles, as necessary.

Apply the formula

$$\mathbf{area = \sqrt{s(s-a)(s-b)(s-c)}}$$

in the following table:

Triangle	Length × of sides (m)			s (m)	(s – x) in m			Area (m²)
	a	b	c		(s – a)	(s – b)	(s – c)	
1	650	860	860	1 185	535	325	325	258 773.25
2	860	980	840	1 340	480	360	500	340 258.66
5	660	420	360	720	60	300	360	68 305.16
Total area of triangles								667 337.07

14. Calculate **the areas of trapeziums** 3 and 4, determining their heights and base lengths, and using the following formula:

area = height × (base 1 + base 2) ÷ 2

Example

Measure the heights and bases of the trapeziums, as necessary.

Apply the formula in the following table:

Lot No.	Height (m)	Base (m) 1	Base (m) 2	$(B^1 + B^2) \div 2$ (m)	Area (m^2)
3	560	980	600	790	442 400
4	460	840	660	750	345 000
Total area of trapeziums					787 400

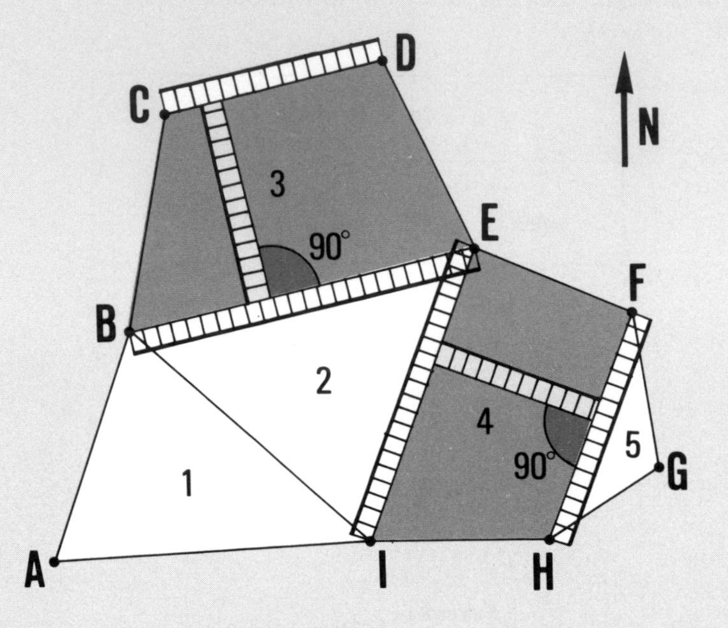

15. Add the total area of the triangles (step 12) to the total area of the trapeziums (step 14) to obtain **total land tract area**.

Example

Total area of triangles	667 337	m^2
Total area of trapeziums	787 400	m^2
Total land tract area	= 1 454 737	m^2
	= **145.47**	**ha**

16. Another way of making the calculations easier is to **measure from a plan the height of each triangle along the perpendicular** laid out from one angle summit to the opposite side (called the base). Then, to calculate each triangle area as:

$$\textbf{area} = (\textbf{height} \times \textbf{base}) \div 2$$

Enter all the data in a single table, as explained in step 11, above.

Example

From a plan, measure heights BJ, BK and LG for triangles 1, 2, and 5, respectively.

Enter all the data in the following table:

Lot No.	Height (m)	Base (m) 1	Base (m) 2	($B^1 + B^2$) ÷ 2 (m)	Area (m²)
1	600	860	–	430	258 000
2	810	840	–	420	340 200
3	560	980	600	790	2 400
4	460	840	660	750	345 000
5	206	660	–	330	67 980
Total area of land tract					**1 453 580**

The total area of the land tract is 145.36 ha, which is slightly different from the previous estimate (see step 15). This was caused by scaling errors when measuring from the plan, which in this case are small enough (0.11 ha or 0.07 percent) to be permissible.

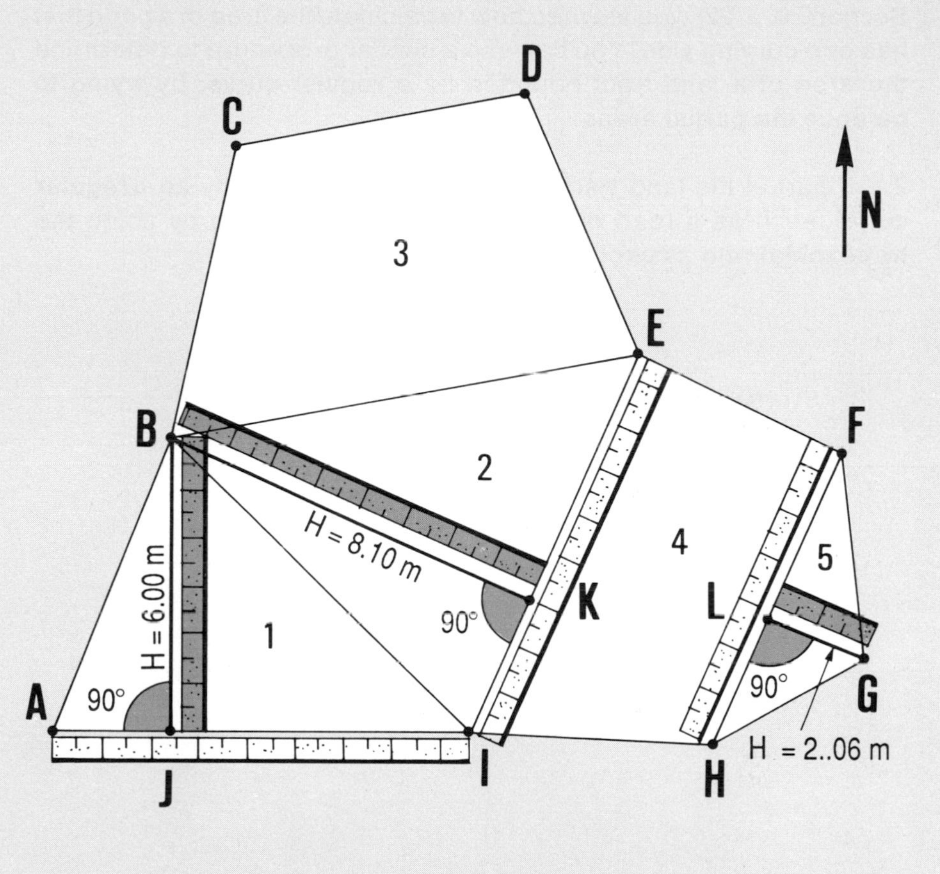

105 How to measure areas bounded by a curve

1. In Volume 4 of this series, **Water for Freshwater Fish Culture** (see Section 20, p. 22), you learned how to calculate the area of a pond that has one curving side. You can use a similar procedure to determine the area of a land tract bounded by **a regular curve**, by trying **to balance the partial areas**.

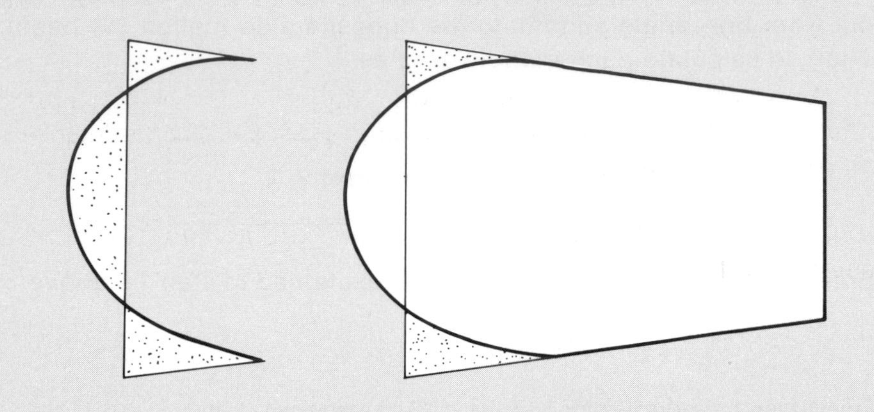

2. If part of the land tract is bounded on one side by **an irregular curve**, such as a road or river, you can find its area by using **the trapezoidal rule** as explained in this section.

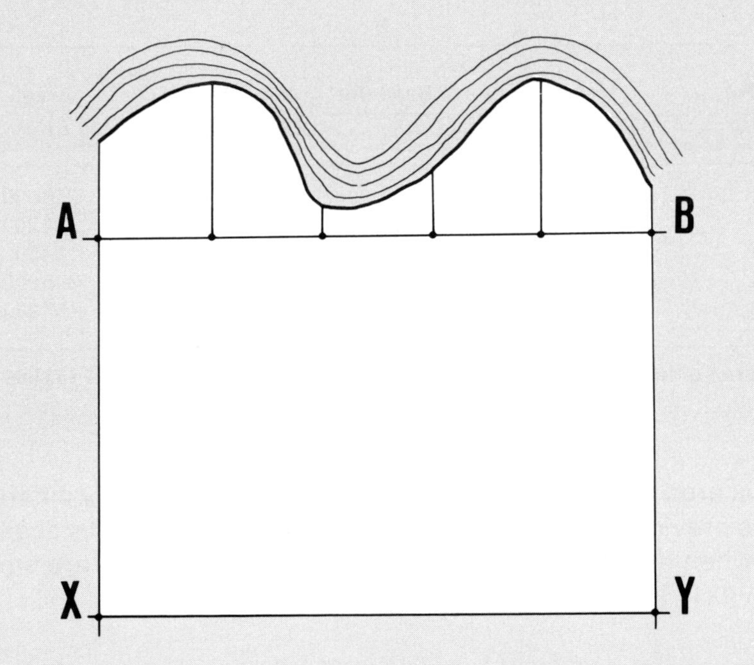

3. Set out **straight line AB** joining the sides of the tract of land and running as closely as possible to the curved boundary. To determine the irregular area ABCDA, proceed as follows.

4. Measure distance AB and subdivide it into a number of **regular intervals**, each, for example, 22.5 m long. Mark each of the intervals on AB with ranging poles.

Note: the shorter these intervals are, the more accurate your area estimate will be.

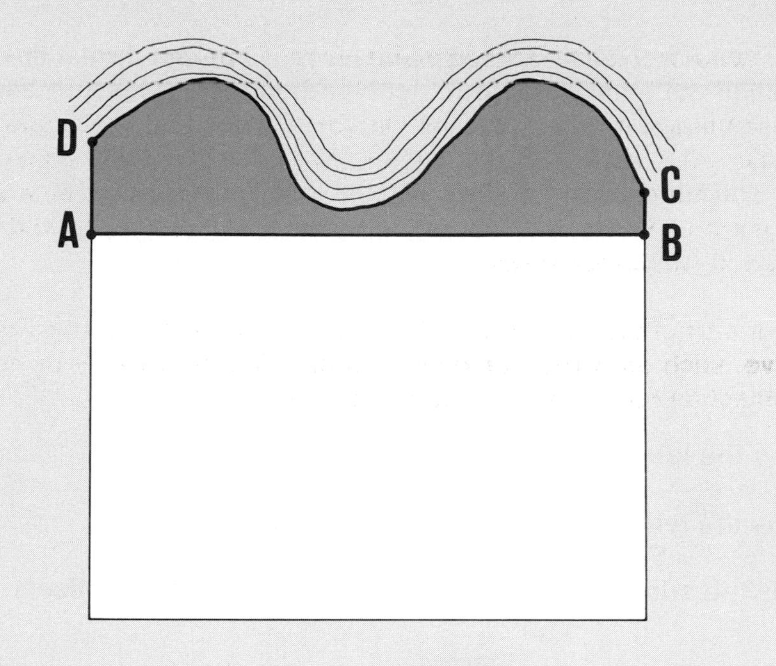

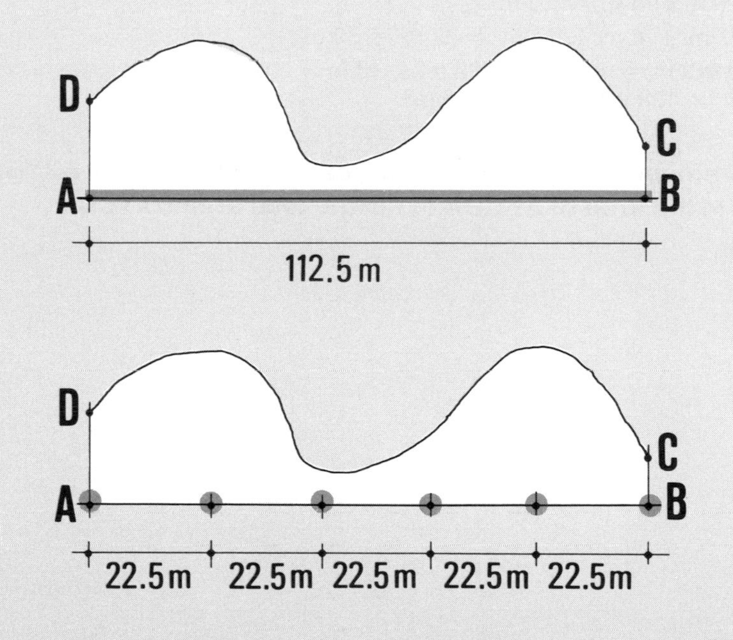

5. At each of these marked points, set out **a perpendicular line** (see Section 36) joining AB to the curved boundary. Measure each of these offsets.

6. Calculate area ABCDA using the following formula:

Area = interval × (h_o + h_n + 2h_i) ÷ 2

where:

h_o is the length of the **first offset**, AD;

h_n is the length of the **last offset**, BC; and

h_i is the sum of the lengths of all the **intermediate offsets**.

Example

Interval = 112.5 m ÷ 5 = 22.5 m
h_o = 20 m and h_n = 10 m
h_i = 27 m + 6 m + 14 m + 32 m = 79 m
Area ABCDA = 22.5 m × (20 m + 10 m + 158 m) ÷ 2 =
(22.5 m × 188 m) ÷ 2 = 2 115 m²

Note: remember that you must still calculate the area of AXYBA and add it to the area of ABCDA to get the **total area DXYCD**.

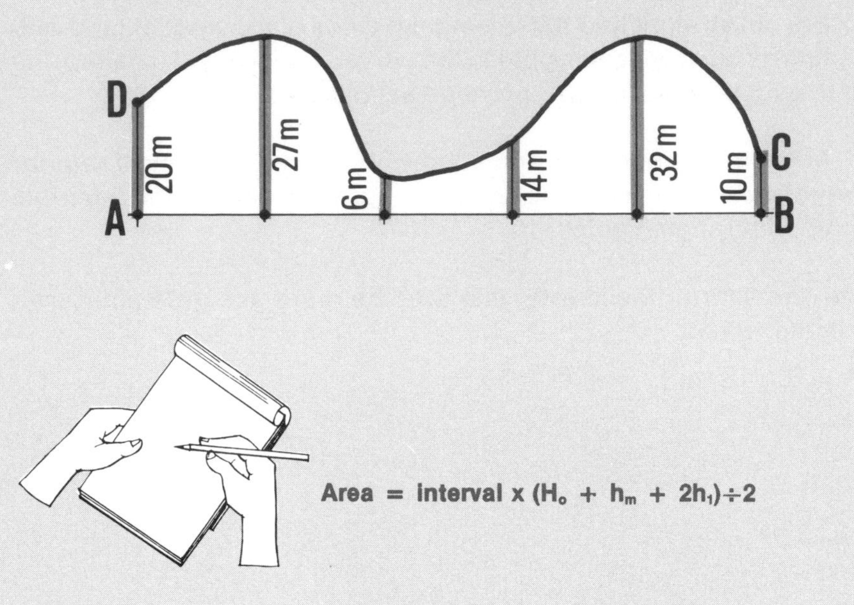

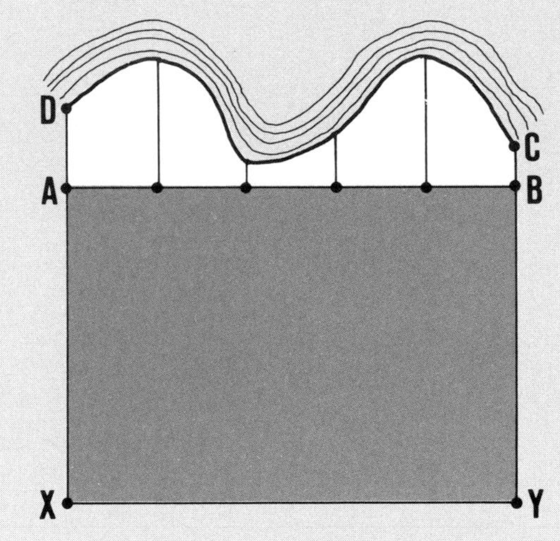

7. If you can lay out **line AB** so that it touches the two ends of the curved boundary, your calculations will be much simpler. In this case, h_o and h_n are both equal to zero, and the formula becomes:

$$\textbf{Area} = \textbf{interval} \times \mathbf{h_i}$$

where h_i is the sum of the lengths of all the intermediate offsets.

Example

Interval = 158 m ÷ 6 = 26.3 m
h_i = 25 m + 27 m + 2 m + 23 m + 24 m = 101 m
Area = 26.3 m × 101 m = 2 656.3 m²

Note: remember that you must still calculate the area of AXYBA and add it to the area of the curved section to get the **total area**.

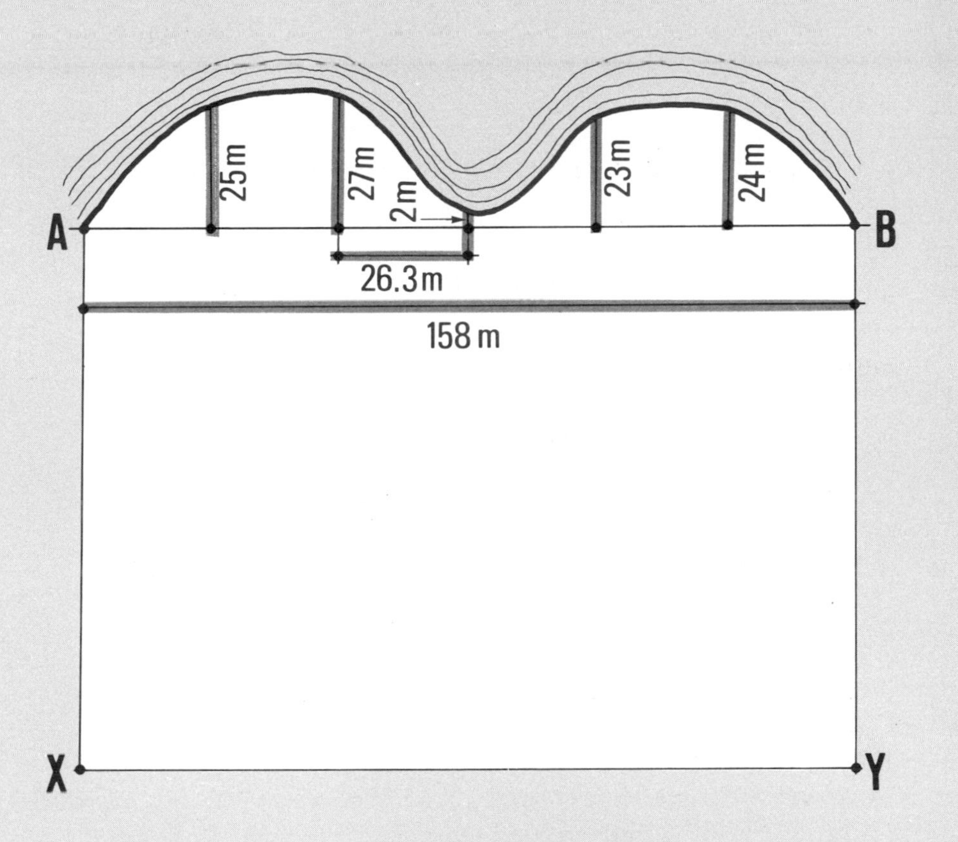

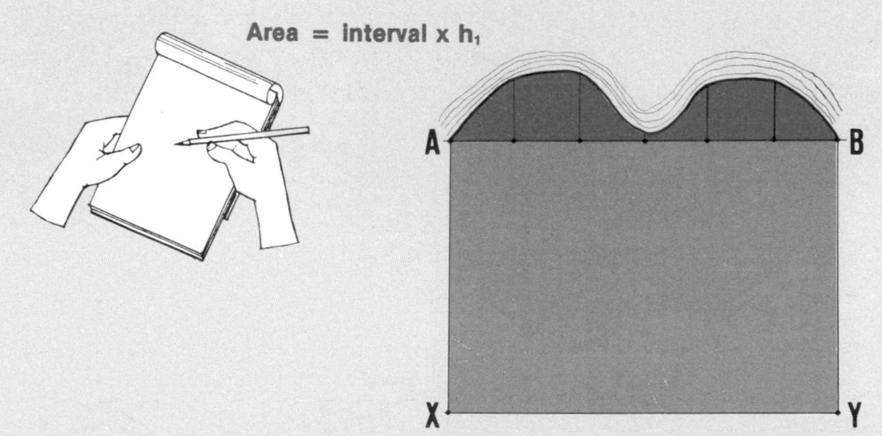

11 TOPOGRAPHY AND FRESHWATER FISH CULTURE

110 What you have learned

1. In the previous chapters of this manual you have learned:

- how to measure the various parameters, or fixed values, which describe the topography of a tract of land, such as distances, horizontal and vertical angles, and differences in elevations;
- how to make a topographical plan survey;
- how to survey the local relief by direct levelling;
- how to determine contours in the field;
- how to prepare topographical plans and maps; and
- how to measure areas, both in the field and from plans or maps.

2. On the basis of this new knowledge, you should now be able to:

- **choose a suitable site** for the construction of a small reservoir or freshwater fish-ponds;
- **design** your fish-farm **and plan** its construction.

3. You will learn more about **site selection** and about **fish-farm design and construction** in the next volumes of **Simple Methods for Aquaculture**. In the following sections, some of this information will be briefly discussed to give you a better idea of how you can use topographical surveys to design and build your fish-farm. You should also refer to some of the examples presented in each of the previous chapters. You can use them as guidelines or suggestions for your own fish-farm project.

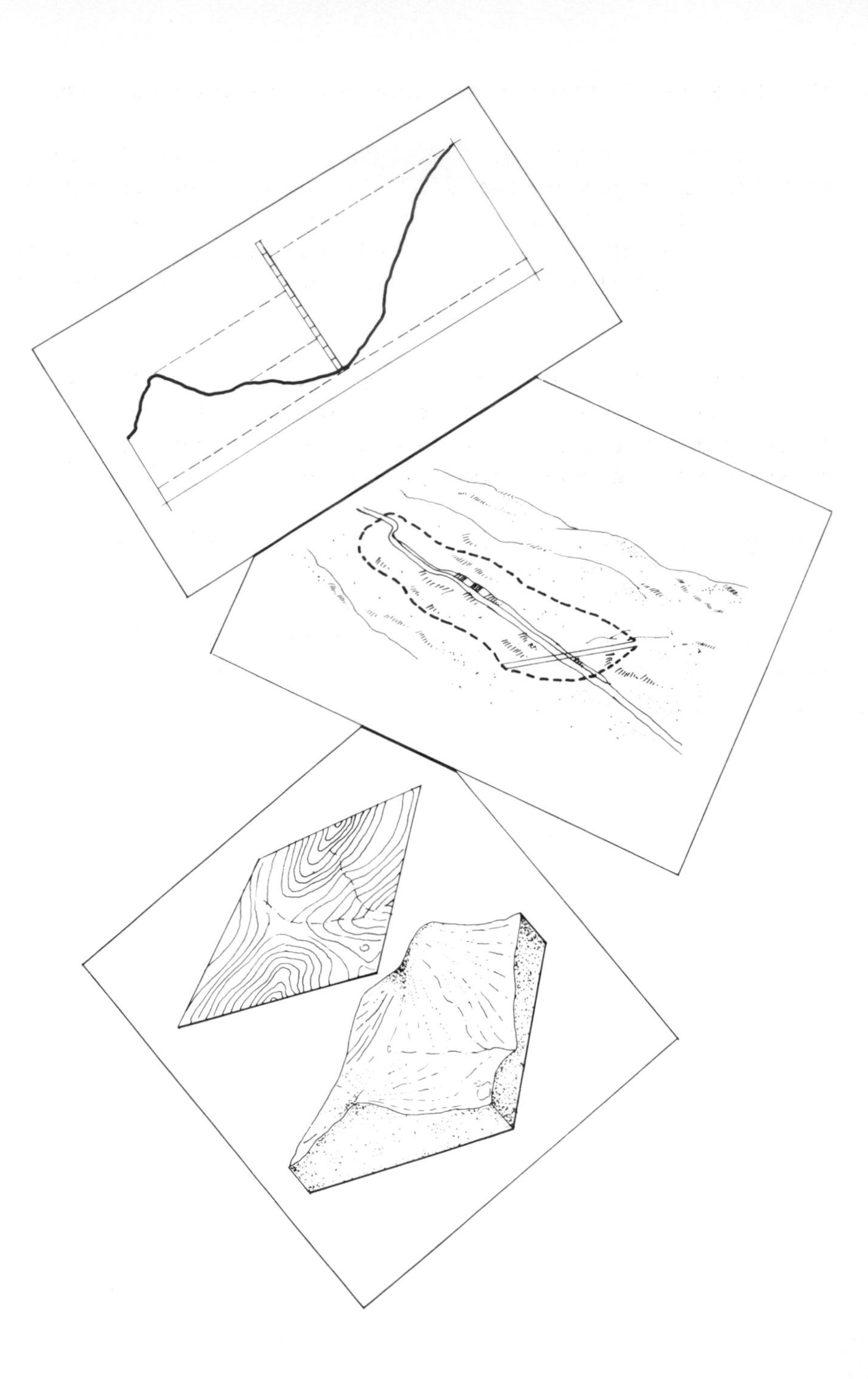

111 How to make preliminary studies from topographical maps

1. Before you begin a field survey, you will usually need to study **available topographical maps** to get information for various projects that you want to do, and to compare these projects and the advantages and disadvantages of each. You can then more easily choose the order in which you will make the reconnaissance field surveys.

2. Some of the most useful things you can learn from a topographical map are the **size of a drainage area**, the **size of a flooded area**, the **characteristics of selected ground profiles** and the **distribution of slope categories** in a given area.

Finding the size of the drainage area and the availability of water for fish culture

3. You have learned (in Volume 4, **Water**, pp. 12-13) that **the catchment basin** of a stream is the total land area which feeds water to that particular stream.

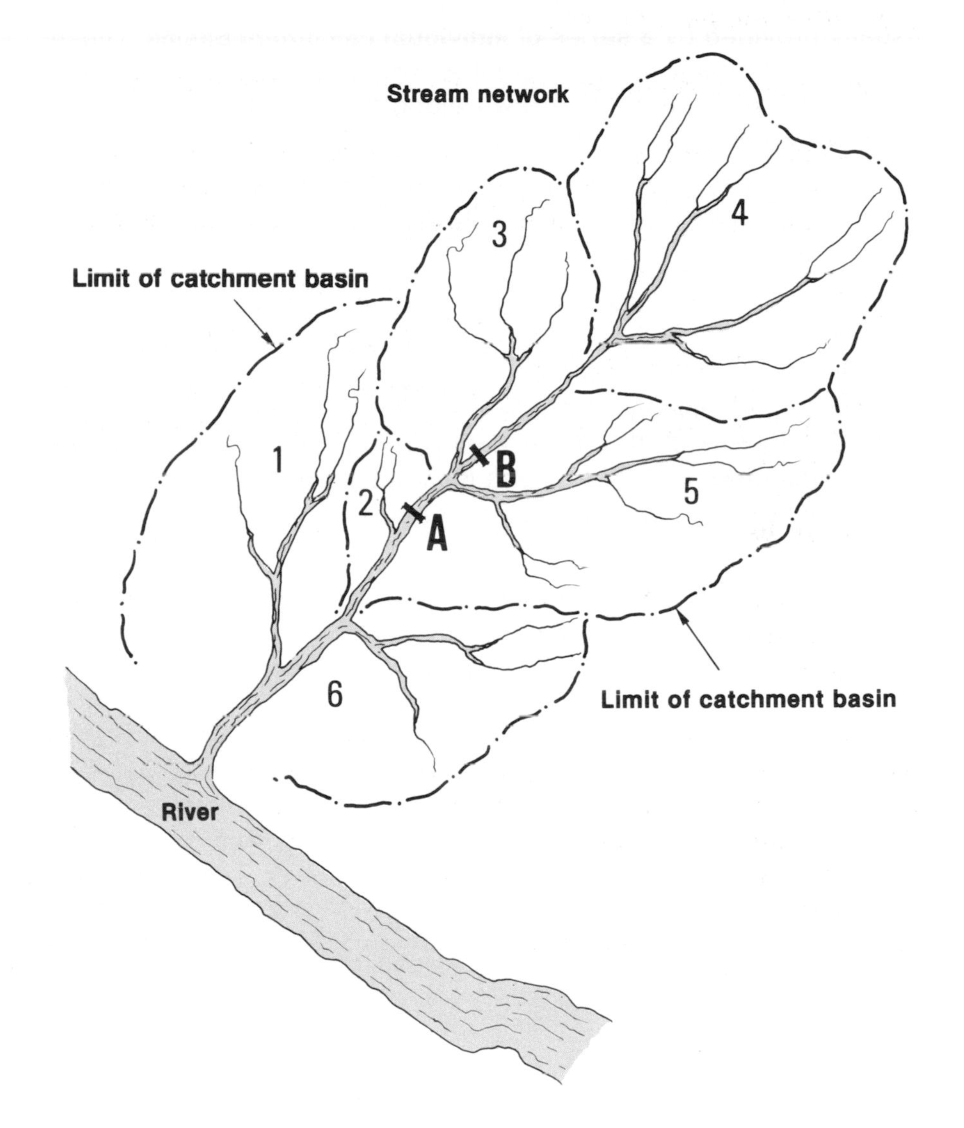

4. At a particular point A on a stream, the total water available is usually provided by **a series of individual catchment basins**. These define **the drainage area** for point A. This area is bounded by **the divide line**, which is a line drawn along the ridges surrounding a drainage area.

5. If you have a topographical map of the region, you can draw **the divide line** and define **the drainage area** for each point you choose along a stream (dotted lines). You may want to obtain the water supply for your fish-farm, which will be built downstream, at point A. Starting from point A, draw a line **perpendicular to the contour lines** on either side of the stream bed, until you reach the points B and C with the **highest altitude**. Then, join B to D, using the same method, and continue from E, F, ... I, etc. until you reach point C on the other bank of the stream. The area you have enclosed by the divide line ABC ... SCA is the drainage area for point A on the stream.

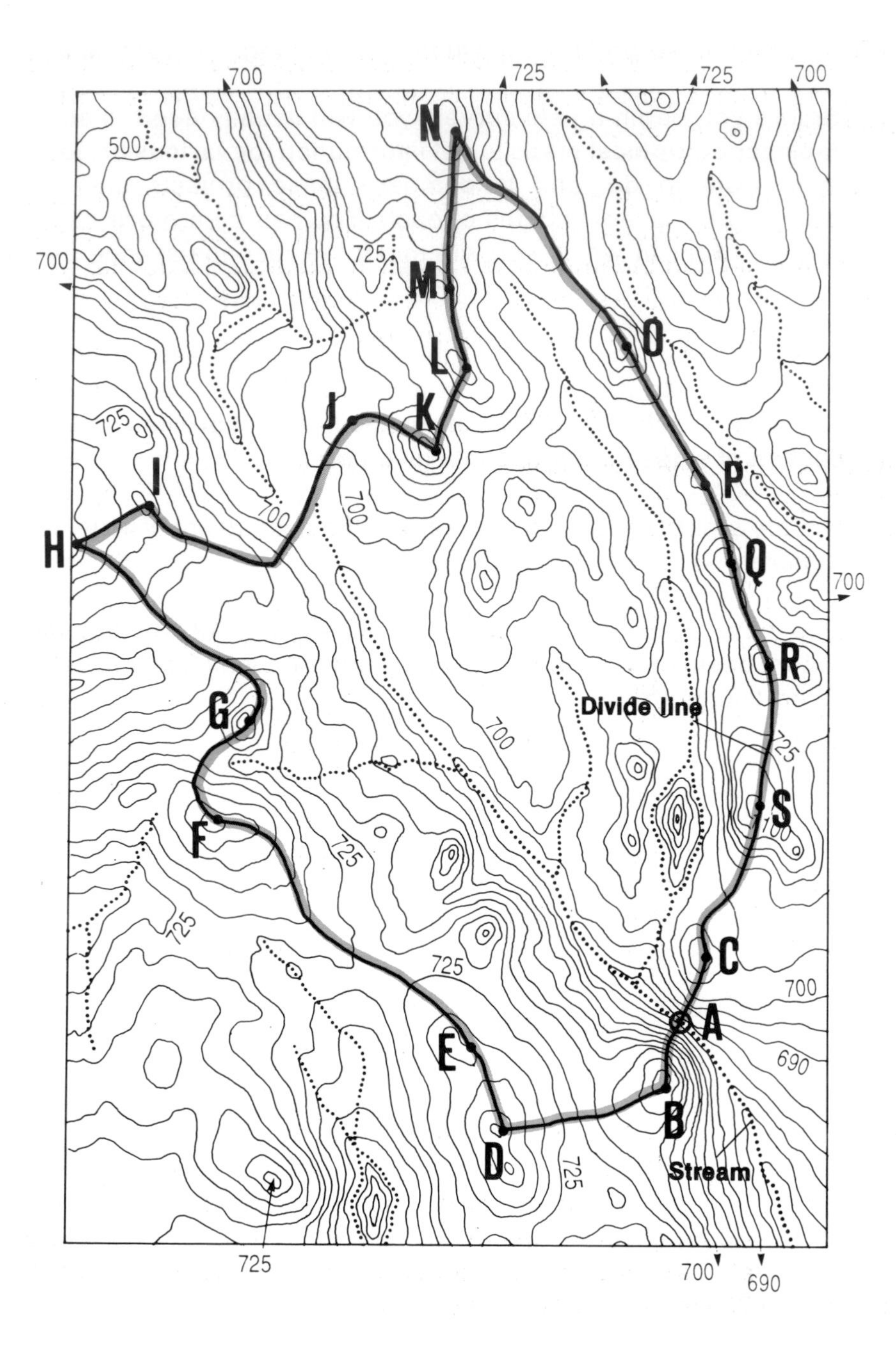

6. Using the square-grid method, preferably (see Section 103), find the area of the region enclosed by the divide line to find **the size of the drainage area** for stream point A.

7. You can estimate **the quantity of water available** at point A from the size of the drainage area, the most common vegetation there, the general relief and the amount of rainfall in the area. To do this, you need to obtain the values of the local **runoff coefficients** from a government office, such as the hydrological service.

Example

Water availability per square kilometre of drainage area in the Bouaké region, Côte d'Ivoire, Western Africa

Annual rainfall (mm)	Runoff coefficient (percent)	Runoff[1] (mm)	Water availability[2] (m^3/km^2)
800	0.9	7	7 000
900	2.8	25	25 000
1 000	4.7	47	47 000
1 100	6.6	72	72 000
1 150	7.5	87	87 000
1 200	8.5	101	101 000
1 300	10.4	135	135 000
1 400	12.3	172	172 000
1 500	14.2	212	212 000
1 600	16.1	257	257 000
1 700	18.0	305	305 000

[1] Runoff (mm) = annual rainfall × (runoff coefficient ÷ 100)

[2] Water availability (m^3/km^2) = runoff (mm) × 1 000

As the **average annual rainfall** in this region amounts to 1 150 mm, you can estimate the **average water availability** to be about 87 000 m^3/km^2 of drainage area for this particular region of the country. If the calculated drainage area upstream from point A is 2.72 km^2, the average water availability at point A can be estimated as 87 000 m^3/km^2 – 2.72 km^2 = 236 640 m^3 per year.

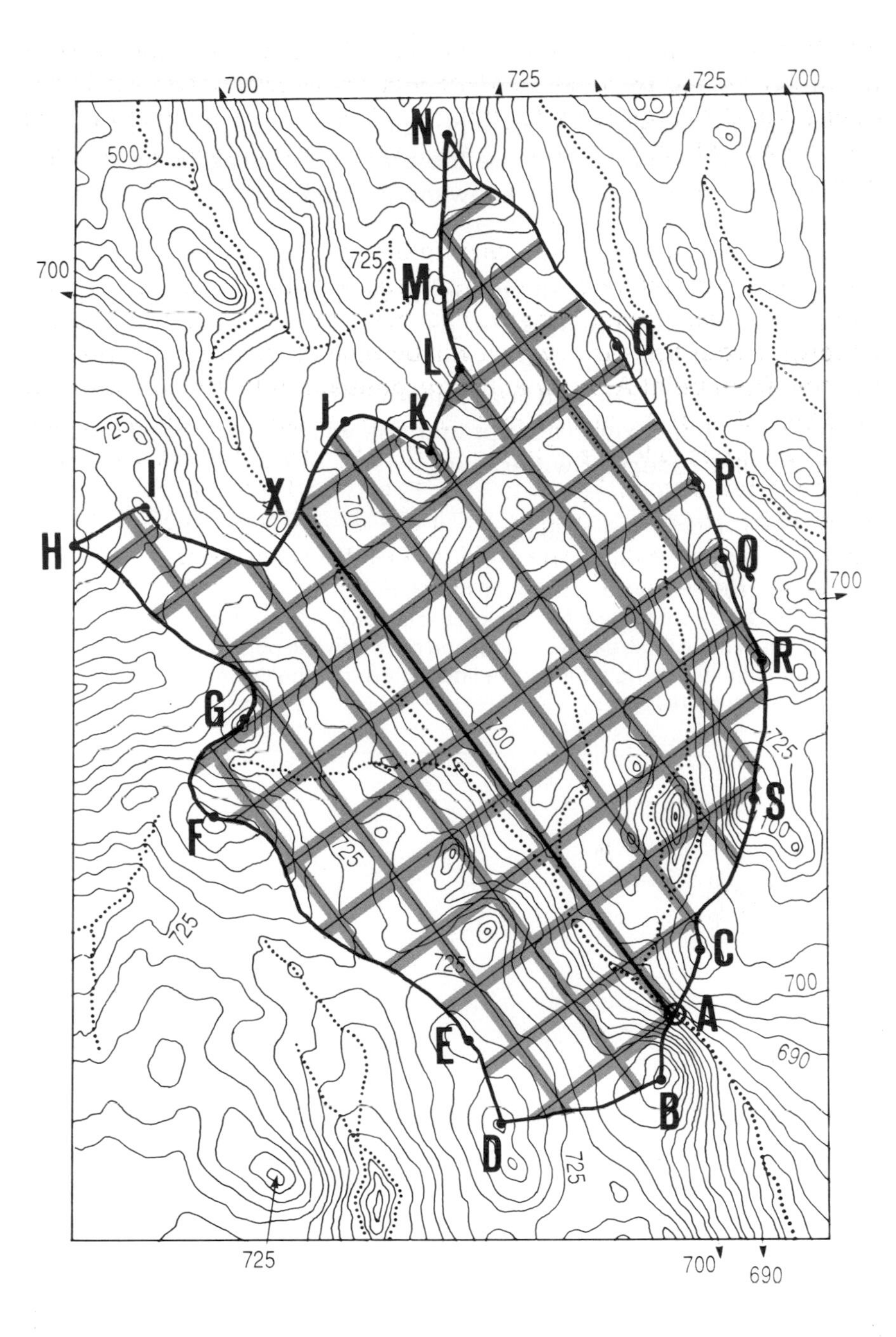

Finding the size of the area to be flooded

8. If you need to build a dam at point A to create a water reservoir, you can fairly easily determine **the size of the area to be flooded** upstream from point A, if you know **the elevation of the water surface** in the reservoir.

9. At the dam site, identify **the contour line** on the topographical map which, on one side of the stream, corresponds to the elevation of the reservoir's water surface. Follow this contour line first upstream, then across the stream, and finally downstream, back to the dam site. The area enclosed within this contour line will be **the flooded area** for that particular reservoir water level.

Example

Elevation of reservoir water surface level: 690 m
Follow the 690 m contour line from point B at the dam site, on the left bank, up to point C (stream crossing) and back down to point D at the dam site, on the right bank. The flooded area at the 690 m water level will be BCDB.

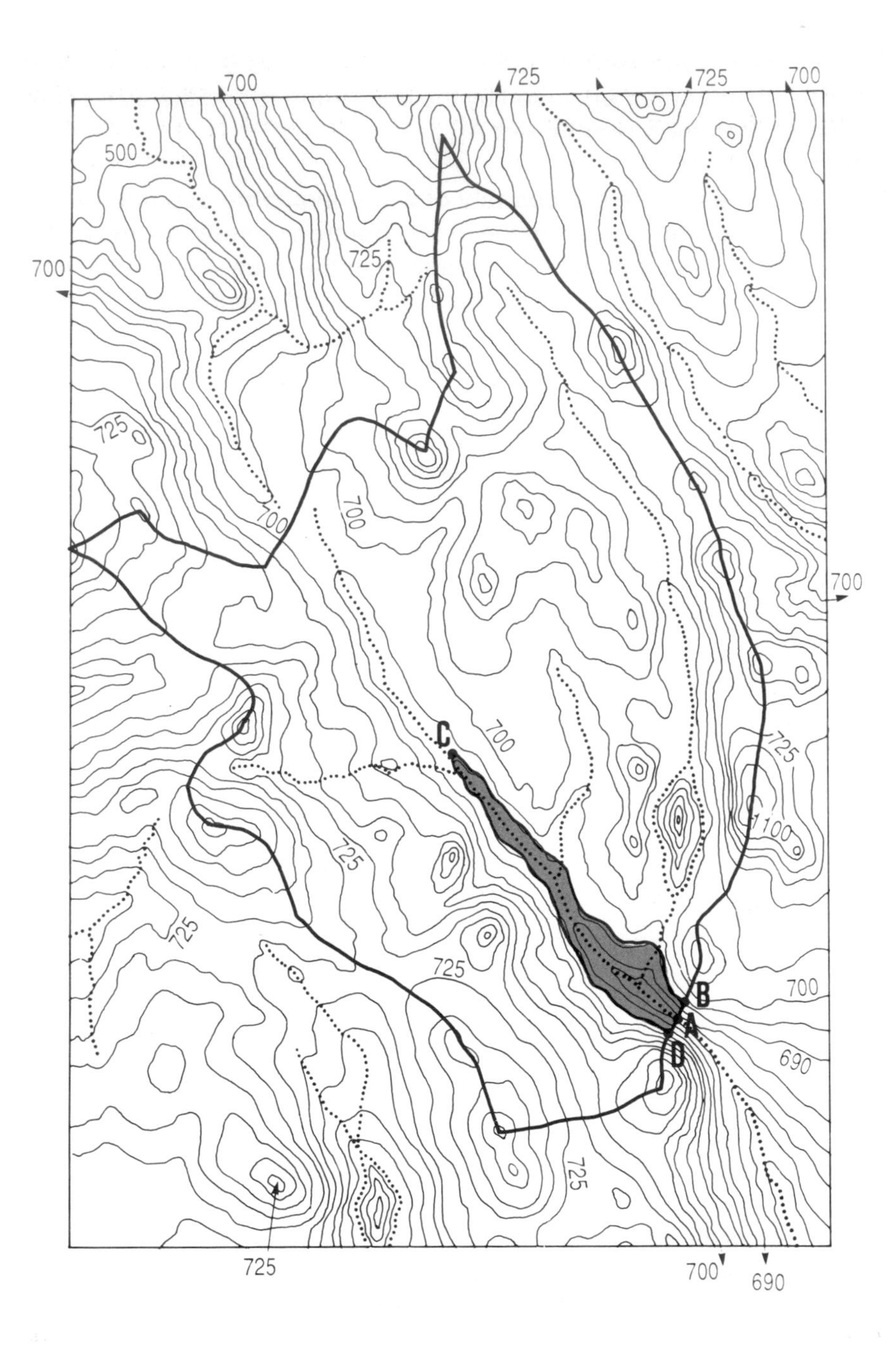

10. Using the square-grid method, preferably (see Section 103), find the area of the zone within the selected contour line to obtain **the size of the flooded area**. If you know the average depth of the reservoir, you can then calculate the volume of water which is stored in it.

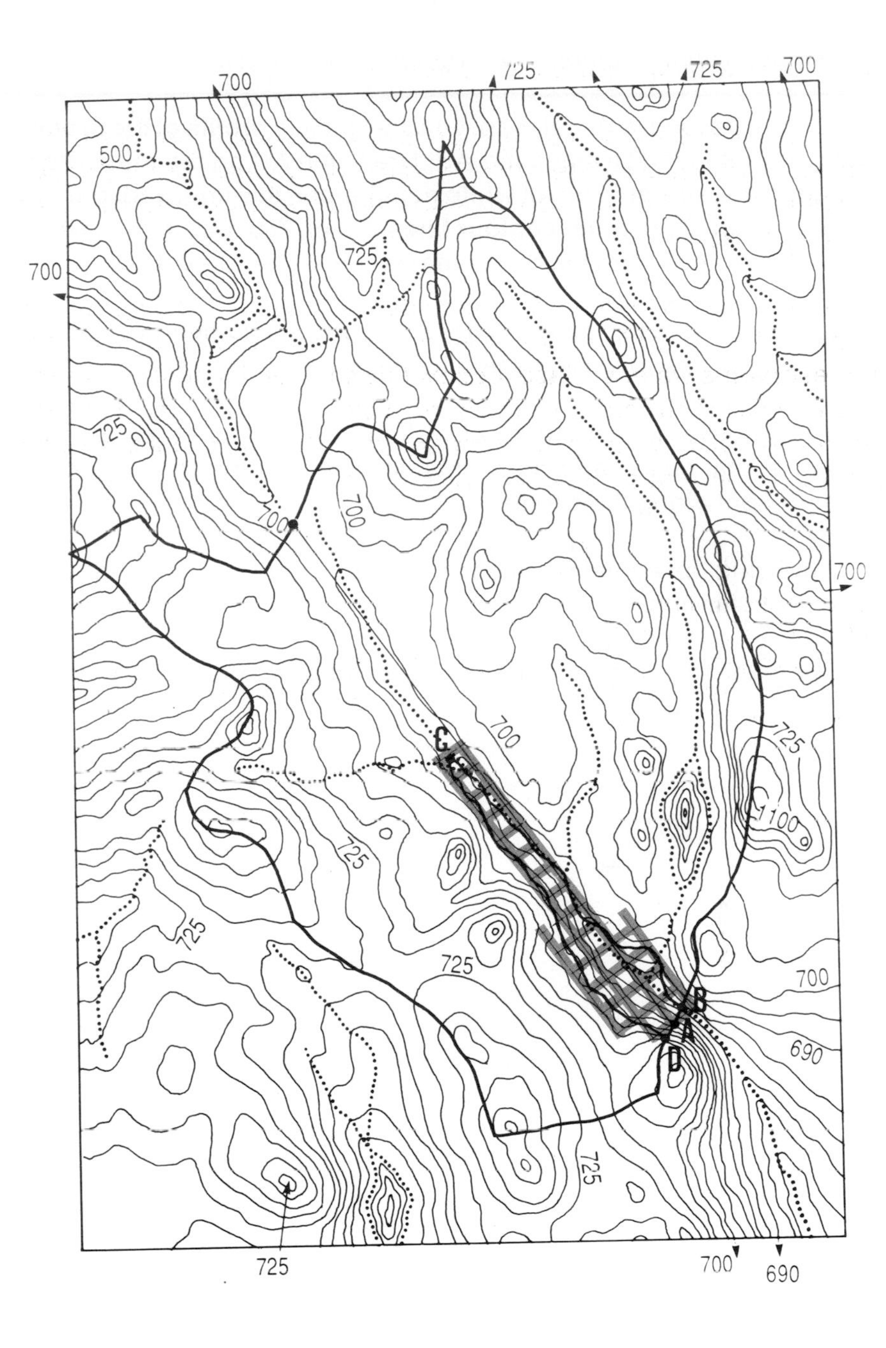

Obtaining ground profiles from topographical maps

11. From a topographical map, you can find the **profile of the ground** along any selected straight line AB. Draw line AB on the map. Place one edge of a straight strip of white paper along this line, and mark on it the position of **the main contour lines** 775 m, 750 m . . . 675 m which line AB intersects. Next to these marks, note their elevations.

12. Transfer these marks onto square-ruled millimetric paper, using a **horizontal distance scale** identical to the map scale.

13. Make **a vertical scale** for elevations 10 to 30 times larger than the horizontal scale, and mark this scale according to the contour lines present along the profile.

14. Indicate the elevation of each distance mark by a point along a perpendicular line.

15. Join these points to obtain **the ground profile** along line AB.

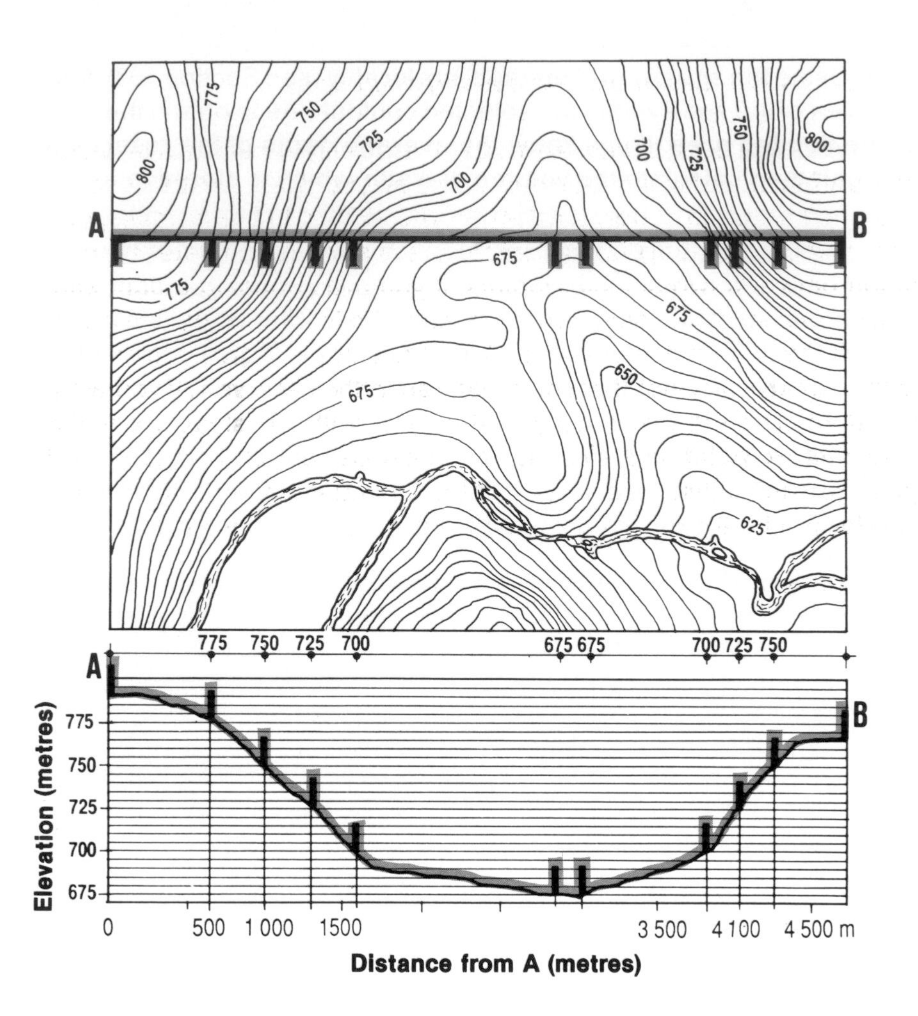

Profile along line AB

Finding differences in elevation along a stream

16. You may need to choose a site on a stream to build a dike which will create a water storage reservoir, or you may want to use an existing stream to supply a fish-farm with water. In these cases, you should study **the longitudinal profile of the stream** and determine its slope between two selected points. If you are choosing a dam site, you should study the slope from the dam site to the highest elevation (upper reach) of the future reservoir. This study will give you an idea of the **volume** of **water that can be stored in the reservoir**. If you want to divert a stream to supply water, you will **the slope from the water-intake point** on the fish-farm to its water outlet. This study will give you the difference in elevation available for you to build a fish-farm on the site which lies between these two points.

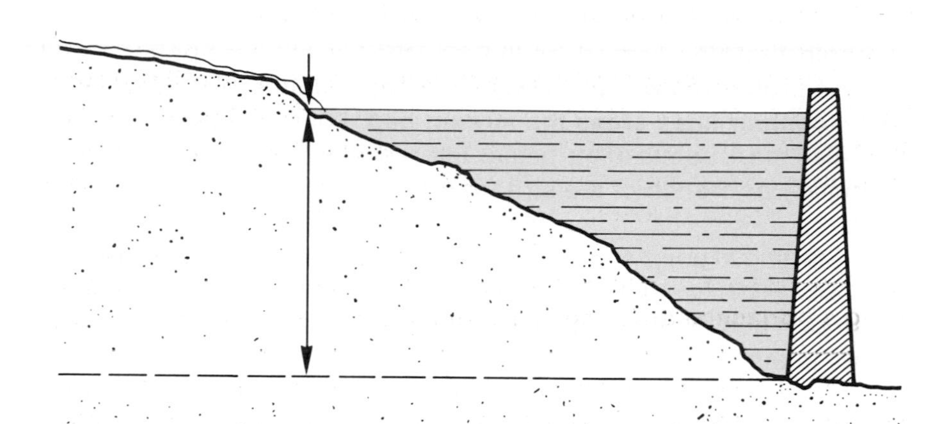

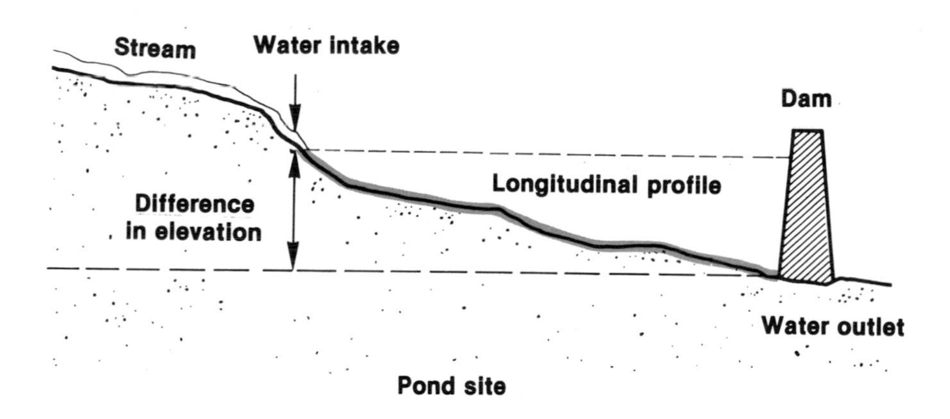

17. To determine the slope of the stream between A and B, for example, first clearly mark on the map the two extreme points A and B of the profile you want to study. Then mark points C, D, E and F, at which **contour lines cross the stream bed**. For later reference, you may also mark points of particular interest along the stream, such as a tributary or branch of the stream (G, H), or a road bridge.

18. Starting at point A, **measure** to the millimetre, **the distances** AG, GC, CD ... FB between these various marked points, closely following the stream bed as you do so. Enter these measurements in a table as shown in the example.

19. Using **the map scale**, transform these map measurements into **ground distances** (in metres), and calculate **the cumulative distances** from point A, as shown in the table in the example.

20. From the contour lines of the map, determine **the elevations** of points A, G, C, D, H, E, F and B, and enter these figures in the table. As you are working downstream, these altitudes should steadily decrease by **a constant value equal to the contour interval of the map**.

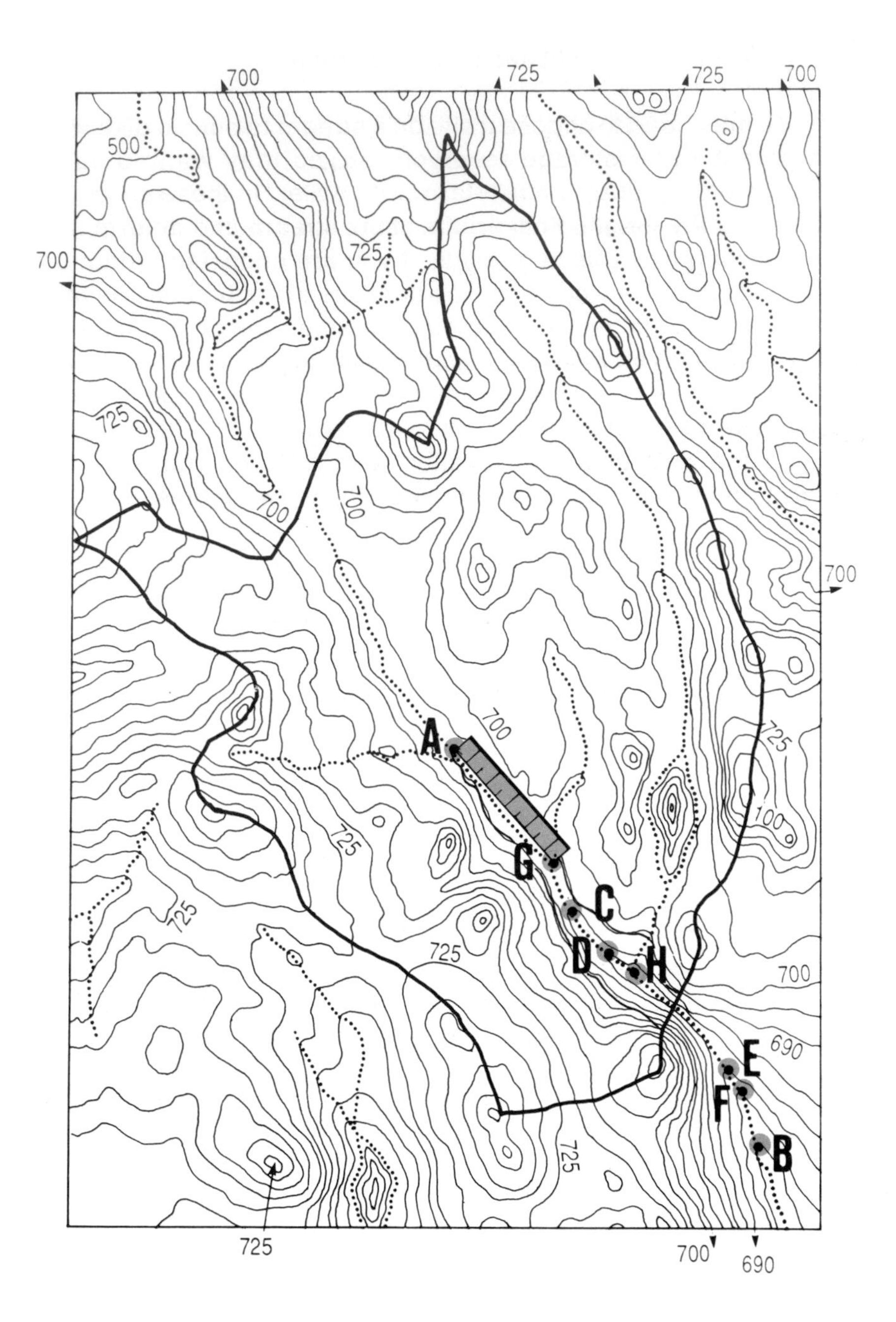

Example

Longitudinal profile of stream section AB

Map scale: 1 cm = 200 m (1:20000); Contour interval = 5 m

Stream point	Map distance (cm)	Ground[1] distance (m)	Cumulative distance (m)	Elevation[2] (m)
A			0	690
	2.9	580		
G			580	(tributary)
	0.7	140		
C			720	685
	1.1	220		
D			940	680
	0.4	80		
H			1 020	(tributary)
	2.1	420		
E			1 440	675
	0.5	100		
F			1 540	670
	1.1	220		
B			1 760	665

[1] Ground distance (m) = map distance (cm) – map scale (m/cm).
[2] From contour lines with contour interval = 5 m.

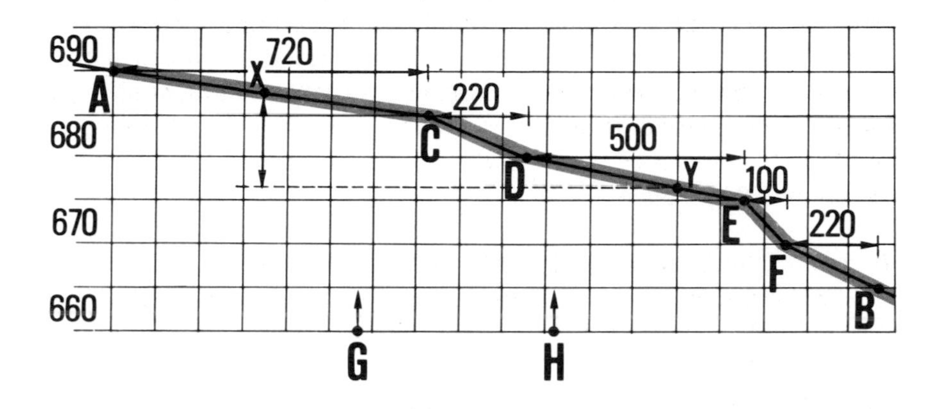

Longitudinal profile AB

21. Draw **the longitudinal profile** of the AB stream section as explained earlier (see Section 95), using the information you have gathered. To its horizontal scale, add the positions of the additional points of interest so that you can refer to them later.

22. From this longitudinal profile, you can now easily determine the difference in elevation existing between any two points X and Y of the stream within this section AB.

Determining the shape of stream valleys

23. Using a method similar to the one described in steps 11-15, you can also determine the general shape of a stream valley. To do this, you will draw **cross-section profiles** perpendicular to the stream bed, at points of interest to you. These points will depend on the purpose of your survey. If you are planning to build **a small dam**, you will draw cross-section AB. If you are looking for a suitable **fish-farm site**, you will draw cross-section CD.

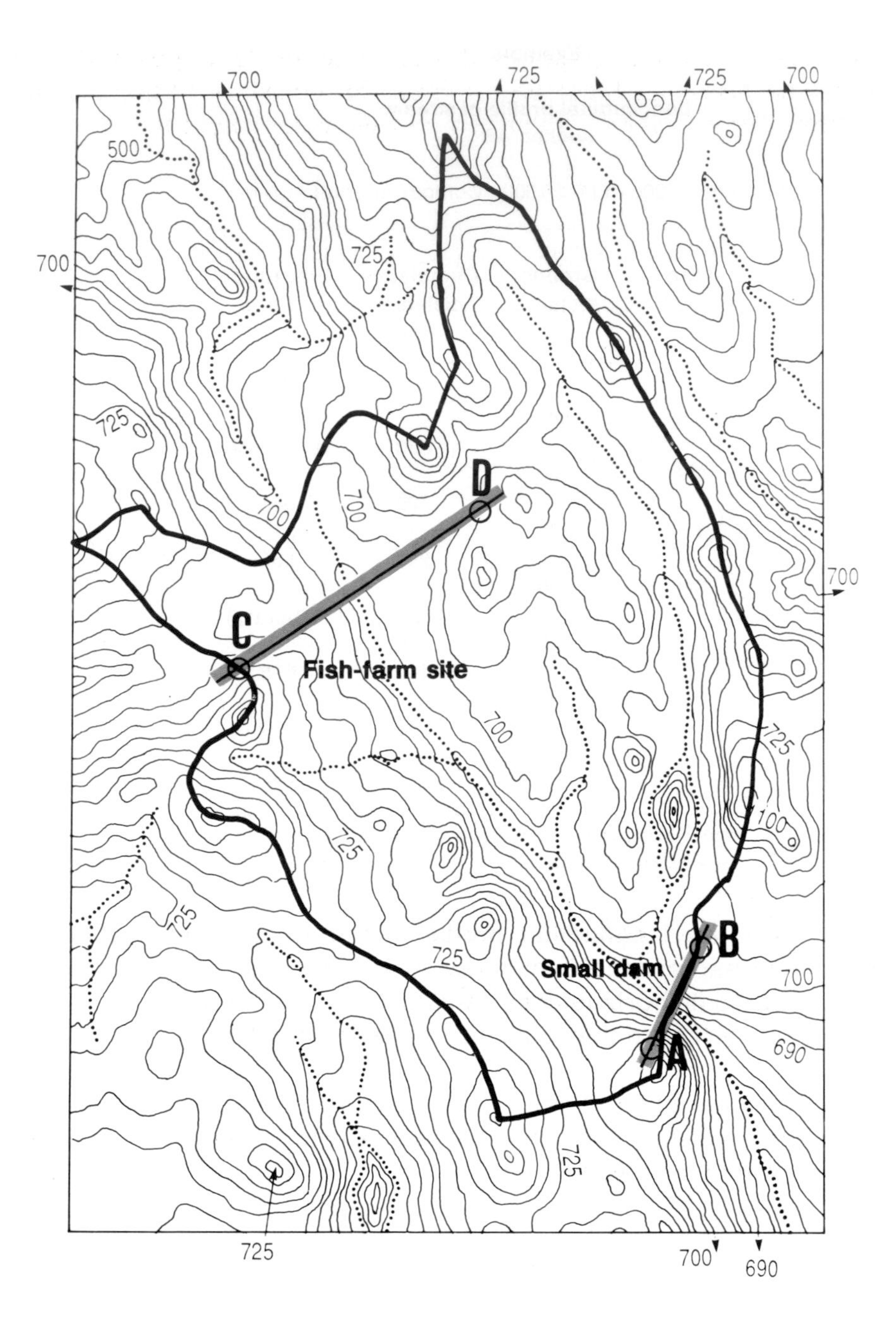

24. Get a strip of paper and align one of its edges with the cross-section line. Mark on the strip the positions of the various **contour lines**, together with a few elevations for reference.

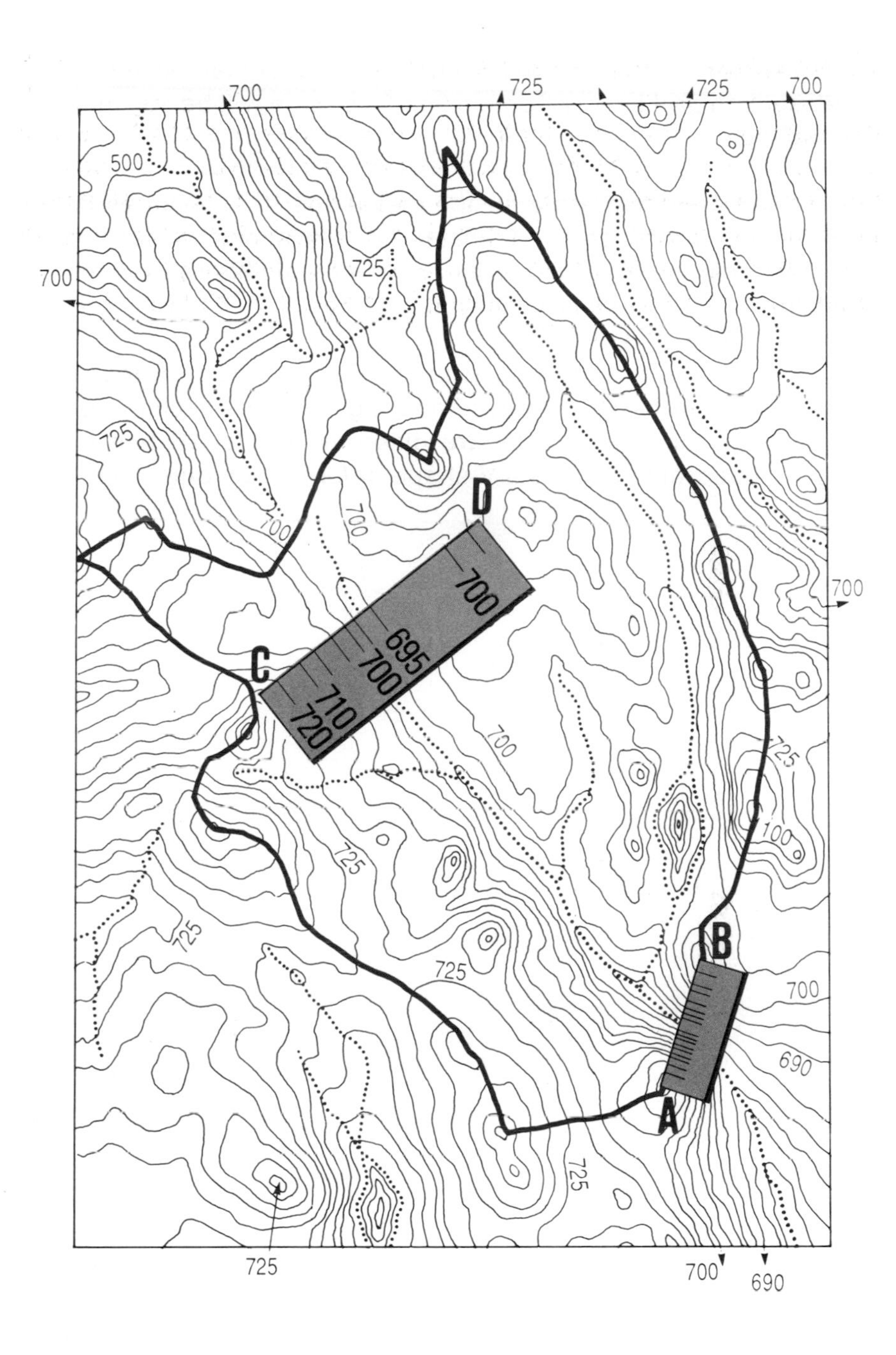

25. Transfer these marks onto the **horizontal scale** of a cross-section profile (see Section 95). This scale will be the same as the map scale.

26. Select a **vertical scale** for the elevations 10 to 30 times larger than the horizontal scale. Transfer the elevations of each mark vertically onto the graph. The line joining the points represents **the cross-section profile of the valley along line AB**.

27. If you repeat this procedure for line CD, you will draw **the valley cross-section profile CD**. When comparing it to cross-section AB, you can see that the two profiles are different. Profile AB has a true V-shape, but profile CD has a V-shape deformed on one side.

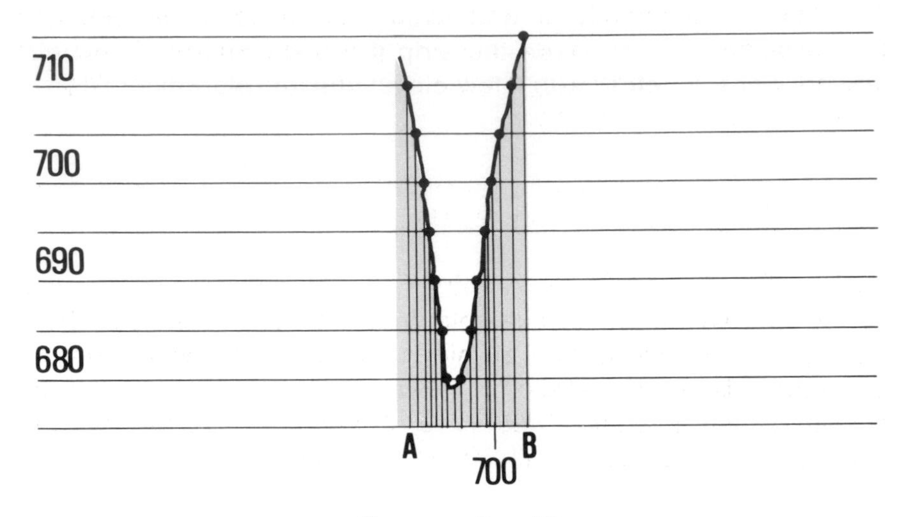

Cross section AB

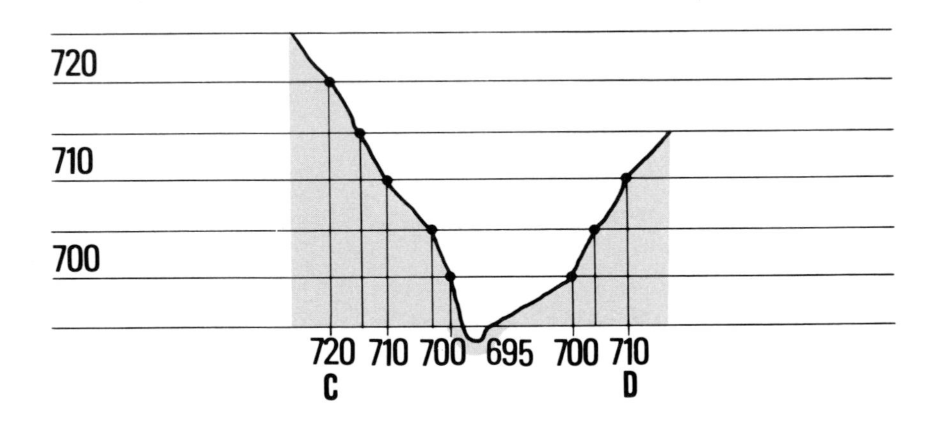

Cross section CD

28. In the context of fish culture, cross-section valley profiles may be classified into **four types**, according to their shape. When you know the shape of the valley in the area where you want to build your fish-farm, you will be able to:

- decide which type of pond to build;
- design your fish-farm better.

This information will be explained to you in the next volume in this series. You will learn, for example, that site AB above would be a good place to build a dam at minimum cost, but it would not be suitable for fish-ponds. At site CD, however, the XYZ side of the valley shows a lateral slope (1.25 to 2.17 percent) suitable for the construction of fish-ponds (see step 27, above).

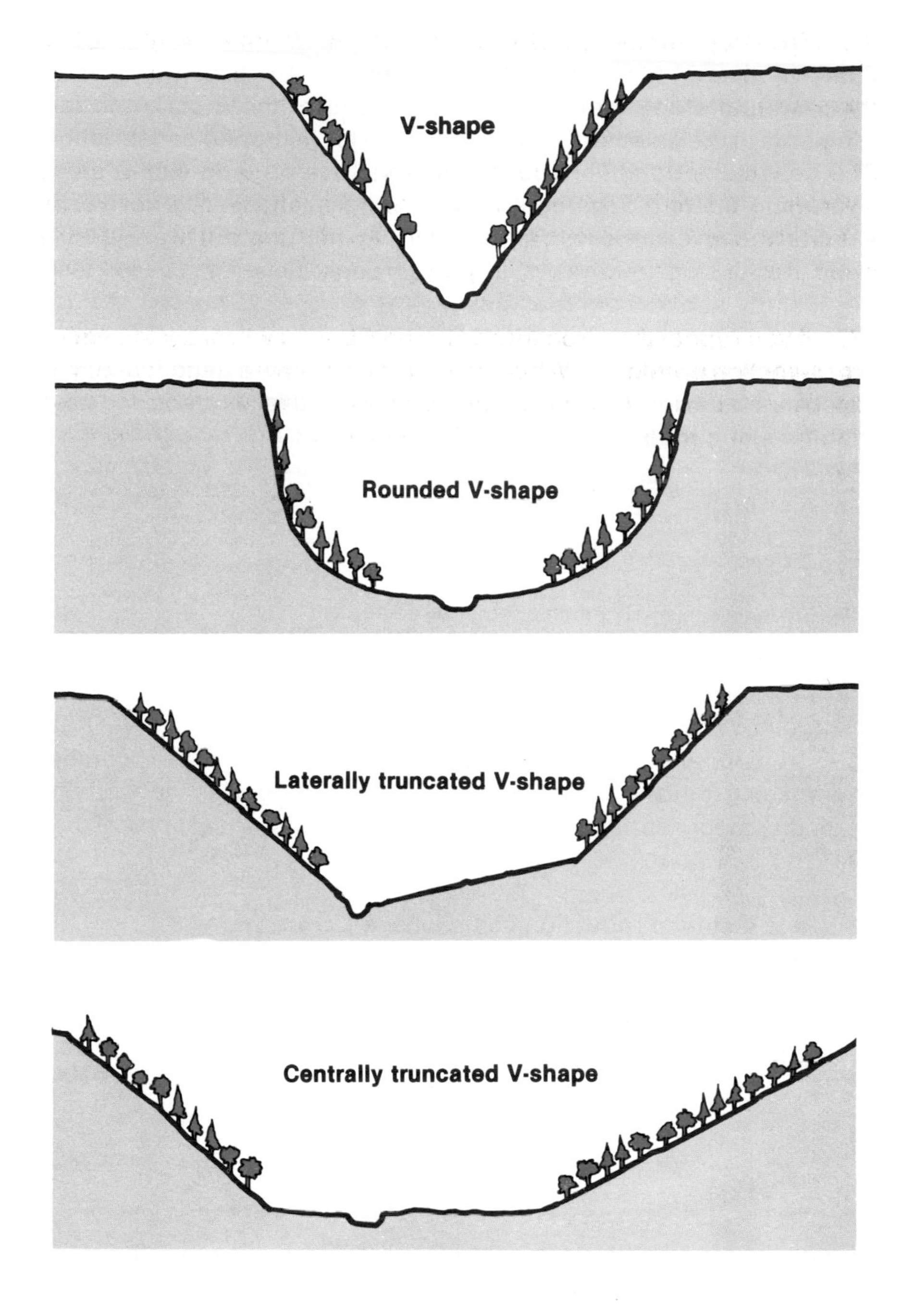

Making a slope map from a topographical map

29. **Ground slope** (see Section 40) is one of the most important elements in the selection of a suitable site for fish-pond construction. The best slope conditions for a fish-pond are on land with a **slope averaging 0.5 to 1.5 percent**, but conditions on near-to-horizontal ground and on ground with slopes from 1.5 to 3 percent are still fairly good. As the slope increases, the cost of construction increases also, particularly above 5 percent slope.

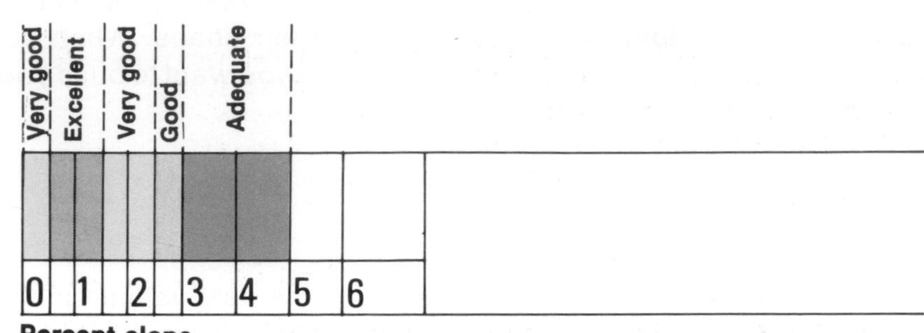

Usefulness of slopes

30. When you study a topographical map, you may find it useful to determine **a series of slope categories** on it. In this way, you will have made a **slope map**.

Example

Ground slope categories useful in fish culture

A — ground slope smaller than or equal to 1.5 percent
B — ground slope between 1.5 and 3 percent
C — ground slope between 3 and 5 percent
D — ground slope greater than 5 percent

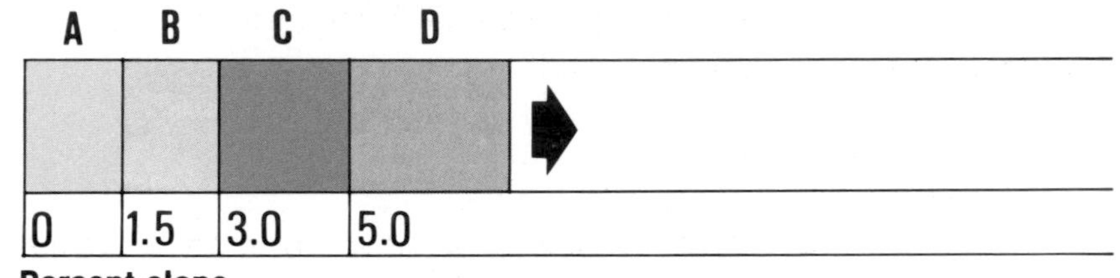

Categories for slope map

31. To prepare a slope map, you must first make **contour line-spacing guides** for each of the slope categories and for the particular topographical map you are using. If this topographical map has:

- a distance reduction scale where 1 cm = **n** (m)
- a contour interval **CI** (m)

the **interval X** (in cm) between the lines of your spacing guide is calculated as:

$$X = (100\ CI) \div (nS)$$

S being the greatest slope in percent characterizing each of the slope categories you want to map.

Example

You have a topographical map on which you want to map the above four slope categories (A, B, C, D). You first need to make three spacing guides for 1.5, 3 and 5 percent slopes, respectively. The topographical map scale is 1 : 50 000 (1 cm = 500 m) and the contour interval is 5 m. Calculate the intervals X between the lines of your spacing guides as follows:

- **1.5 percent slope**: X = (100 – 5) ÷ (500 – 1.5) = 500 ÷ 750 = 0.67 cm or 6.7 mm
- **3 percent slope**: X = (100 – 5) ÷ (500 – 3) = 500 ÷ 1 500 = 0.33 cm or 3.3 mm
- **5 percent slope**: X = (100 – 5) ÷ (500 – 5) = 500 ÷ 2 500 = 0.2 cm or 2 mm

32. Get several sheets of squared-ruled millimetric paper and prepare your **spacing guides** as follows:

- using a sharp pencil with a hard lead, draw a line across the sheet of paper close to the bottom;
- for a vertical distance of about 15 cm, draw a series of lines parallel to this bottom line, at a distance equal to the interval X (calculated as explained above), for example, 6.7 mm for the 1.5 percent slope. 1.5 percent is your **spacing guide**;
- up to a further vertical distance of about 10 cm, draw a second series of lines parallel to the first series, at a distance equal to the interval X calculated for the second slope value, for example, 3.3 mm for the 3 percent slope. 3 percent is your **spacing guide**;
- repeat the procedure to obtain the **5 percent spacing guide**.

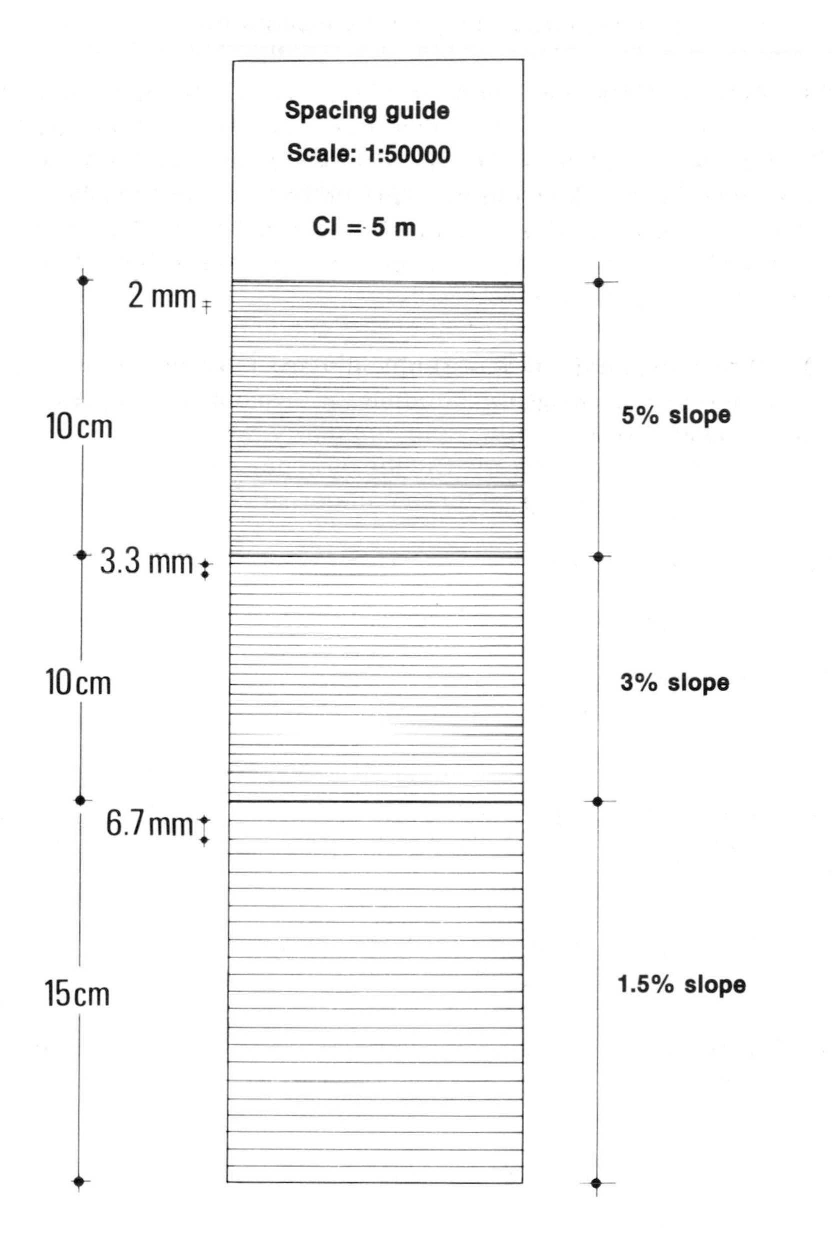

33. You are now ready to make your **slope map**, identifying, for example, the four slope categories on the topographical map. Proceed in the following way:

(a) Get several differently coloured crayons or coloured pencils. Select **one colour to represent each slope category**. The lightest colour can represent the least sloping ground (0 to 1.5 percent), while the darkest colour can be for the most sloping ground (more than 5 percent).

(b) Cut a 2 cm strip off your contour-line spacing guide (see step 32) which gives **the minimum interval between contour lines** on your topographical map for the "1.5 to 3 percent" category, the "3 to 5 percent" category, and the "greater than 5 percent" category. Cut the strip perpendicular to the spacing lines so that all three categories are shown on it.

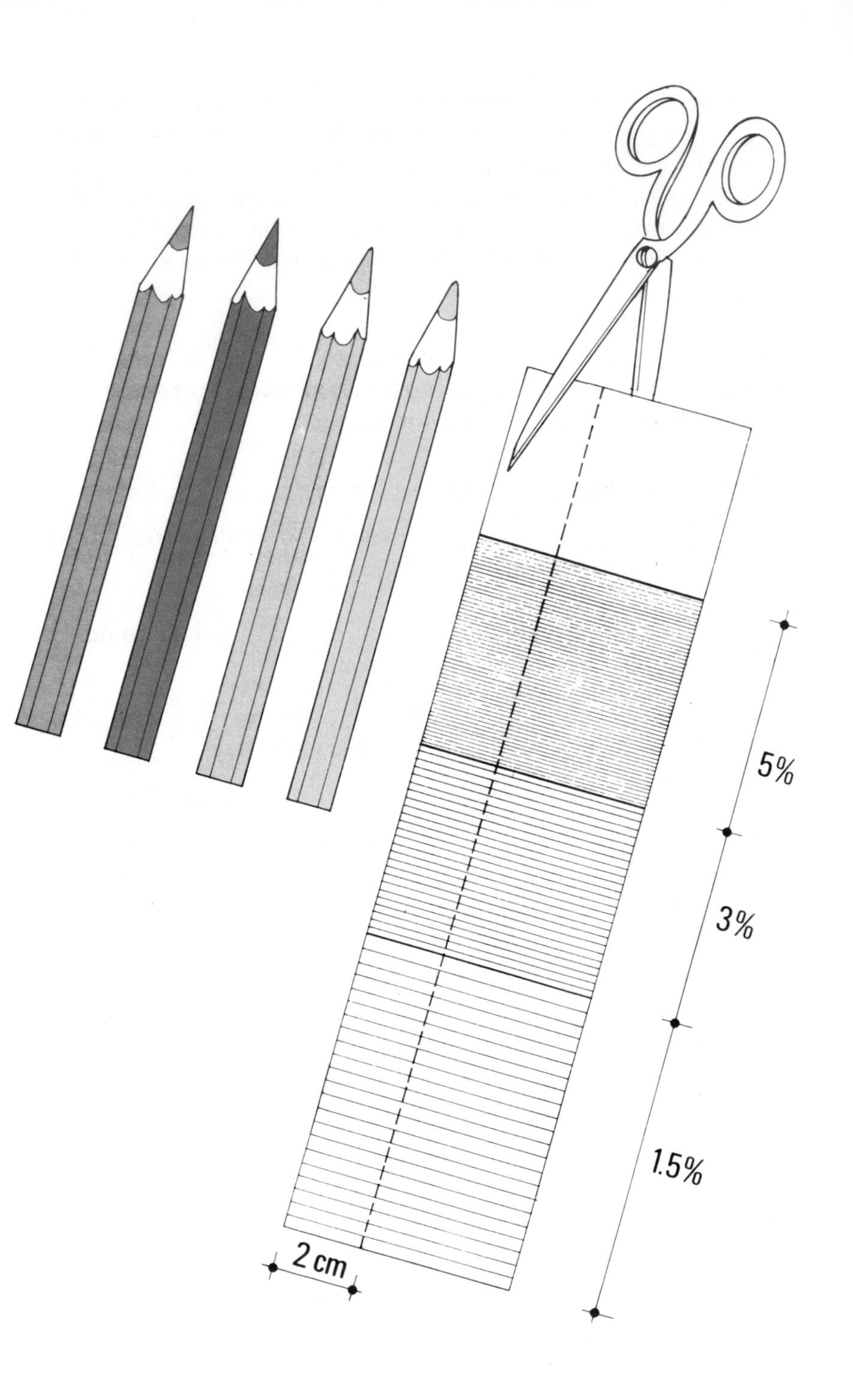

(c) Place this strip on the map. Then, going through them one by one, determine which sections have **intervals between the contour lines smaller than or equal to** the interval on the strip section that corresponds to the **5 percent spacing guide**. Since such sections have slopes equal to or greater than 5 percent, you should colour them with the darkest colour.

Note: you should compare the interval of the strip with the intervals between contour lines **along a line perpendicular to the contour lines**, which is the direction of maximum slope.

(d) Repeat the same procedure with the section of the strip corresponding to the **3 percent spacing guide**. Then determine which **uncoloured sections** of the map have intervals between contour lines **smaller than or equal** to the strip interval. Such sections have slopes from 3 to 5 percent, and you should colour them with the next lightest colour.

(e) Repeat this procedure with the section of the strip that corresponds to **the 1.5 percent spacing guide**, and determine **the uncoloured sections** of the map which have slopes from 1.5 to 3 percent. Colour them with the next lightest colour.

(f) Finally, using the same procedure, check that the **uncoloured sections** of the map have intervals between the contour lines **greater than the interval on the 1.5 percent section of the strip**. Colour these sections with slopes less than 1.5 percent with the lightest colour of all.

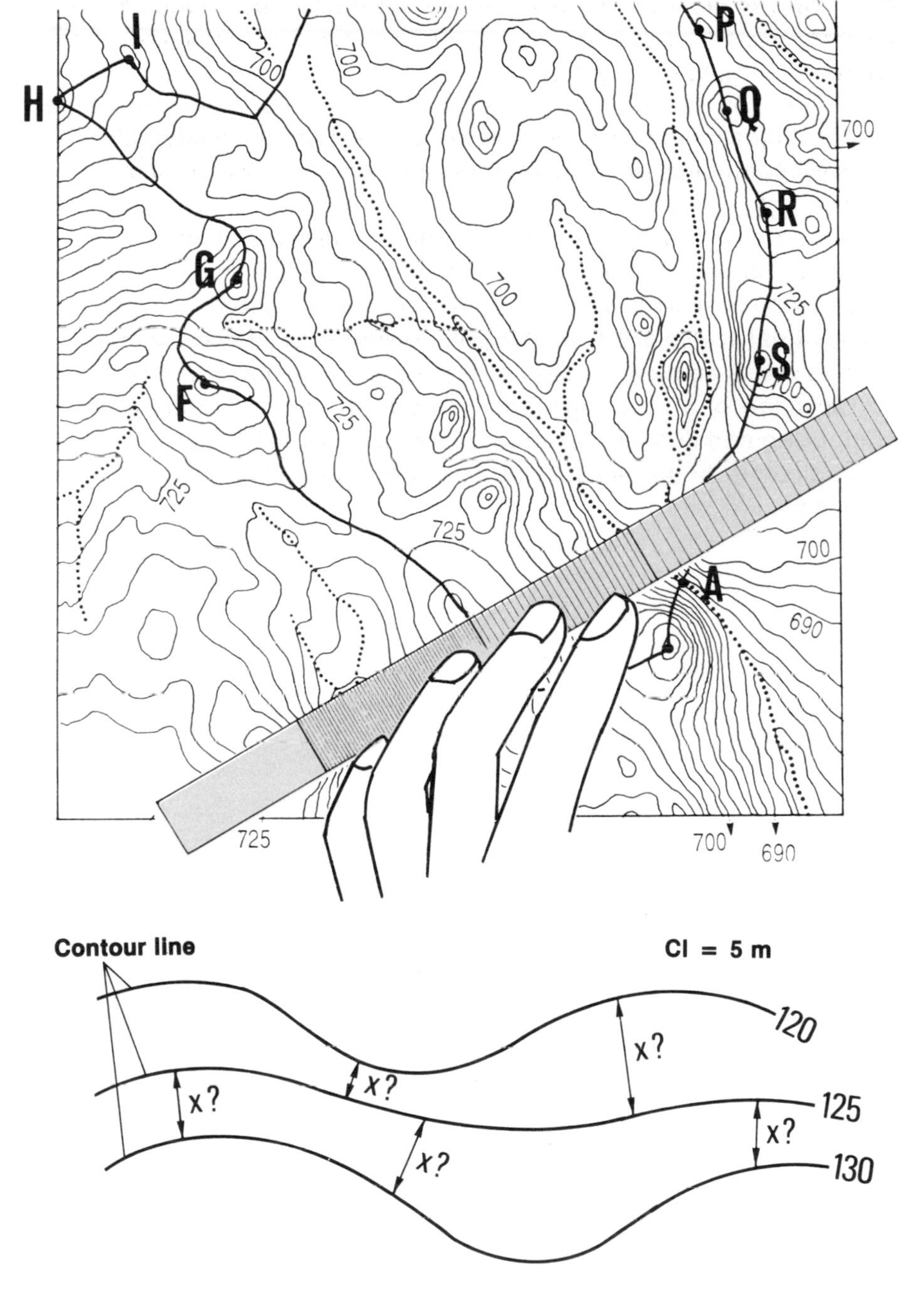

112 How to make a reconnaissance survey of a possible site

1. After you have made preliminary studies from available topographical maps, you can choose **the best potential site(s)** for the construction of fish-ponds (see the next volume in this series).

2. You should now organize a **reconnaissance survey** of the selected site(s) in the field in order to obtain more detailed topographical information. This survey should include, at least, **a longitudinal profile of the stream valley and/or the selected site(s)**, as well as **cross-section profiles of the site(s)**. If you are planning the construction of a reservoir, you can survey its maximum area by contouring and find its maximum water level by levelling. Then, you can calculate both the surface area and the water volume of the reservoir.

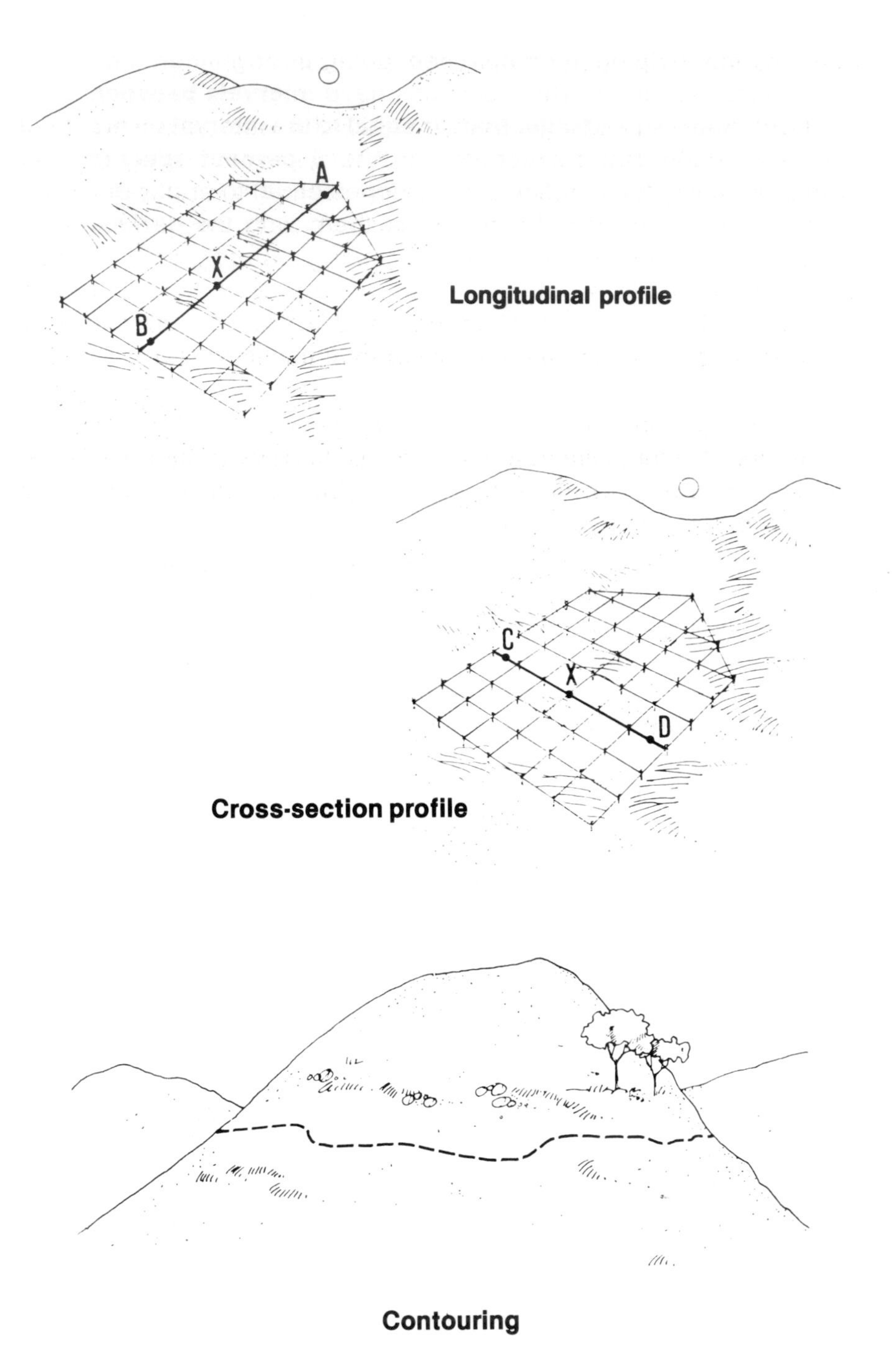

Longitudinal profile

Cross-section profile

Contouring

Studying the longitudinal profile of the stream valley

3. First, make a levelling survey to obtain **the longitudinal profile of the stream valley** (see Section 82). To do this, traverse along a series of straight lines closely following the stream. Plot the profile on graph paper (see Section 95). Then you can calculate the **difference in elevation** between any two points of the longitudinal profile, such as the future water intake F and the future water outlet A. This difference in elevation E(F)-E(A) should be great enough to allow you to build the proposed fish-farm (see next volume in this series).

4. If you are planning **to build a reservoir**, you can use a similar procedure to easily estimate the approximate height of the dam at point A, together with the maximum water depth of the future reservoir.

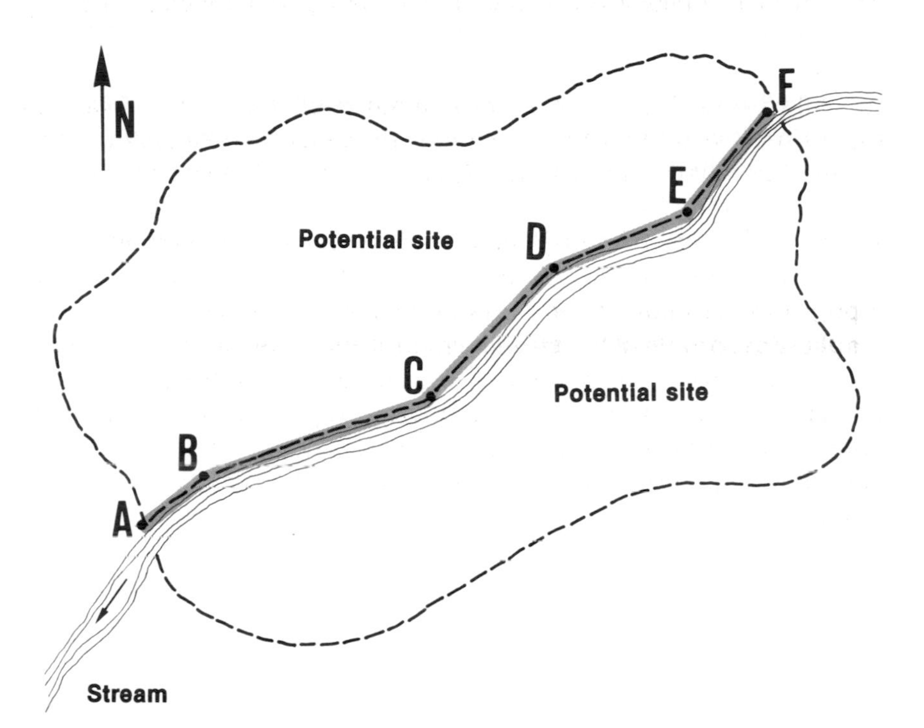

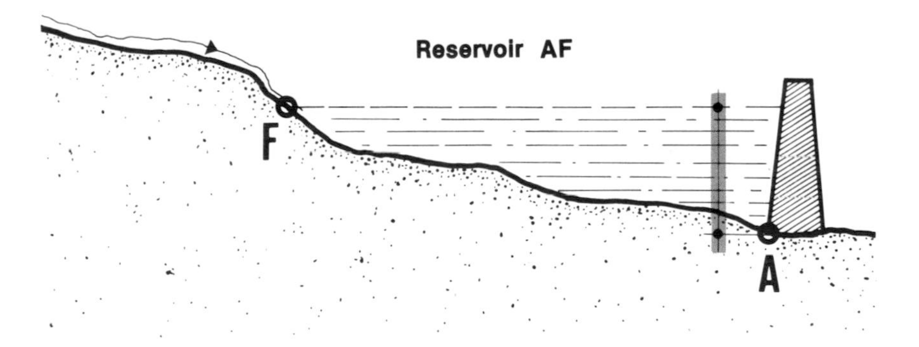

5. You survey a series of **cross-section profiles** GH, IJ, ..., based on the above longitudinal profile, across a valley at intervals of 20 to 50 m, covering the entire area of the potential site (see Section 82). Plot the profiles on graph paper (see Section 96). This makes it possible to calculate **the ground slopes** and, on this basis, design the fish-farm.

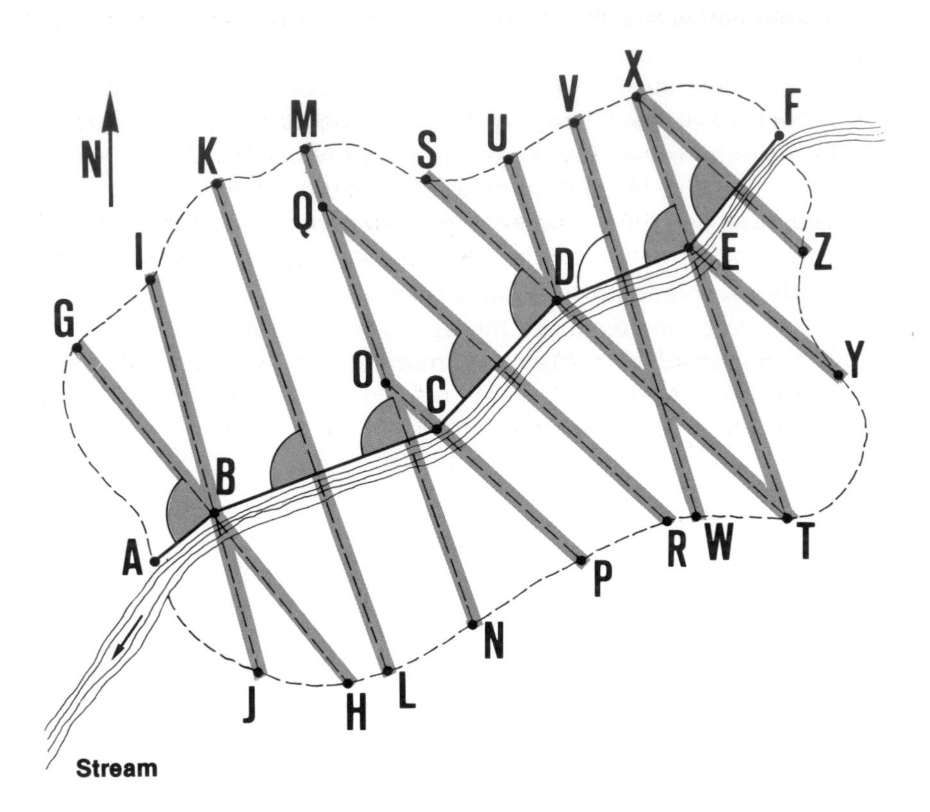

6. From the results of the above survey, you can also prepare **a plan of the potential site**, and **plot the contours** on it (see Section 94, step 10).

7. In another contour mapping method, you first choose **a reference point** on the potential site, such as the water intake point A at the stream. Point A will then be used as a **bench-mark** with a given elevation E(A) = 100 m, for example: from point A, you will identify **the E(A) contour** ABC ... H in the potential site, by contouring (see Section 83).

8. At intervals of 20 to 50 m on this E(A) contour, you will survey perpendicular **cross-section profiles** from the contour to the stream. From the results of this survey you can identify other contours and prepare a contour map.

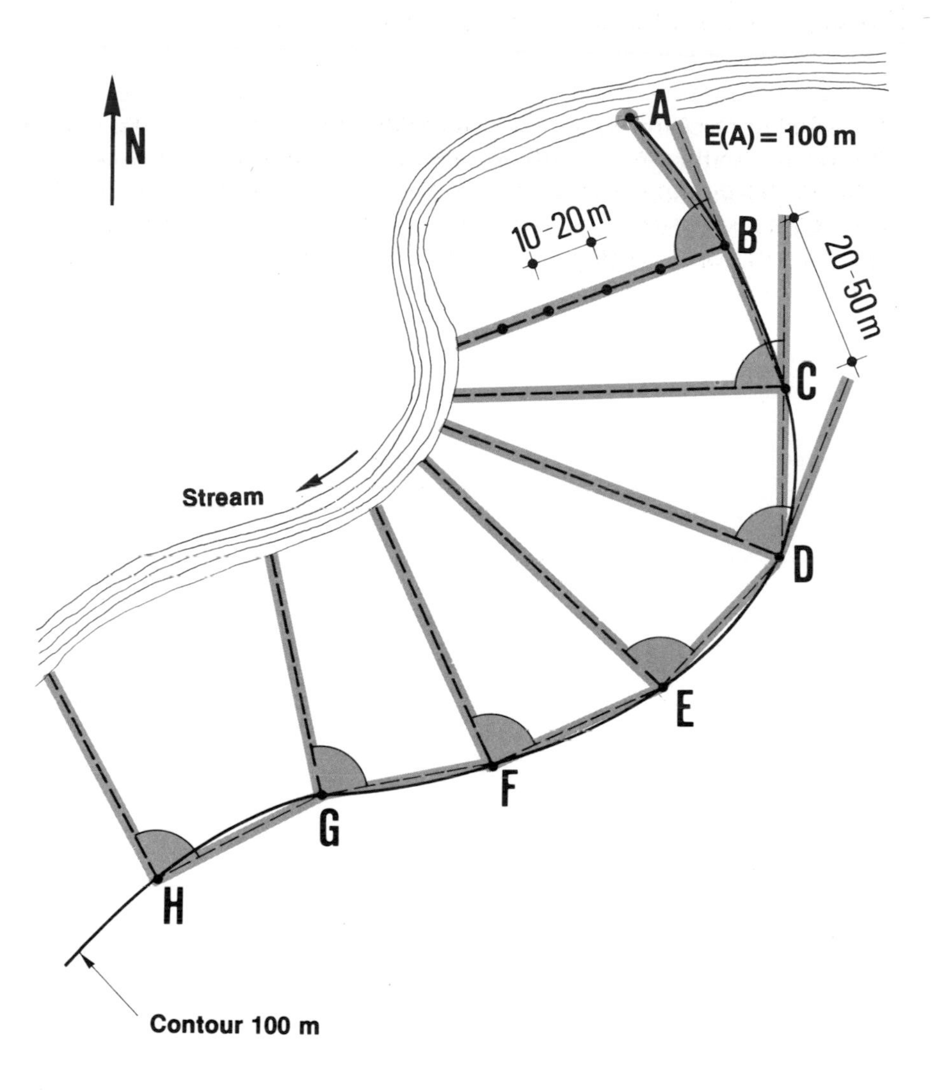

Finding the maximum area of a reservoir

9. When you have fixed the positions of both the water inlet A and the dam centre-line XY of a proposed reservoir, you can easily determine, on the ground, **the maximum area of this future reservoir**. From the water inlet A, find the position of contour E(A) by contouring from A on both stream banks to line XY.

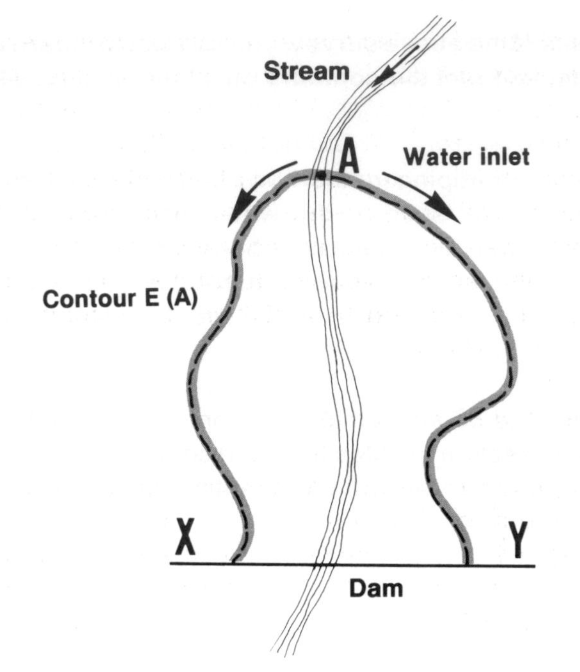

113 How to make a feasibility study of a potential site

1. From the data obtained during the reconnaissance surveys, you can draw **a topographical plan** to scale, showing the various distances and elevations. You can study the design of the fish-farm or, if you need to build a dam, you can obtain additional information on the characteristics of the water reservoir that will result from the dam. (Fish-farm design will be discussed further in the next volume in this series, and you will learn more about the characteristics of reservoirs there.) In this section you will learn how to estimate the volume of a dike built from earth, and how to calculate height differences for pumping stations.

Estimating the area of the future reservoir

2. On your topographical plan, transfer the position of **the E(A) contour** which encloses the maximum area of the future reservoir (see step 9 above). You may then estimate **the size of the area** enclosed by the E(A) contour line, using measurements on the map (see Sections 104 and 105).

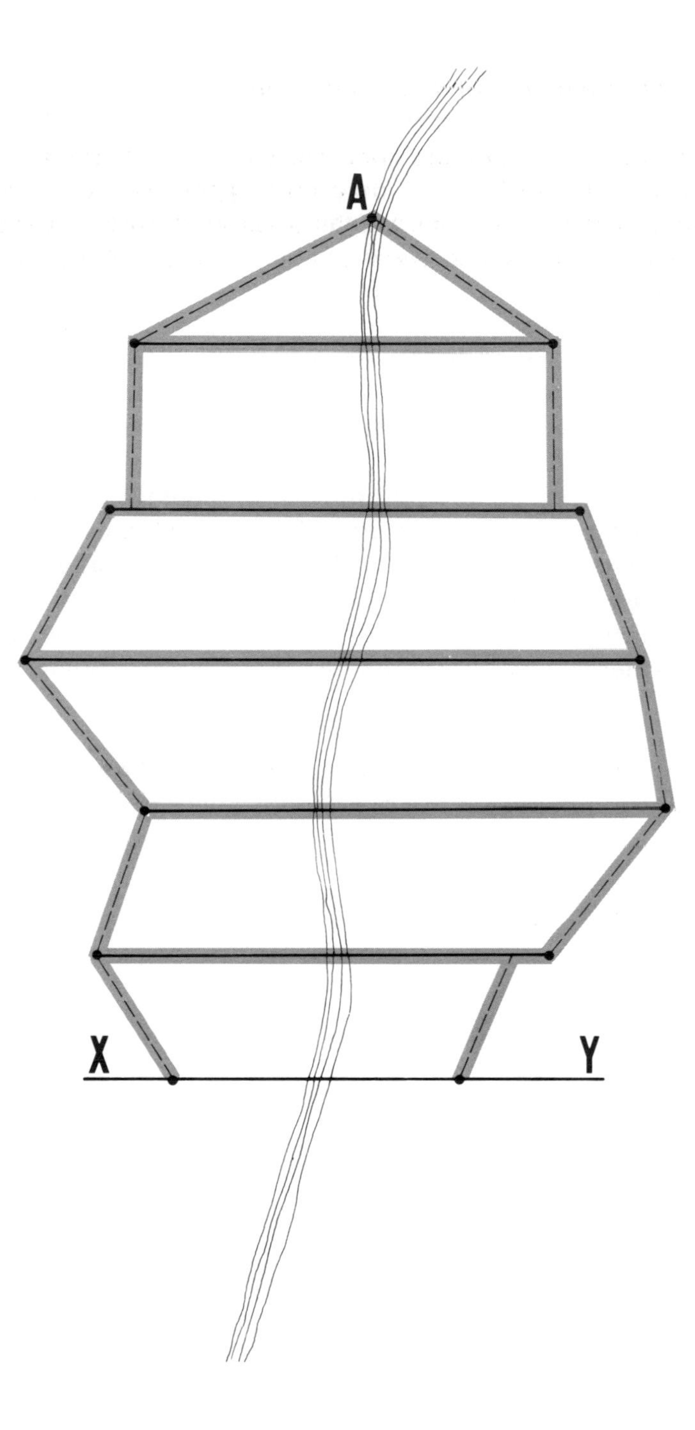

Estimating the volume of the future reservoir

3. In **Water for Freshwater Fish Culture** (Volume 4, Section 42), you learned simple methods for **determining the volume of a future reservoir**. Now, you will learn a fairly accurate, rapid field method based on topographical surveys.

4. Every 20 to 50 m along the longitudinal axis AZ of the future reservoir, mark **perpendicular lines** BC, DE ... QR within the area enclosed by contour E(A). Use wooden pegs at 25 m intervals along the perpendiculars, and mark them **on each side of the axis AZ**.

5. Start from end-point B of line BC, on **contour E(A)** where the maximum water level will line up with the ground elevation. Using a target levelling staff and a sighting level, **transfer elevation E(A) to the top of the pegs a, b and c**. Drive them into the ground until they are at the correct elevation. **Pegs** a, b and c now clearly show the **maximum depth of water** which will be present at each of these points when the reservoir is full.

6. Repeat this procedure for each **transversal line** DE, FG ... QR successively; similarly, stake out XY, the centre line of the proposed dike.

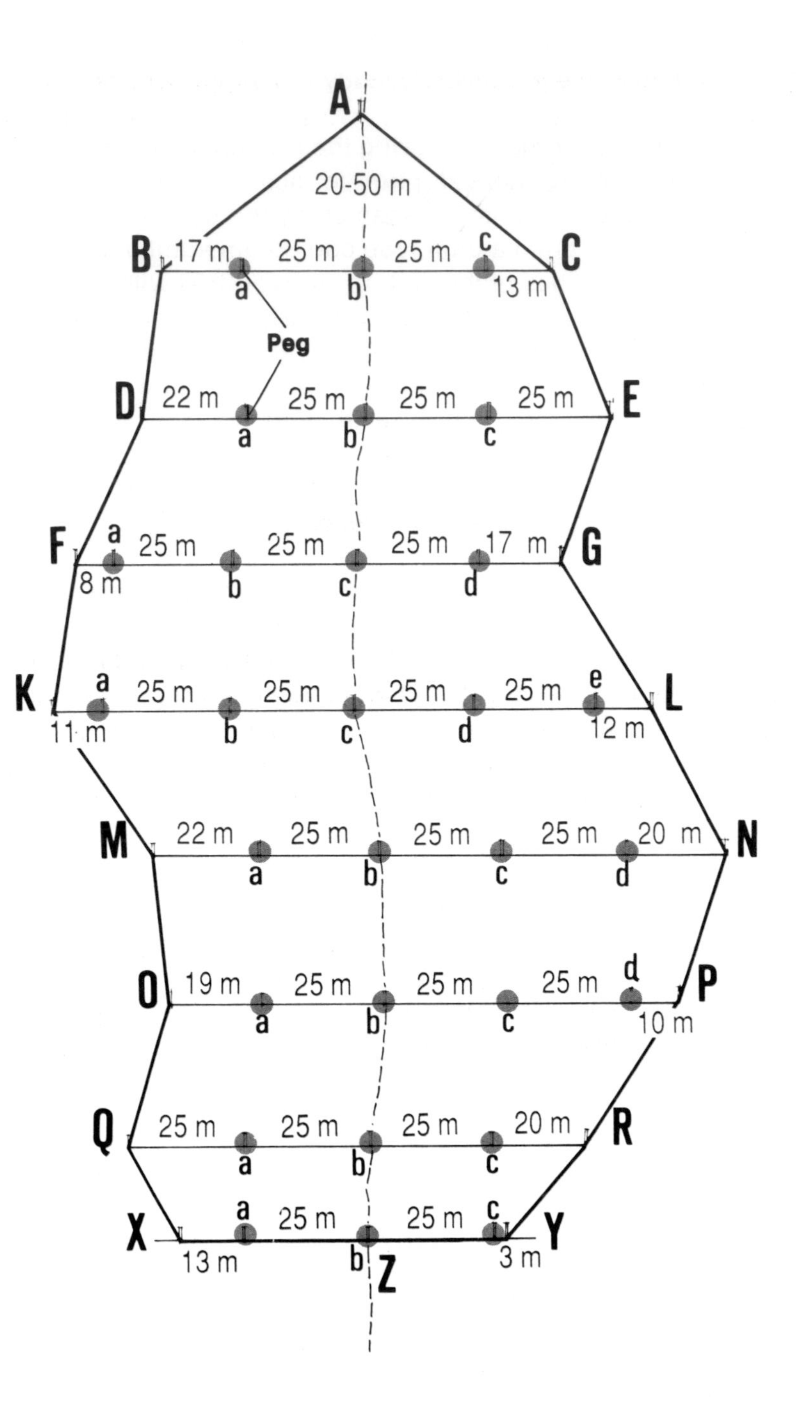

7. In **a simple table**, measure and record **the height of each peg above ground level** for each transversal line, including line XY, as shown in the example.

Example

Line	Peg height, m				
	a	b	c	d	e
BC	0.45	0.87	0.38	–	–
DE	0.85	1.42	0.73	–	–
FG	0.22	0.87	1.63	0.79	–
KL	0.49	0.98	1.89	0.91	0.58
...	...	...	...	...	...
XY	0.82	2.42	0.84	–	–

8. Using the correct scale, draw **the cross-sections BC, DE ... QR, and XY of the completely filled reservoir** on square-ruled millimetric paper. Use a vertical scale 10 times larger, for example, than the horizontal distance scale. Remember that the end-points of each cross-section have, by definition, a zero depth of water.

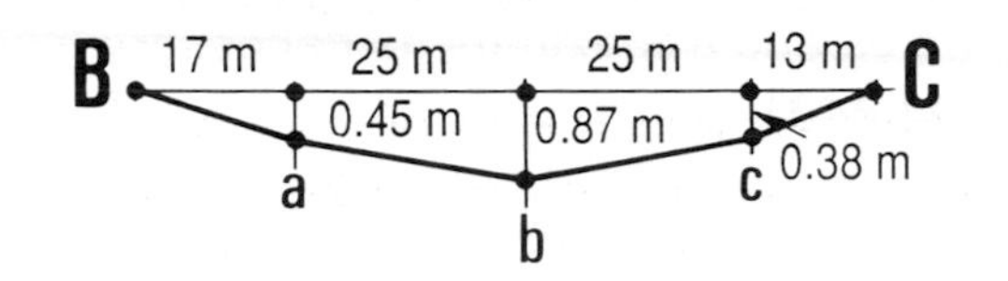

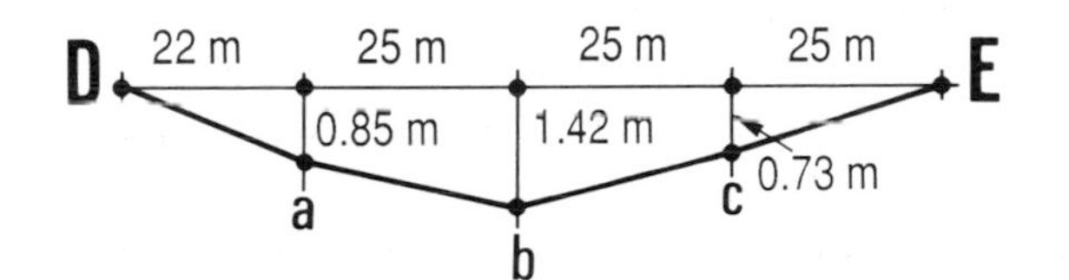

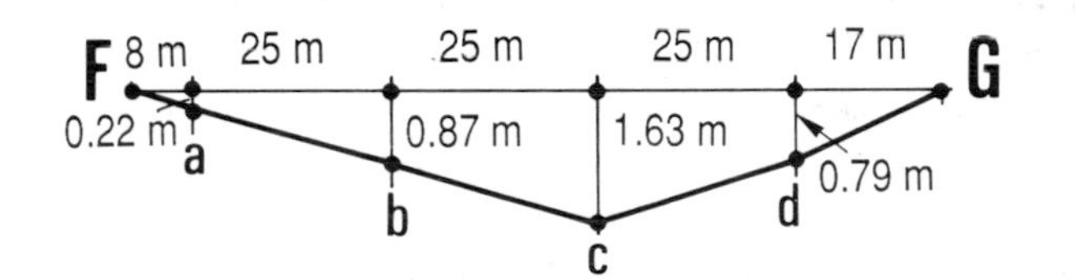

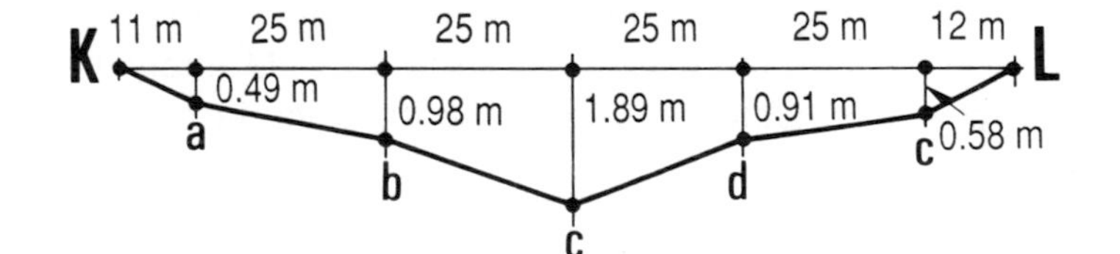

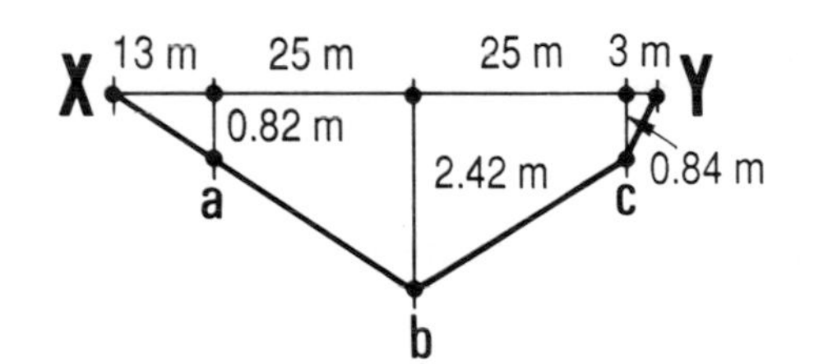

9. Calculate **the area of each cross-section,** adding the partial areas of triangles and trapeziums as necessary (see mathematical formulas in **Annex 1**).

Example

Area BC = triangle 1 + trapezium 2 + trapezium 3 + triangle 4

— Triangle 1 = (17 m × 0.45 m) ÷ 2	=	3.825 m²
— Trapezium 2 = [(0.45 m + 0.87 m) ÷ 2] × 25 m	=	16.500 m²
— Trapezium 3 = [(0.87 m + 0.38 m) ÷ 2] × 25 m	=	15.625 m²
— Triangle 4 = (13 m × 0.38 m) ÷ 2	=	2.470 m²
Area BC	=	38.420 m²

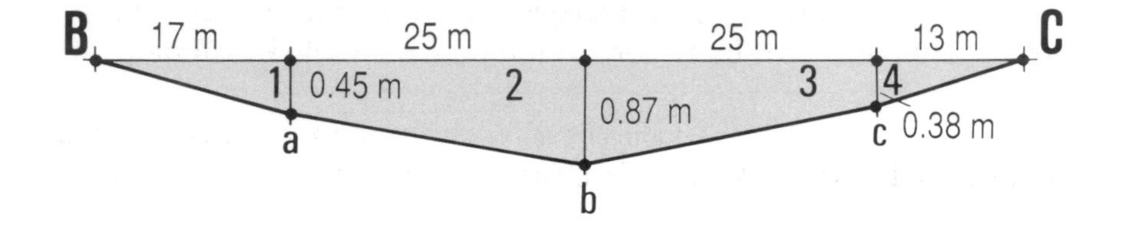

Area BC

10. Add the areas of cross-sections BC, DE ... QR, and **multiply this sum by the fixed interval between cross-sections** (in this case, 25 m) to obtain an estimate of **the volume of the reservoir upstream from the last cross-section QR**.

Example

Volume of reservoir from point A to line QR = (area BC + area DE + ... + area QR) × 25 m

11. Estimate **the volume of the last section of the reservoir**, between cross-section QR and the dike's centre-line XY. Multiply the area of cross-section XY (see step 9) by **half the distance between previous cross-sections**.

Example

Volume section QR/Z = (area XY) × (25 m ÷ 2)

12. Calculate **the volume of the entire reservoir** by adding:

- the volume A/QR obtained in step 10; and
- the volume of QR/Z obtained in step 11.

Example

Volume A/Z = volume A/QR + volume QR/Z

Estimating the volume of an earth dam

13. To rapidly estimate **the volume of an earth dike XY** to be built across a given valley, use the following method. It will provide a volume estimate **about 10 percent smaller** than the actual volume. But this level of accuracy is good enough for the initial estimate.

14. Using the information from the table in step 7, calculate **the heights h of the dam XY** at mid-points between consecutive pegs.

Example

For the dam XY there are only two mid-points to consider: one between pegs a and b, and one between pegs b and c.

- Dike height a/b = (0.82 m + 2.42 m) ÷ 2 = 1.62 m
- Dike height b/c = (2.42 m + 0.84 m) ÷ 2 = 1.63 m

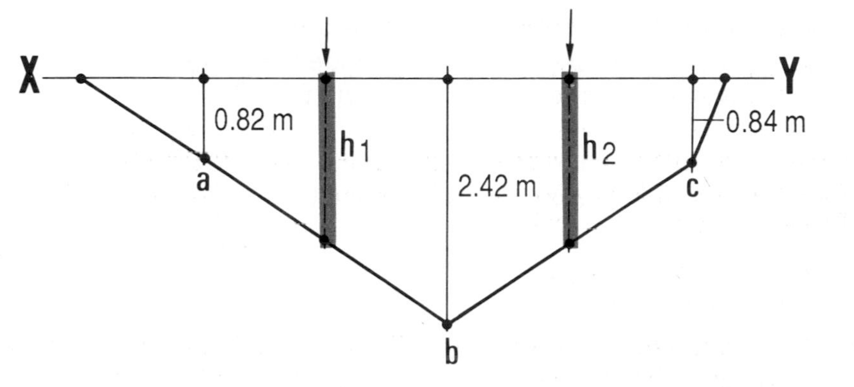

Mid-points to calculate the volume of dam XY

15. Using the correct scale, draw **the transversal section of the type of dike** you plan to build (see next volume in this series). There are three measurements, in particular, which you must determine:

- **the width C** of the highest point, or crown, of the dike;
- **the dry-side slope** of the wall outside the reservoir, D : 1;
- **the wet-side slope** of the wall inside the reservoir, W : 1.

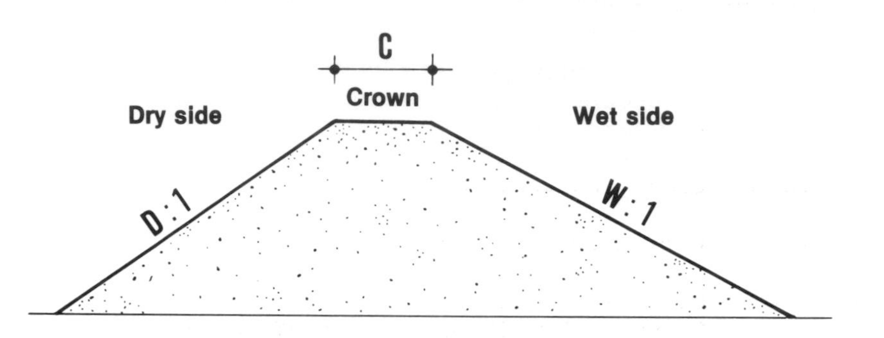

16. If you know these characteristics of the dike, you can calculate **the area of any transversal section of this dike** by adding:

- rectangle 1 area = C × h;
- triangle 2 area = (D × h) × (h ÷ 2);
- triangle 3 area = (W × h) × (h ÷ 2).

Therefore, **the area A of any transversal section of the dike** equals:

$$A = (Ch) + (Dh^2 \div 2) + (Wh^2 \div 2)$$

where **C** is the crown width of the dike, **h** the height, **D** the dry slope, and **W** the wet slope.

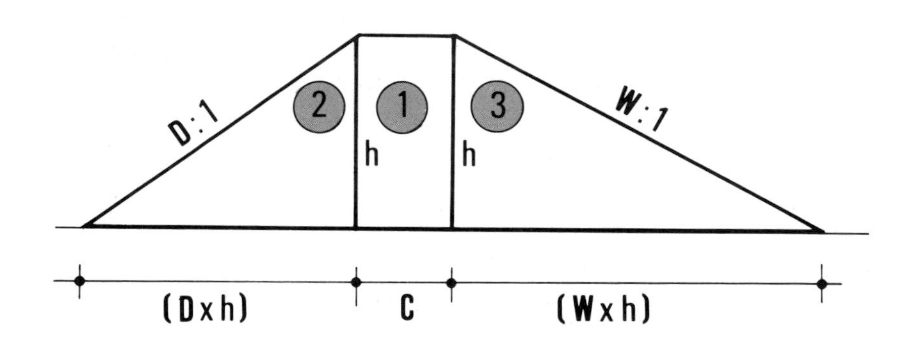

17. Apply the above formula to successively calculate **the section area of the dike at each of the XY mid-points**, using the h-values obtained in step 14.

Example

If the dike characteristics are fixed as follows:

C = 4 m; dry slope = 1.5 : 1; wet slope = 2 : 1;

the dike's transversal section areas are:

- *At mid-point a/b* where h_1 = 1.62 m
  $A_1 = (4\,m \times 1.62\,m) + (1.5 \times 1.62^2\,m) \div 2 + (2 \times 1.62^2\,m) \div 2$
  $= 6.48\,m^2 + 1.97\,m^2 + 2.62\,m^2 = 11.07\,m^2$
- *At mid-point b/c* where h_2 = 1.63 m
  $A_2 = (4\,m \times 1.63\,m) + (1.5 \times 1.63^2\,m) \div 2 + (2 \times 1.63^2\,m) \div 2$
  $= 6.52\,m^2 + 1.99\,m^2 + 2.66\,m^2 = 11.17\,m^2$

18. Calculate **the partial volumes** of each portion of the dike marked by pegs a, b, c, etc. To do this, multiply each corresponding mid-point section area by the length of the dam portion.

Example

Partial volumes of the dike with pegs a, b and c at 25 m intervals:

- *For portion a/b* $V_1\ A_1 \times 25\,m = 11.07\,m^2 \times 25\,m = 276.75\,m^3$
- *For portion b/c* $V_2 = A_2 \times 25\,m = 11.17\,m^2 \times 25\,m = 279.25\,m^3$

19. Obtain the estimate of the **total volume of the dike** by adding these partial volumes.

Example

Total volume of dike XY $= 276.75\,m^3 + 279.25\,m^3 = 556\,m^3$

Using what you know about topography to instal a pumping station

20. You may be planning to pump water for your reservoir either from a well or from an existing body of water. If so, **the kind of pump you choose** will greatly depend on the difference in elevation between the two extremities of your water pipeline and the pump. Usually, you will site the pump at an intermediate elevation, where it can bring the water up by suction from a lower elevation (the source of water) and force it on to a higher elevation (a reservoir tank, for example).

21. When you are choosing a site for your pumping station, two differences in elevation are particularly important:

- From the water-source surface to the pump, the **suction head** (in metres);
- From the pump to the reservoir, the **delivery head** (in metres).

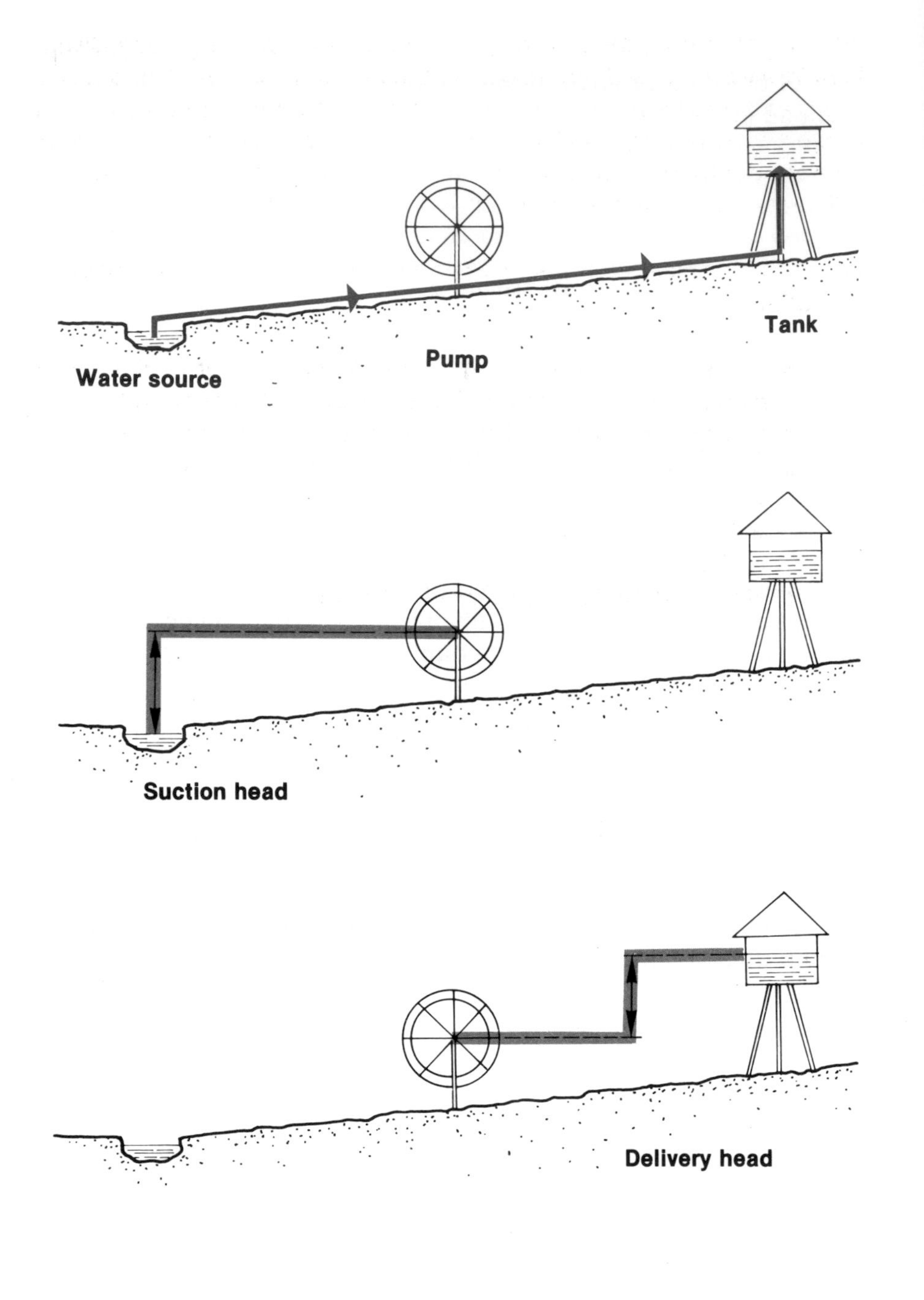

22. To obtain these differences in elevation, you may use **direct levelling** from the water source A to the pumping station B, and to the site C of the reservoir tower. Then, knowing **the ground elevations** at these points and the various vertical distances (for example, water source pump axis and pump axis/tank water), you can easily calculate the suction head and the delivery head.

23. You can instead use **an indirect method** to find the differences in elevation:

- measure the vertical angles made with the horizontal plane by successively sighting lines AB and BD (see Chapter 4);
- measure horizontal distances AE and BC (see Chapter 2);
- calculate the differences in elevation (see Section 50, step 14), as follows:
  - EB = AE tan BAE
  - CD = BC tan DBC,
- obtaining the tangent values from **Table 3**.

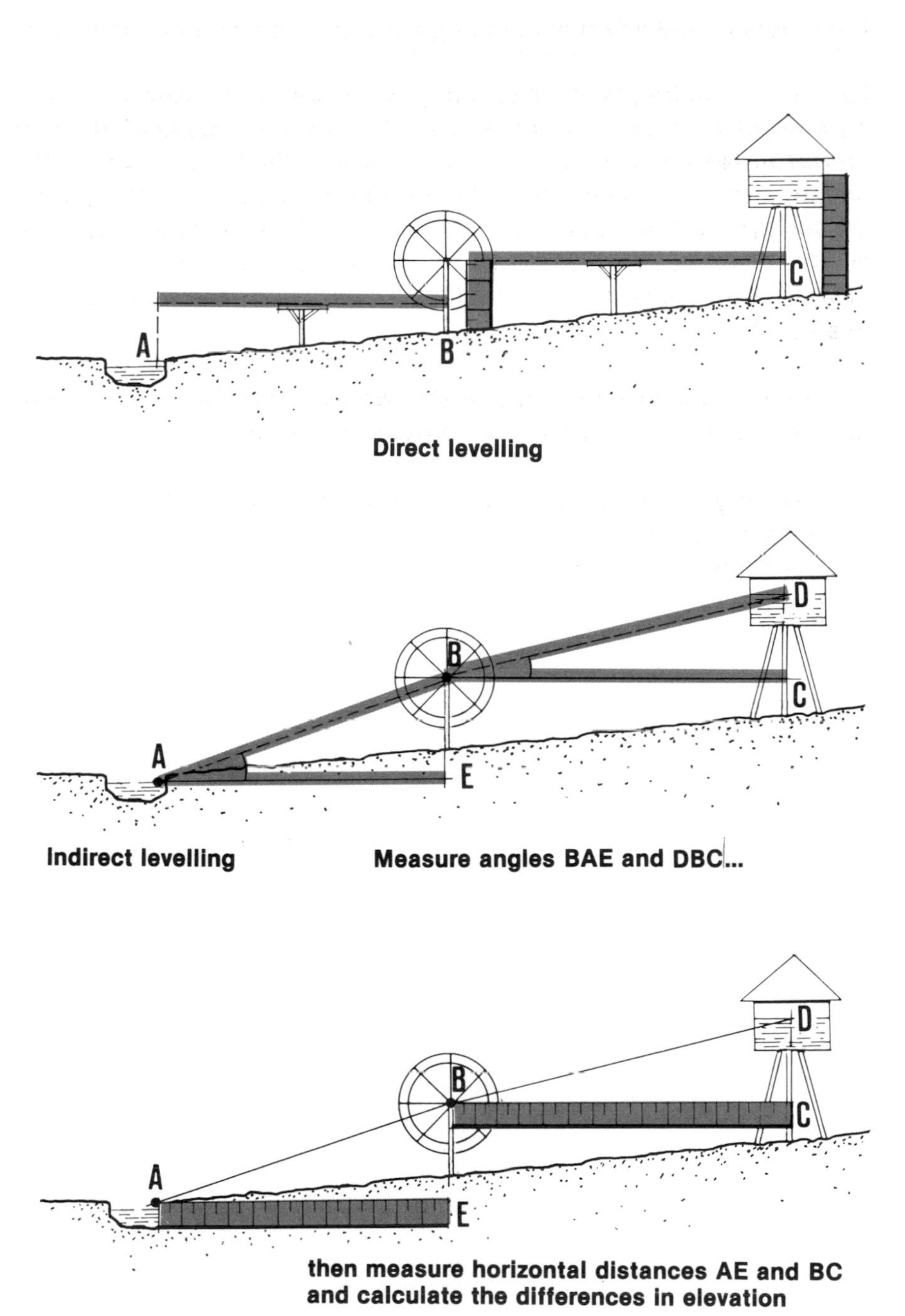

114 How to make levelling surveys for fish-farm construction

1. You have already learned that you need to know how to use the right topographical methods during the construction of your fish-farm. Now you will learn about two additional topographical methods, one for the construction of a water-supply canal and one for the construction of a pond.

Staking out a water supply canal for construction

2. You learned, in earlier chapters, how to first survey **the centre-line of a water-supply canal** (see Sections 71 and 82), then how to draw its longitudinal profile (see Section 95) and its cross-section profile (see Section 96). You have also learned how you can contour (see Section 83) to rapidly **identify the route a canal can take** between the water intake point and the fish-farm water inlets.

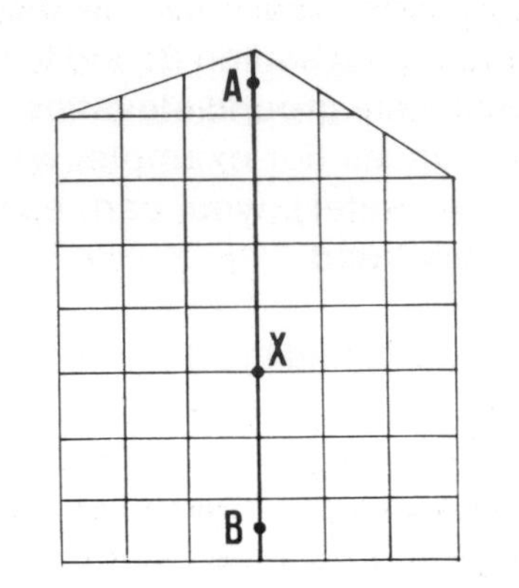

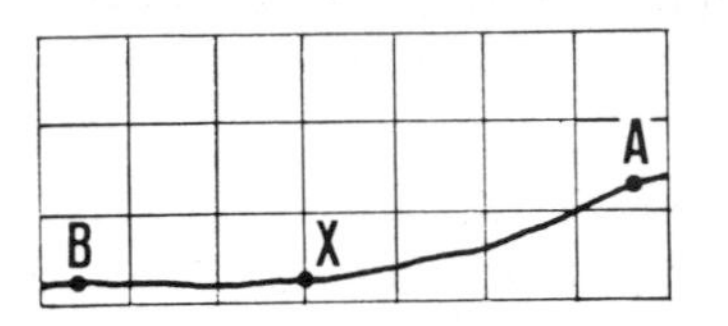

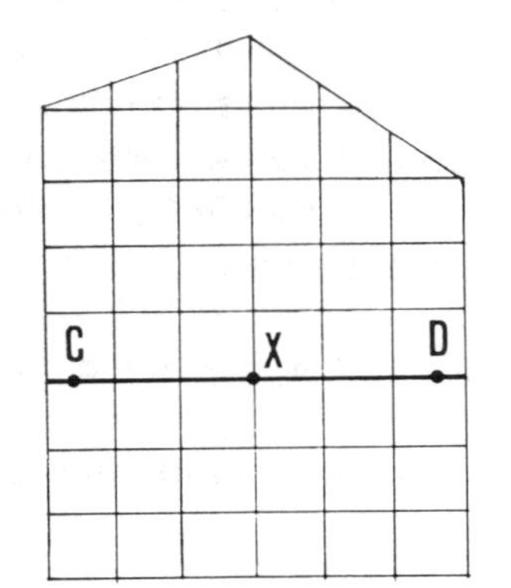

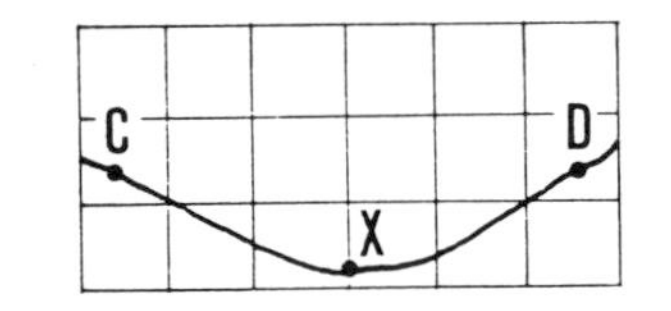

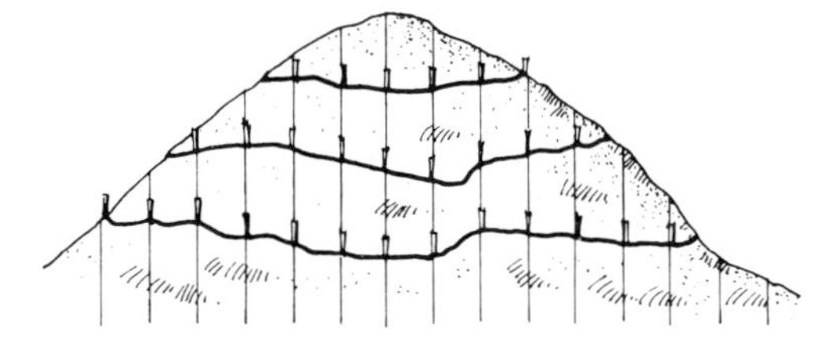

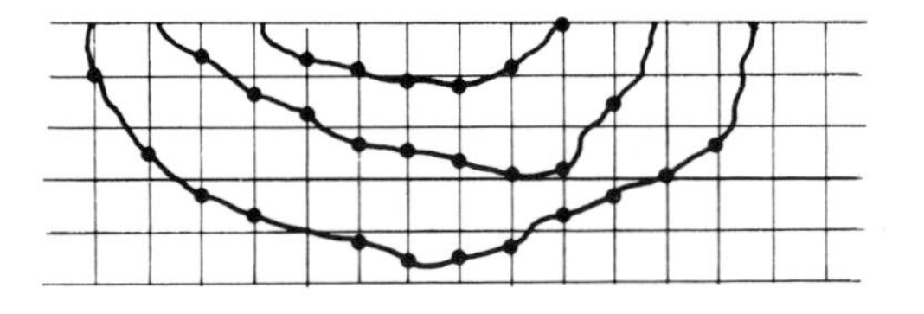

3. Once the route of the water-supply canal has been carefully defined and identified in the field, you need to stake it out before you can begin building it. To do this, you will first have to clear a 1 to 2 m stretch of land along the canal centre-line. Then you will set out **a series of short pegs** along this centre-line. **The summits** of the pegs must correspond to **a horizontal reference level**, that is, the tops must all be at the same height.

4. The distance between these reference pegs depends on the levelling method you are using. Usually, the simplest method to use is to proceed from the starting point A with **a 4 m straight edge and a mason's level** (see Section 66). But you could also use **a clisimeter** (see Section 45) and a target levelling staff. In this case, the pegs are placed at 5 to 10 m intervals.

5. Then, define **the cross-section profile** of the canal (see next volumes in this series). Add pegs as you need them, to direct the workers in their digging.

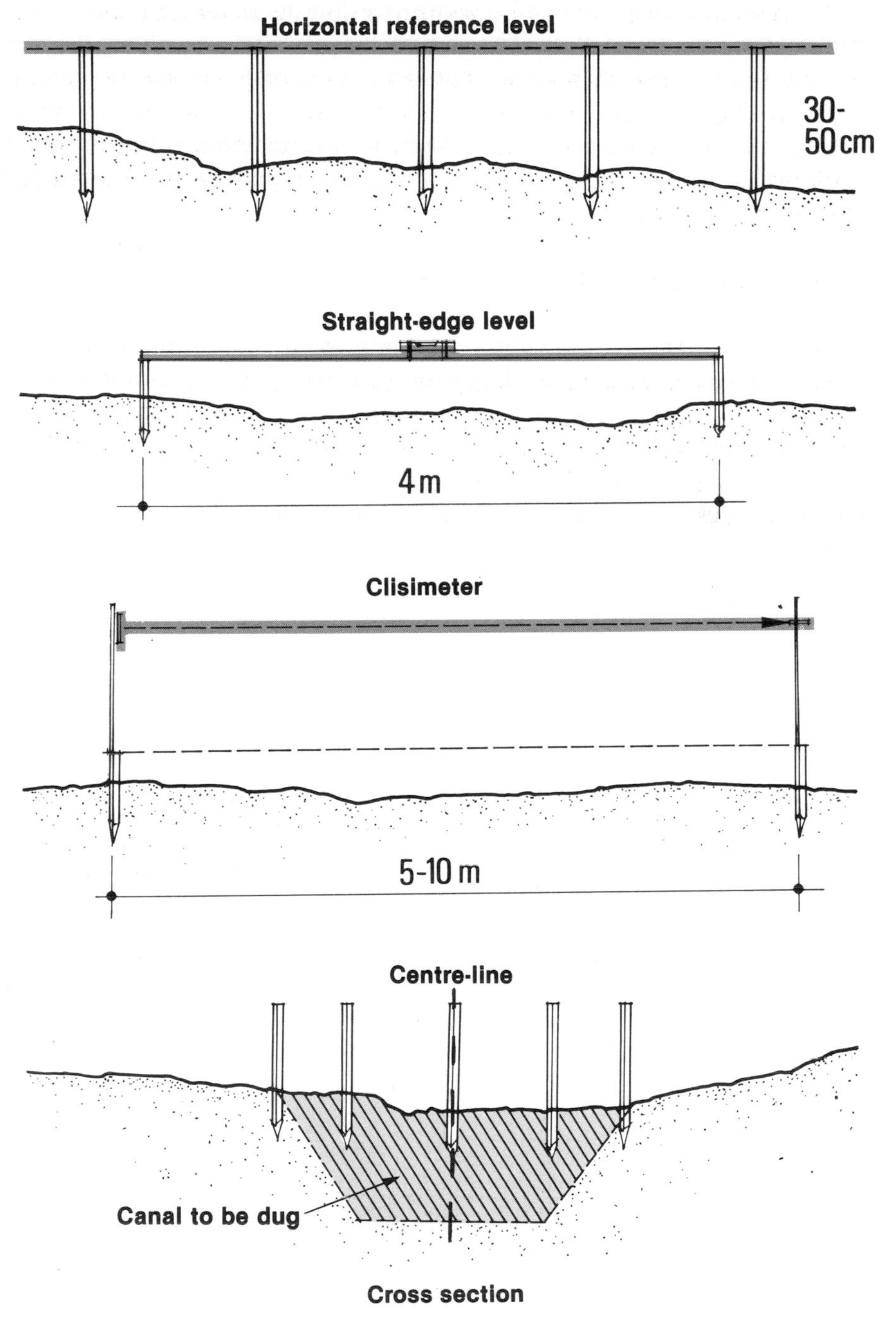

6. If you are building a canal **without a bottom slope**, you can show the workers **the fixed distance to** which they must dig by giving them a wooden stake the same length as the top level of the centre-line pegs.

7. If you are building a canal **with a bottom slope**, the simplest method you can use is to give a similar **slope to the horizontal reference level** which joins the top of the centre-line pegs. To do this, proceed in the following way:

(a) From the value of the slope and the distance between consecutive stakes, calculate how much **difference in elevation** should exist from one peg to the next one.

Example

Using a straight-edge and mason's level for levelling, place your pegs at 4 m intervals. If the slope of the canal bottom is to be, for example, 0.1 percent, the difference in elevation between one peg and the next equals (0.1 m × 4 m) ÷ 100 m = 0.004 m = 4 mm.

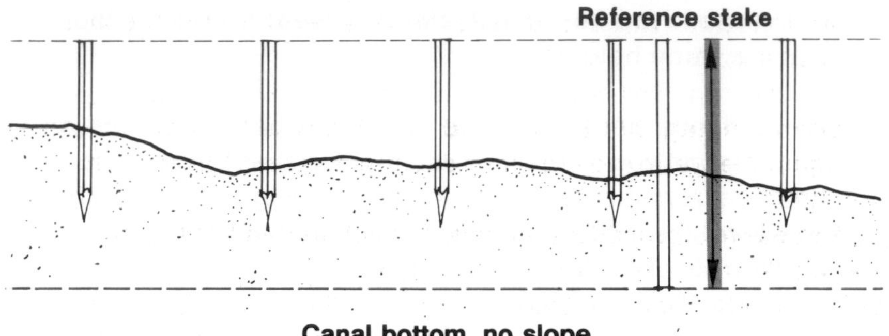

Canal bottom, no slope

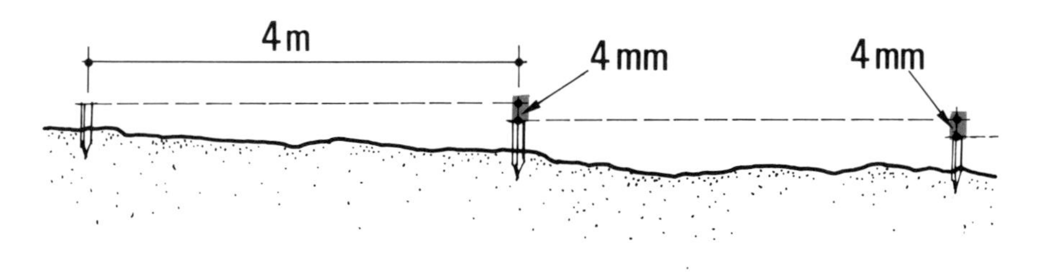

Setting slope

(b) Cut **a small piece of wood** of the same thickness as this difference in elevation.

(c) Put this piece of wood **on top of the second peg, and drive this peg deeper** into the ground until the top of the first peg and the top of this second one are again horizontal. Use a straight-edge and mason's level, for example, to guide you in doing this.

(d) Move to the third peg, put the piece of wood on top of it, and **drive the peg into the ground** until the wood is horizontal again, between the top of the second peg and the top of this one.

(e) Repeat this procedure until you reach the end of the canal centre-line.

(f) The line joining the top of the centre-line pegs now has a slope equal to the required slope of the canal bottom. You can dig the canal as in step 6, using **a constant reference length** from the top of each peg.

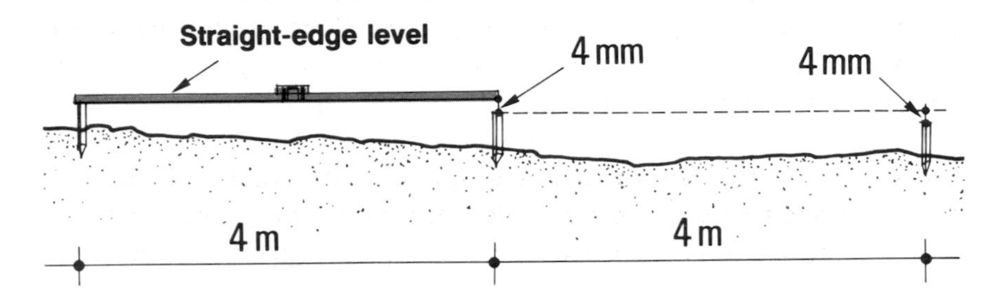

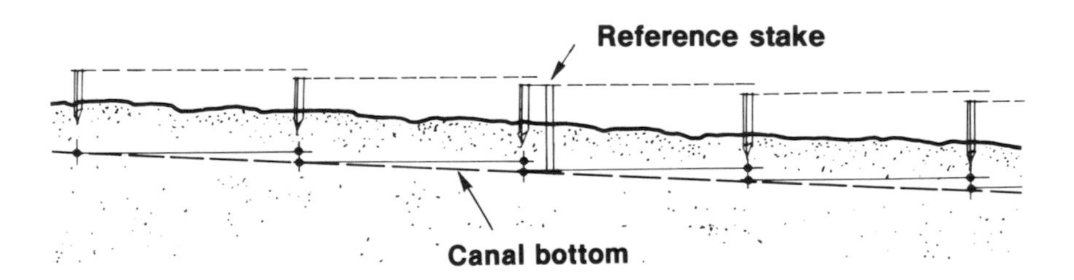

Staking out a pond bottom before construction

8. You have already learned how to **survey the periphery, or boundary line, of your ponds** by setting out a series of rectangular areas and the centre-lines of the dikes you need to build (see Section 37) or by contouring at the maximum water level (see Section 110).

9. When you have found the area of your ponds, but before you start building them, you will usually need to know how much earth you will have to remove at the various points of the pond. You will also need to indicate this clearly on the site so that the workers can proceed with the pond construction correctly. To do this, you will need to stake out the bottom of each pond.

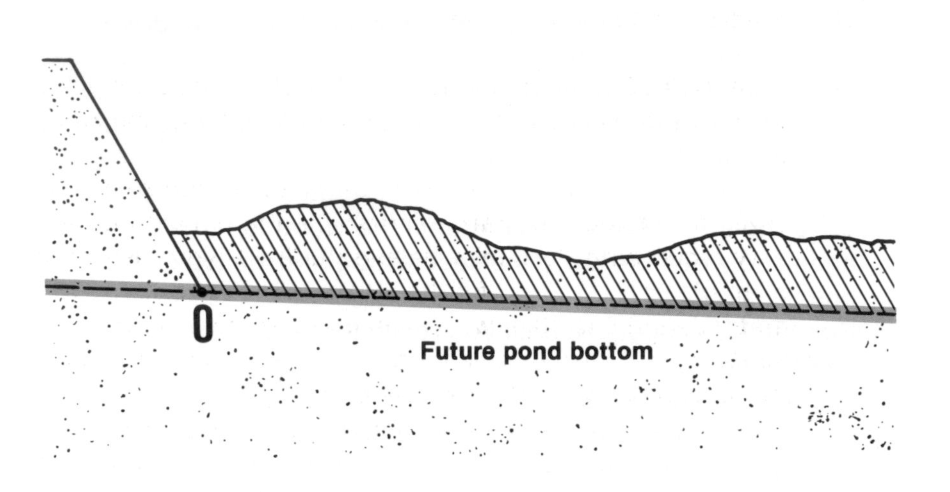

10. One simple way of staking out a pond bottom is to make **a survey by radiation from the lowest point** of the future pond, its water outlet. Proceed in the following way.

(a) From **point 0** (the lowest point of the future pond), mark **a series of lines** radiating over the major part of the pond bottom area, using pegs.

(b) Starting with line OX, for example, place **a series of pegs along it at fixed intervals**. The size of these intervals will vary according to the levelling method you use. For example, it will be 4 m for the straight-edge/mason's level method, and 5 to 10 m when you use a sighting level.

(c) Starting from point 0, survey line OX and put **the top of all of the pegs** at a horizontal reference level.

(d) Repeat the same procedure for **all the marked radiating lines**; in this way, you will obtain a series of pegs whose tops are all at the same reference level, and which are dispersed all over the pond bottom area.

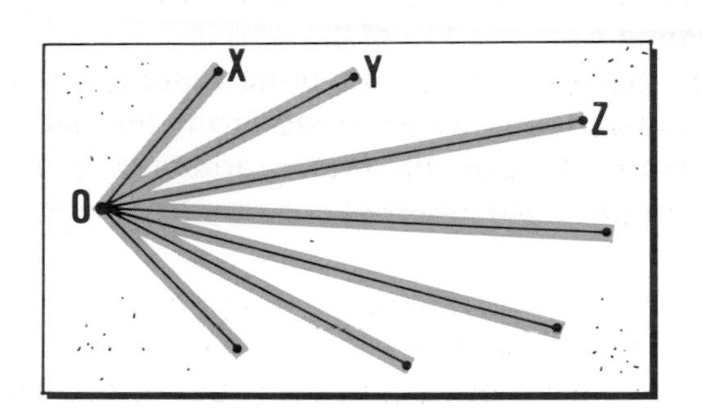

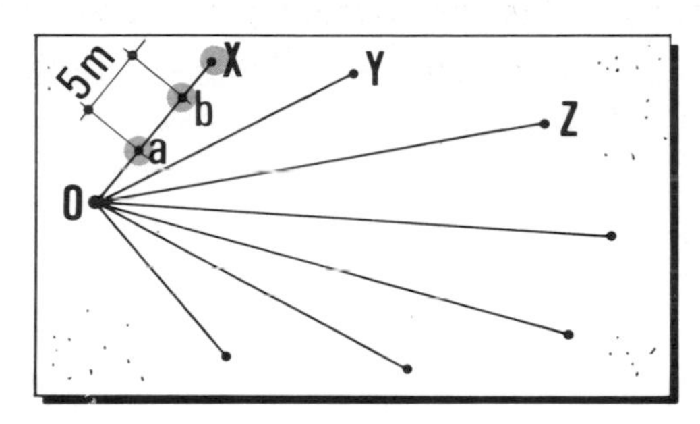

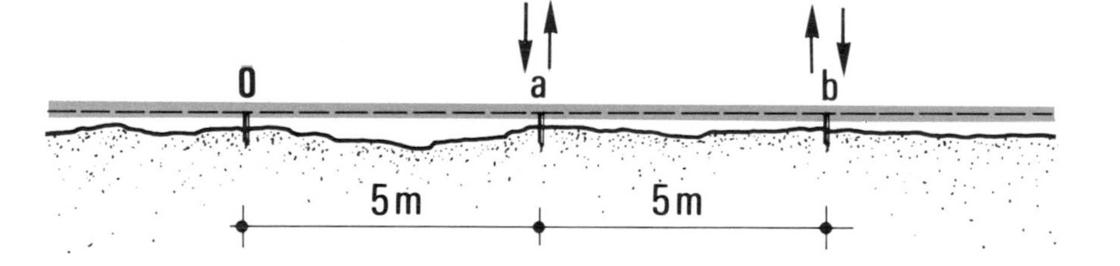

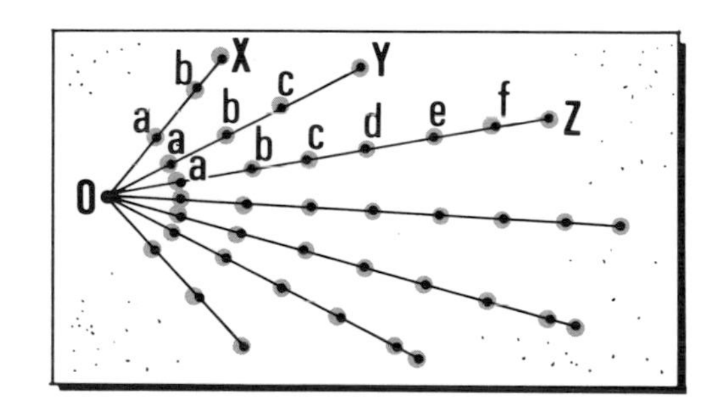

(e) On each peg you must now clearly indicate **the vertical distance** from the top of the peg to the bottom of the pond.

(f) From the slope which the pond bottom will have (see next volumes in this series) and from **the distance** between consecutive pegs on each radiating line, calculate **the difference in elevation** necessary from one peg to the next one in each line.

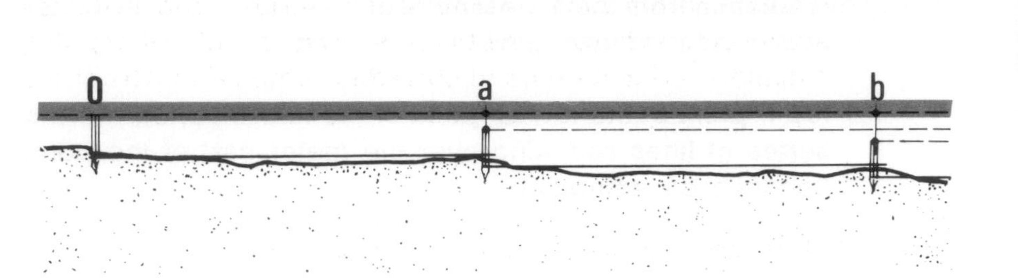

Example

- Slope of pond bottom is 1 percent.
- Distance between stakes is 5 m.
- Difference in elevation should be (1 m × 5 m) ÷ 100 m = 0.05 m = 5 cm.

Calculate the difference in elevation for the desired slope

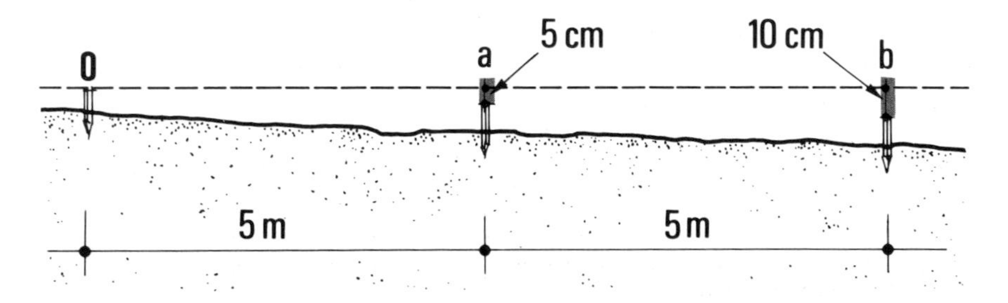

(g) From the fixed elevation to be given to **point 0 at the pond outlet** (see the next volume in this series), and from the surveyed elevation of **the top of the peg** set out at that point, calculate **the difference in elevation** required at 0 from the top of the peg. Mark this difference clearly on stake 0.

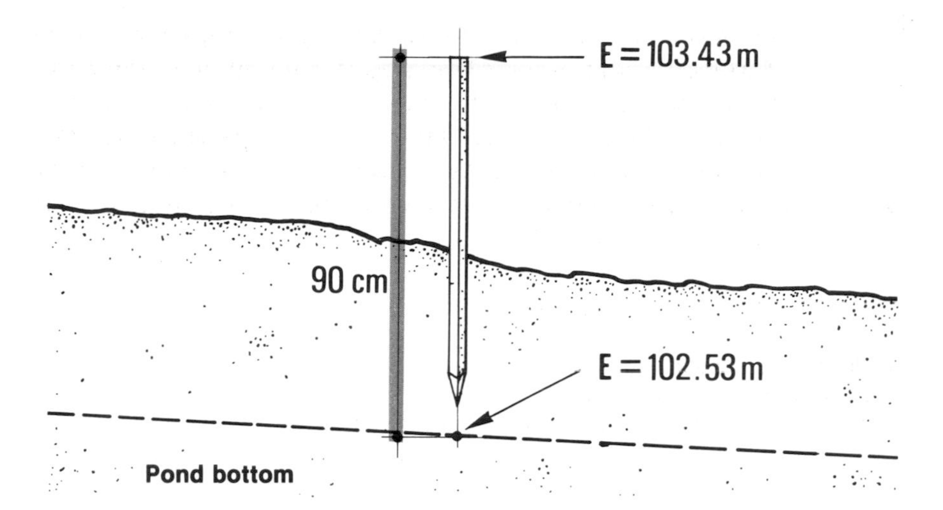

Example

- Elevation to be given according to construction plans to point 0 = 102.53 m.
- Actual elevation of the top of peg 0 obtained by levelling from bench-mark = 103.43 m.
- Difference in elevation required at 0 = 103.43 m – 102.53 m = 0.90 m = 90 cm.

(h) Starting with line OX at point 0, calculate the difference in elevation between the peg top and the pond bottom for each successive peg. **Subtract** the difference obtained above in (f) each time. Repeat the same procedure for all the other radiating lines. To make your measurements clearer, you can **use a simple table** as shown in the example. Clearly mark these values on the pegs. When the workers dig the pond, they will use the values as a reference guide.

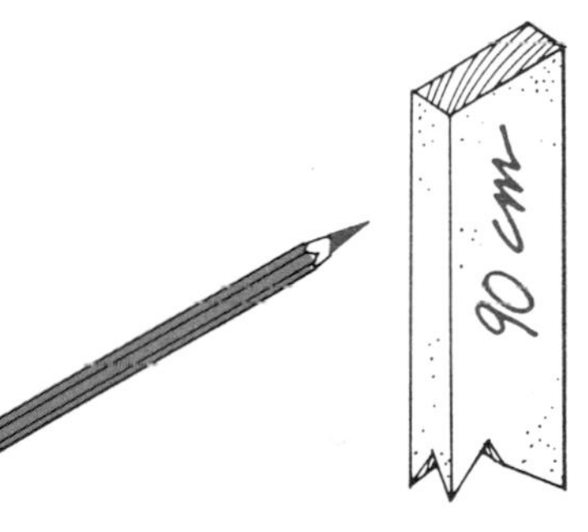

Example

Difference in elevation between peg tops and pond bottom, in centimetres

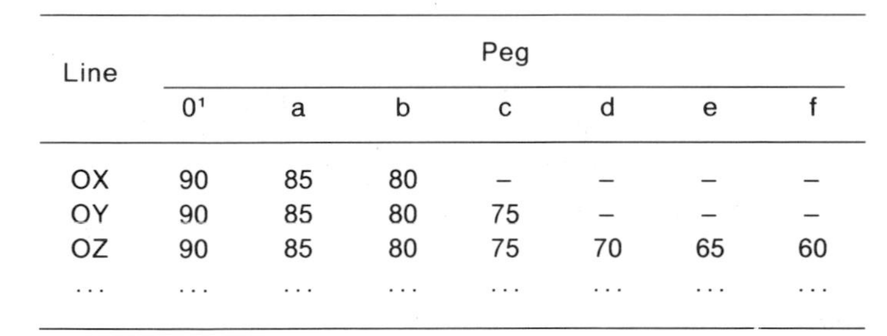

Line	Peg						
	0[1]	a	b	c	d	e	f
OX	90	85	80	–	–	–	–
OY	90	85	80	75	–	–	–
OZ	90	85	80	75	70	65	60
...	...	...	...	...	...	...	...

[1] At point 0, the difference in elevation is 90 cm from plans. The fixed difference to be subtracted is 5 cm, see above.

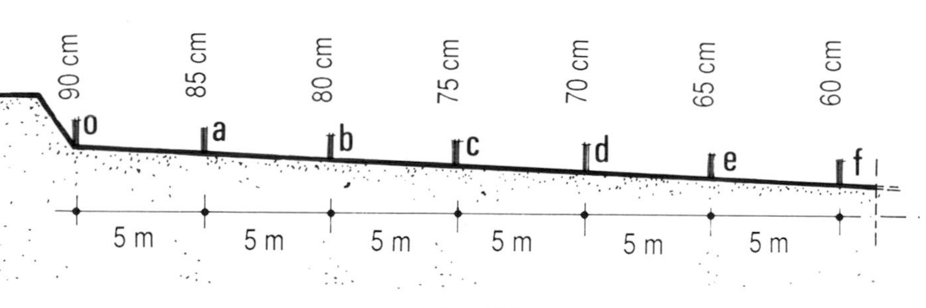

Line OZ

ANNEX I

Some useful mathematical formulae for regular geometrical figures

PERIMETER LENGTH AND SURFACE AREA

Figure	*Perimeter (P)*	*Area (A)*
Square a	$P = 4a$	$A = a^2$
Rectangle ab	$P = 2a + 2b$	$A = ab$
Right-angle triangle abc	$P = a + b + c$	$A = ab \div 2$
Trapezium abcdh (a parallel to c)	$P = a + b + c + d$	$A = (a + b)(h \div 2)$
Circle rd	$P = 6.28r$ $P = 3.14d$	$A = 3.14^2$ $A = 3.14d^2 \div 4$

SIDE LENGTH OF RIGHT-ANGLE TRIANGLE

ABCabc	$c = a \cos ABC$
	$b = a \sin ABC$
	$c = a^2 - b^2$
	$b = c \tan ABC$

Note: *tan* – see Table 3; *cos* – see Table 5; *sin* – see Table 14

ANNEX II

Tangents and values of angles
(*Tan* = angles expressed in degrees *d* and minutes *m*)

Tan	*d*	*m*	*Tan*	*d*	*m*	*Tan*	*d*	*m*	*Tan*	*d*	*m*	*Tan*	*d*	*m*	*Tan*	*d*	*m*	*Tan*	*d*	*m*
0	0	0	0.0875	5	0	0.1763	10	0	0.2679	15	0	0.3640	20	0	0.4663	25	0	0.5774	30	0
0.0029		10	0.0904		10	0.1793		10	0.2711		10	0.3673		10	0.4699		10	0.5812		10
0.0058		20	0.0934		20	0.1823		20	0.2742		20	0.3706		20	0.4734		20	0.5851		20
0.0087		30	0.0963		30	0.1853		30	0.2773		30	0.3739		30	0.4770		30	0.5890		30
0.0116		40	0.0992		40	0.1883		40	0.2805		40	0.3772		40	0.4806		40	0.5930		40
0.0145		50	0.1022		50	0.1914		50	0.2836		50	0.3805		50	0.4841		50	0.5969		50
0.0175	1	0	0.1051	6	0	0.1944	11	0	0.2867	16	0	0.3839	21	0	0.4877	26	0	0.6009	31	0
0.0204		10	0.1080		10	0.1974		10	0.2899		10	0.3872		10	0.4913		10	0.6048		10
0.0233		20	0.1110		20	0.2004		20	0.2931		20	0.3906		20	0.4950		20	0.6088		20
0.0262		30	0.1139		30	0.2035		30	0.2962		30	0.3939		30	0.4986		30	0.6128		30
0.0291		40	0.1169		40	0.2065		40	0.2994		40	0.3973		40	0.5022		40	0.6168		40
0.0320		50	0.1198		50	0.2095		50	0.3026		50	0.4006		50	0.5059		50	0.6208		50
0.0349	2	0	0.1228	7	0	0.2126	12	0	0.3057	17	0	0.4040	22	0	0.5095	27	0	0.6249	32	0
0.0378		10	0.1257		10	0.2156		10	0.3089		10	0.4074		10	0.5132		10	0.6289		10
0.0407		20	0.1287		20	0.2186		20	0.3121		20	0.4108		20	0.5169		20	0.6330		20
0.0437		30	0.1317		30	0.2217		30	0.3153		30	0.4142		30	0.5206		30	0.6371		30
0.0466		40	0.1346		40	0.2247		40	0.3185		40	0.4176		40	0.5234		40	0.6412		40
0.0495		50	0.1376		50	0.2278		50	0.3217		50	0.4210		50	0.5280		50	0.6453		50
0.0524	3	0	0.1405	8	0	0.2309	13	0	0.3249	18	0	0.4245	23	0	0.5317	28	0	0.6494	33	0
0.0553		10	0.1435		10	0.2339		10	0.3281		10	0.4279		10	0.5354		10	0.6536		10
0.0582		20	0.1465		20	0.2370		20	0.3314		20	0.4314		20	0.5392		20	0.6577		20
0.0612		30	0.1495		30	0.2401		30	0.3346		30	0.4348		30	0.5430		30	0.6619		30
0.0641		40	0.1524		40	0.2432		40	0.3378		40	0.4383		40	0.5467		40	0.6661		40
0.0670		50	0.1554		50	0.2462		50	0.3411		50	0.4417		50	0.5505		50	0.6703		50
0.0699	4	0	0.1584	9	0	0.2493	14	0	0.3443	19	0	0.4452	24	0	0.5543	29	0	0.6745	34	0
0.0729		10	0.1614		10	0.2524		10	0.3476		10	0.4487		10	0.5581		10	0.6787		10
0.0758		20	0.1644		20	0.2555		20	0.3508		20	0.4522		20	0.5619		20	0.6830		20
0.0787		30	0.1673		30	0.2586		30	0.3541		30	0.4557		30	0.5658		30	0.6873		30
0.0816		40	0.1703		40	0.2617		40	0.3574		40	0.4592		40	0.5696		40	0.6916		40
0.0846		50	0.1733		50	0.2648		50	0.3607		50	0.4628		50	0.5735		50	0.6959		50

ANNEX III

Cosine values of angles
(*d* = degrees, *m* = minutes, *cos* = cosine, *x* = difference)

MAIN TABLE

d	m	cos	x
3	0	0.9986	1
	10	0.9985	2
	20	0.9983	3
	30	0.9981	1
	40	0.9980	2
	50	0.9978	2
4	0	0.9976	2
	10	0.9974	3
	20	0.9971	2
	30	0.9969	2
	40	0.9967	3
	50	0.9964	2
5	0	0.9962	3
	10	0.9959	2
	20	0.9957	3
	30	0.9954	3
	40	0.9951	3
	50	0.9948	3
6	0	0.9945	3
	10	0.9942	3
	20	0.9939	3
	30	0.9936	4
	40	0.9932	3
	50	0.9929	4
7	0	0.9925	3
	10	0.9922	4
	20	0.9918	4
	30	0.9914	3
	40	0.9911	4
	50	0.9907	4
8	0	0.9903	4
	10	0.9899	5
	20	0.9894	4
	30	0.9890	4
	40	0.9886	5
	50	0.9881	4
9	0	0.9877	5
	10	0.9872	4
	20	0.9868	5
	30	0.9863	5
	40	0.9858	5
	50	0.9853	5
10	0	0.9848	5
	10	0.9843	5
	20	0.9838	5
	30	0.9833	6
	40	0.9827	5
	50	0.9822	6
11	0	0.9816	5
	10	0.9811	6
	20	0.9805	6
	30	0.9799	6
	40	0.9793	6
	50	0.9787	6
12	0	0.9781	6
	10	0.9775	6
	20	0.9769	6
	30	0.9763	6
	40	0.9757	7
	50	0.9750	6
13	0	0.9744	7
	10	0.9737	7
	20	0.9730	6
	30	0.9724	7
	40	0.9717	7
	50	0.9710	7
14	0	0.9703	7
	10	0.9696	7
	20	0.9689	8
	30	0.9681	7
	40	0.9674	7
	50	0.9667	8
15	0	0.9659	7
	10	0.9652	8
	20	0.9644	8
	30	0.9636	8
	40	0.9628	7
	50	0.9621	8
16	0	0.9613	8
	10	0.9605	9
	20	0.9596	8
	30	0.9588	8
	40	0.9580	8
	50	0.9572	9
17	0	0.9563	8
	10	0.9555	9
	20	0.9546	9
	30	0.9537	9
	40	0.9528	8
	50	0.9520	9
18	0	0.9511	9
	10	0.9502	10
	20	0.9492	9
	30	0.9483	9
	40	0.9474	9
	50	0.9465	10
19	0	0.9455	9
	10	0.9446	10
	20	0.9436	10
	30	0.9426	9
	40	0.9417	10
	50	0.9407	10
20	0	0.9397	

TABLE OF PROPORTIONAL PARTS, P

	Cos difference, x										
m	1	2	3	4	5	6	7	8	9	10	*m*
1	0.1	0.2	0.3	0.4	0.5	0.6	0.7	0.8	0.9	1.0	1
2	0.2	0.4	0.6	0.8	1.0	1.2	1.4	1.6	1.8	2.0	2
3	0.3	0.6	0.9	1.2	1.5	1.8	2.1	2.4	2.7	3.0	3
4	0.4	0.8	1.2	1.6	2.0	2.4	2.8	3.2	3.6	4.0	4
5	0.5	1.0	1.5	2.0	2.5	3.0	3.5	4.0	4.5	5.0	5
6	0.6	1.2	1.8	2.4	3.0	3.6	4.2	4.8	5.4	6.0	6
7	0.7	1.4	2.1	2.8	3.5	4.2	4.9	5.6	6.3	7.0	7
8	0.8	1.6	2.4	3.2	4.0	4.8	5.6	6.4	7.2	8.0	8
9	0.9	1.8	2.7	3.6	4.5	5.4	6.3	7.2	8.1	9.0	9

Example

To calculate intermediate cosine values using the proportional parts, for cos 7°38′ for example, proceed as follows:

- from the Main Table, calculate cos 7°30′ = 0.9914;
- obtain the difference between this value and the next, x = 3;
- find column 3 in Table of Proportional Parts, P;
- move down this column to line m = 8, to find P = 2.4;
- subtract P from the last number (4) of the value read from the Main Table, 0.9914 – 0.00024 = 0.99116. This is cos 7°38′.

COMMON ABBREVIATIONS

Az	=	azimuth (magnetic)
BS	=	backsight
BM	=	bench-mark
Cos	=	cosine (angle)
CI	=	contour interval
E(A)	=	elevation of point A
FS	=	foresight
HI	=	height of the instrument
LS	=	levelling station
Sin	=	sine (angle)
Tan	=	tangent (angle)
TP	=	turning point
TBM	=	temporary bench-mark

MEASUREMENT UNITS

Lengths/distances	**km**	= kilometre = 1 000 m
Height differences	**m**	= metre
Elevations	**cm**	= centimetre = 0.01 m
	mm	= millimetre = 0.001 m
Surface areas	**km^2**	= square kilometre = 100 ha
	ha	= hectare = 10 000 m^2
	m^2	= square metre
	cm^2	= square centimetre = 100 mm^2
	mm^2	= square millimetre
Volumes	**m^3**	= cubic metre
Angles	**sec**	= ″ = second
	min	= ′ = minute = 60 sec
		° = degree = 60 min = 3 600 sec
Slopes/gradients	**%**	= percent
	‰	= per thousand

GLOSSARY OF TECHNICAL TERMS[1]

ALIDADE
: Sighting ruler used together with a plane-table.

ALTITUDE
: Vertical distance or height above the mean sea level which is in this case the reference **horizontal plane***; see also **"elevation"** and **"reference"**.

AZIMUTH
: **Horizontal*** angle formed by **the magnetic north*** and a straight line or a direction; always measured clockwise from the magnetic north to the line/direction.

BACKSIGHT
: (a) Direction of a line measured when looking back to a previous survey point from a new point, whose direction has been defined as a foresight from the previous survey point. Commonly used in traversing.
(b) Measurement of the height above ground of a point of known **elevation***, for example in direct levelling; in this case, also known as a **plus sight*.**

BENCH-MARK
: Permanent, well-defined ground point of known or assumed **elevation*** used for example as the starting point of a topographical survey or as a reference point during constructions. **A temporary bench-mark** is used for a short period of time only, and is not permanently marked as a reference point.

CONTOUR
: Imaginary line which joins all ground points of an equal **elevation*** above a given reference plane.

CONTOUR INTERVAL
: Difference in **elevation*** between two adjacent **contours***.

CONTOUR LINE
: A drawn line which joins points of equal elevation on a plan or a map; it represents a contour as it is found in the field.

[1] This glossary contains definitions of the technical terms marked with an asterisk (*) in the text.

GLOSSARY OF TECHNICAL TERMS, continued

CUMULATIVE DISTANCE — Total distance from the starting point of a survey line.

CUT — Area where it is necessary to lower the land to a required **elevation***, by digging soil away.

ELEVATION — **Vertical*** distance or height above a given **reference horizontal plane***; see also **"altitude"** and **"reference"**.

FILL — Area where it is necessary to raise the land to a required **elevation*** by bringing soil in.

FORESIGHT —
(a) Direction of a line measured ahead (forward), from the line's initial point, for example in traversing.
(b) Measurement of the height above ground of a point of unknown **elevation***, for example in direct levelling; in this case, also known as a **minus sight***.

GRAVITY — Attractive force by which all bodies (including water) tend to move toward the centre of the earth, for example when moving or falling from a higher elevation to a lower elevation.

GROUND PROFILE — A drawn representation of the ground surface which shows change in **elevation*** (along the vertical axis) with distance (along the horizontal axis).

HEIGHT OF THE INSTRUMENT — Height above ground of the sighting line of a **levelling*** instrument.

HORIZONTAL — Line or **plane*, parallel*** to the plane of the horizon and at right angles to the **vertical plane***; flat, level.

LEVELLING — Operation of measuring differences in **elevation*** at several ground points through a topographical survey.

LEVELLING STATION — Ground point where a levelling instrument is set up for a topographical survey.

LINE OF SIGHT — Imaginary line which begins at the eye of the surveyor and runs toward a fixed point. It is always a straight line, also called the **"sighting line"**.

MAGNETIC BEARING — Direction in which any point lies from a point of reference as measured from the **magnetic north*** with a compass.

MAGNETIC NORTH — Direction in which the magnetized end of a compass needle points, namely toward the north magnetic pole of the earth. Note: the position of the north magnetic pole is affected by local variations, and corrections may be required for detailed use.

MINUS SIGHT — An elevation which is always subtracted, see **foresight***, definition (b).

OBLIQUE LINE — For a given **horizontal** and **vertical plane***, an oblique line is:

- within the **horizontal plane*** but not **perpendicular*** to the **vertical plane***, or
- within the **vertical plane***, but not **perpendicular*** to the **horizontal plane***, or
- within neither **plane***.

PARALLEL (LINE) — A line equally distant from another line at every point along its length.

PERPENDICULAR — A line/plane having a direction at right angles to a given line/plane.

PLANE — An imaginary flat surface: every straight line joining any two points in it lies totally in it.

PLUS SIGHT — An elevation which is always added, see **backsight***, definition (b).

GLOSSARY OF TECHNICAL TERMS, continued

POINT OF REFERENCE	Fixed point, usually identified in the field by a marker at the end of a line of sight.
POLYGON	A geometrical figure or an area of land having more than three straight sides.
RECTANGLE	A four-sided **polygon*** with four **right angles***.
REDUCED LEVEL	Vertical distance to a **common reference plane***, such as the mean sea level (see **"altitude"**) or an assumed horizontal plane (see **"elevation"**). It is calculated from survey data.
REFERENCE LEVEL/ PLANE	**Elevation*** or **plane*** which is repeatedly used during a particular survey, and to which survey points or lines are referred.
RIGHT ANGLE	A 90-degree angle.
RIGHT-ANGLED	A **triangle*** with one 90-degree angle.
SCALE	Relationship existing between the distance shown on a drawing and the actual distance across the ground.
SIGHTING LINE	Synonym for **"line of sight"**.
TANGENT	Mathematical function for angles.
TRAPEZIUM	A four-sided **polygon*** with two **parallel*** sides.
TRAVERSE	A set of straight lines connecting established points around or along the route of a plan survey.
TRIANGLE	A three-sided geometrical figure or land area.
TURNING POINT	Temporary intermediate or reference point being surveyed between two established points; it is no longer needed after the necessary reading has been taken.

VERTICAL	Line or **plane*** which is perpendicular to a horizontal line or plane; in practice defined by the position of a freely suspended weighted line.

FURTHER READING

CLENDINNING, J. and J.G. OLLIVER, 1966. *The principles of surveying.* London, Blackie and Son Ltd., 463 p. 3rd. edition.

STERN, P. *et al.* (eds), 1983. *Field engineering. An introduction to development work and construction in rural areas.* London, Intermediate Technology Publications Ltd., 251 p.

NOTES

NOTES

NOTES

NOTES